Nuclear Physics in Astrophysics VIII (NPA8 2017)

EPJ Web of Conferences Volume 165 (2017)

Catania, Italy
18 - 23 June 2017

Editors:

M. La Cognata
M. Lattuada
S. Palmerini
R.G. Pizzone
C. Spitaleri

ISBN: 978-1-5108-5571-7

Printed from e-media with permission by:

Curran Associates, Inc.
57 Morehouse Lane
Red Hook, NY 12571

CURRAN ASSOCIATES INC. proceedings.com

Some format issues inherent in the e-media version may also appear in this print version.

Printed by Curran Associates, Inc. (2018)

For additional information, please contact EDP Sciences – Web of Conferences
at the address below.

EDP Sciences – Web of Conferences
17, Avenue du Hoggar
Parc d'Activité de Courtabœuf
BP 112
F-91944 Les Ulis Cedex A
France

Phone: +33 (0) 1 69 18 75 75
Fax: +33 (0) 1 69 28 84 91

contact-edps@webofconferences.org

Additional copies of this publication are available from:

Curran Associates, Inc.
57 Morehouse Lane
Red Hook, NY 12571 USA
Phone: 845-758-0400
Fax: 845-758-2633
Email: curran@proceedings.com
Web: www.proceedings.com

TABLE OF CONTENTS

EPJ Web of Conferences **165**, 00001 (2017) DOI: 10.1051/epjconf/201716500001
NPA8 2017

VIII Nuclear Physics in Astrophysics International Conference (NPA8): Preface

Marco La Cognata[1], *Marcello* Lattuada[2,1], *Sara* Palmerini[3,4], *Rosario Gianluca* Pizzone[1], and *Claudio* Spitaleri[2,1]

[1] *INFN - Laboratori Nazionali del Sud, Via S. Sofia 62, 95123 Catania, Italy*
[2] *Dipartimento di Fisica e Astronomia, University of Catania, Via S. Sofia 64, 95123 Catania, Italy*
[3] *Dipartimento di Fisica e Geologia, University of Perugia, Via A. Pascoli, 06123 Perugia, Italy*
[4] *INFN - Sezione di Perugia, Via A. Pascoli, 06123 Perugia, Italy*

NPA8 — Nuclear Physics in Astrophysics VIII
18-23 June 2017 Laboratori Nazionali del Sud

NPA8 is the successor of earlier events held in Eilat, Israel, in 2001 and, with a 2-year frequency, in Debrecen (Hungary, 2 times), in Frascati (Italy), Eilat (Israel) and in York (UK). In the same spirit as the earlier editions, NPA8 was the showcase for the most recent developments in the field of nuclear astrophysics covering a wide range of topics from fundamental aspects to instrumentation and astrophysical applications. Room was given not only to the traditional aspects of experimental nuclear astrophysics, such as direct measurements above and under ground and indirect approaches, but also to emerging fields such as the measurement of fusion reactions in laser induced plasmas and of photodissociation reactions. Also significant room has been given to novel experimental techniques that might have important impact on future nuclear astrophysics experiments and on astrophysical modelling and observations, which constitute the necessary motivations and background of the proposed measurements. The event was attended by a large number of researchers at the start of their scientific career together with well-established senior scientists. Such a diversity in experience, nationality and research expertise made NPA8 an ideal platform for cross fertilisation between the disciplines, stimulating new ideas and scientific networks.

During NPA8 a special event was organised, to celebrate the scientific achievements of the conference chairman, C. Spitaleri, on the occasion of his retirement.

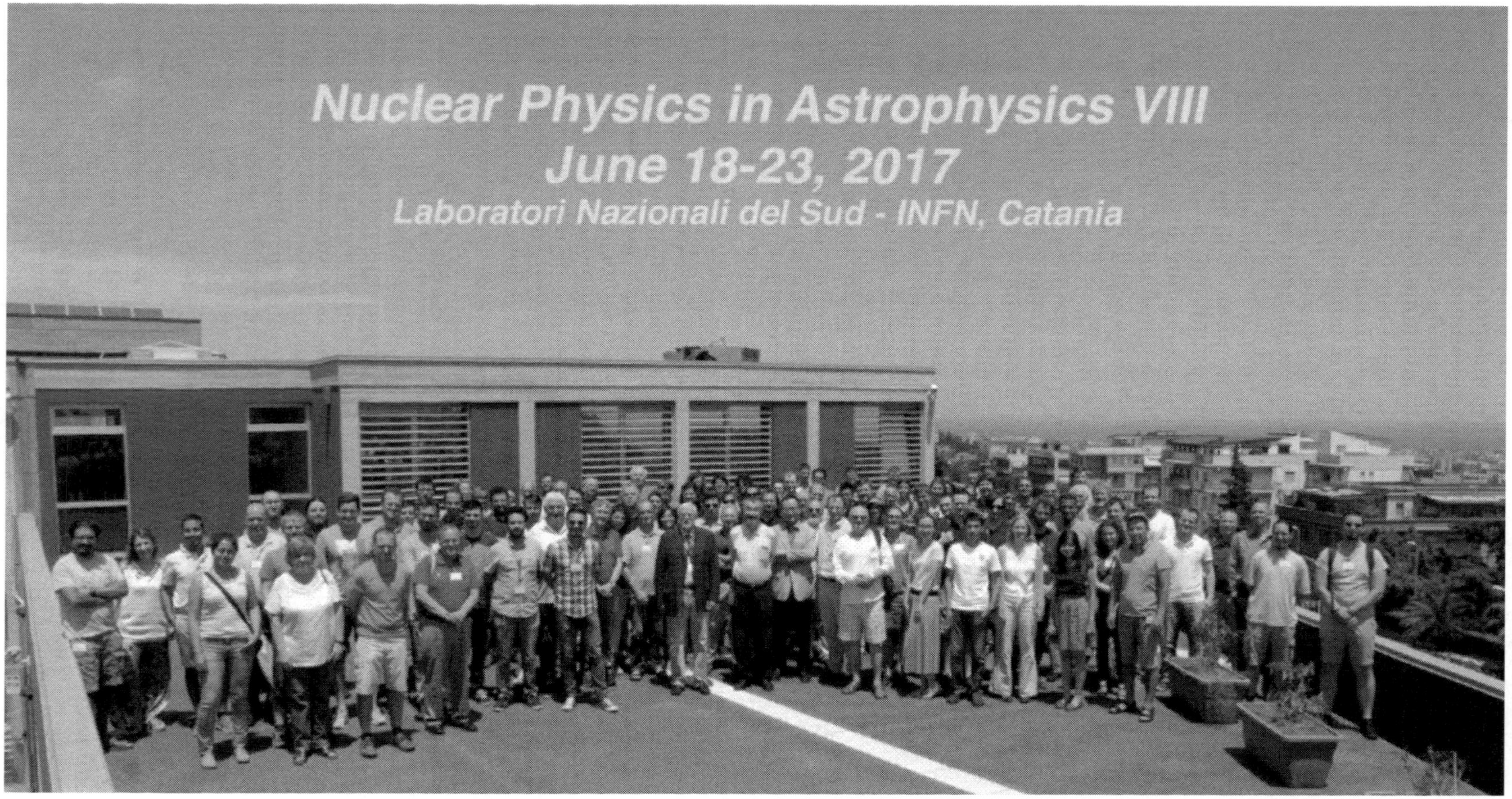
Nuclear Physics in Astrophysics VIII
June 18-23, 2017
Laboratori Nazionali del Sud - INFN, Catania

edp sciences

EPJ Web of Conferences

Statement of Peer review

In submitting conference proceedings to EPJ Web of Conferences, I certify to the Publisher that I adhere to the **Policy on Publishing Integrity** of the journal in order to safeguard good scientific practice in publishing.

1. All articles have been subjected to peer review administered by the proceedings editors.
2. Reviews have been conducted by expert referees, who have been requested to provide unbiased and constructive comments aimed, whenever possible, at improving the work.
3. Proceedings editors have taken all reasonable steps to ensure the quality of the materials they publish and their decision to accept or reject a paper for publication has been based only on the merits of the work and the relevance to the journal.

Title, date and place of the conference

Nuclear Physics in Astrophysics 8, 18-23 june 2017 , Catania

Proceedings editor(s):

La Cognata M., Lattuada M., Palmerini S., Pizzone R.G. , Spitaleri C.

Date and editor's signature

25.11.2017

EPJ Web of Conferences **165**, 01001 (2017) DOI: 10.1051/epjconf/201716501001
NPA8 2017

Reaction production + AMS: An alternative method to study low energy reactions. ^{26}Al as a test case

L. Acosta[1,2,*], *V.* Araujo-Escalona[3], *E.* Chávez[1], *E.* Andrade[1], *L.* Barrón-Palos[1], *F.* Favela[1,2], *M.A.* Flores[1], *J.* García-Ramírez[1], *A.* Huerta[1], *O.* de Lucio[1], C. Méndez-García[1], *M.E.* Ortiz[1], S. Padilla[1], *A.M.* Sánchez-Benítez[4], *P.* Santa Rita[1], and *C.* Solís[1]

[1]Instituto de Física, Universidad Nacional Autónoma de México, A.P. 20-364, Cd.Mx. 01000, Mexico
[2]Istituto Nazionale di Fisica Nucleare, Sezione di Catania, I-95123 Catania, Italy
[3]KU Leuven, Instituut voor Kern-en Stralingsfysica, 3001 Leuven, Belgium
[4]Facultad de Ciencias Experimentales, Universidad de Huelva, E-21071, Huelva, Spain

Abstract. Considering the importance of the ^{26}Al nuclei in Astrophysics, in this work, preliminary results regarding a campaign of measurements related with this radioisotope production, are presented. We have taken advantage of two different facilities: first, the radio-nucleus is produced by means of irradiation of targets selected in correlation with particular reactions; once the enrichment with ^{26}Al was made, the targets are analyzed in an AMS machine to obtain the concentration of ^{26}Al produced during the irradiation. With this off-line method, it is possible to measure acceptable small cross sections of a selected low energy reaction. In this work, our preliminary results for three different energies of ^{28}Si(d,α)^{26}Al reaction cross sections are shown, as well as our first considerations to commence with measurements of ^{25}Mg(p,γ)^{26}Al reaction cross sections below 1 MeV.

1 Introduction

The ^{26}Al is a long-lived radioisotope ($T_{1/2}$ = 0.716 My). The characteristic gamma at E_γ = 1.809 MeV, following its β+ decay to ^{26}Mg, has been observed for decades in the galactic plane, giving evidence of presently undergoing nucleosynthesis. The detection of the radiation generated by this decay allows to trace the ^{26}Al production sources in the Milky Way, and impose with this, constrains to the existing nucleosynthesis models [1]. This radioisotope is present for instance, in the stars where there is H, C and Ne fusion at high temperatures; besides, indications of extinct ^{26}Al have been found in meteorites where it could be deposited or created *in situ* [2].

Though many studies regarding ^{26}Al have been carried out, some discrepancies still remain about its production in explosions of massive stars. There are important gaps in knowledge regarding the kind of events or environments which most contribute to its production (Wolf-Rayet, novae, asymptotic giant branch, supernovae explosion, etc.) [3].

* Corresponding author: acosta@fisica.unam.mx

Many important reactions to produce ^{26}Al happen at very low energies, making difficult to get an approach of the cross section by using standard spectroscopic methods to measure them. Several γ resonances, in case of ^{25}Mg(p,γ)^{26}Al appear for instance, below 500 keV, a region where real γ-ray signal may be coupled with background of a conventional detection system.

In order to overcome the technical difficulties afore mentioned, an alternative technique proposed by Arazi et al. [1] has been identified. This technique is particularly useful for measuring extremely low cross sections of reactions that produce long-lived radionuclides, like ^{26}Al. In [1], authors used ^{25}Mg(p,γ)^{26}Al reaction to produce the radio-nucleus by means of a low energy proton beam interacting with a MgO target. The irradiated target is subsequently analyzed for its ^{26}Al content by using accelerator mass spectrometry technique (AMS). This technique allow us to yield information about the number of ^{26}Al nuclei which were produced previously, during the irradiation. To achieve this goal, the [1] authors used a chemical separation of the Al from the MgO target including ^{27}Al as carrier, in order to normalize a ratio of events (^{26}Al/^{27}Al) by using AMS. The final composed to be employed as material for the negative-ion sputter source of the AMS system was Al_2O_3. The ^{26}Al/^{27}Al ratio obtained with AMS can be later traduced as a cross section of the reaction to be studied.

In the following section the main characteristics of our measurements are described, which were based on the technique mentioned above, in order to obtain cross section values for the (d,α) reaction related as well with the ^{26}Al production. In section 3 our preliminary results are shown and after that, at the end of such section, our perspectives to develop other measurements regarding (p,γ) reaction (always measuring ^{26}Al nuclei with AMS), are also described. We dedicated the section 4 to present the main conclusions extracted from the exposed measurements.

Fig. 1. CN-Van de Graaff Accelerator 5.5 MV, Carlos Graef Laboratory (left picture) and Accelerator Mass Spectrometry National Laboratory, LEMA (right picture) at Physics Institute, UNAM, Mexico City. The Van de Graaff was used for the irradiation of the samples, later analysed in the AMS facility (see text for details).

2 Experimental procedure

In the CN-Van de Graaff Carlos Graef Laboratory of IF-UNAM at Mexico City (left picture, Fig. 1), 300 nA beams of protons, deuterons, ^{3}He and ^{4}He can be currently produced between 0.8-3.0 MeV energy range. Considering this, the performances of this facility are particularly appropriate to study the production of ^{26}Al by means of ^{28}Si(d,α)^{26}Al reaction at low energies (Q = 1.43 MeV). The study of (d,α) reaction at low energies could contribute to explain the fraction of ^{26}Mg found in meteorites, considering

DOI: 10.1051/epjconf/201716501001

that not all ^{26}Mg found, for instance in Allende meteorite, may be related to the previous presence of ^{26}Al [2]. Even if the cosmic rays flux of deuterons compared with cosmic rays flux of protons is really low (2 to 5 % [4, 5]), we consider interesting to know with precision the cross sections related with the (d,α) reaction. Moreover, the high purity Silicon powder (used as target for this reaction) shows a very stable behaviour when interacts with stable beams (the case of the irradiation in the first step) and allows the production of a stable Al current in the sputter negative ion cesium source (which is the case of AMS ion source to be used for the second part of the measurement).

2.1 Target preparation and irradiation

As a first step, we decided to carry out the irradiation directly on the container capsule of the sample to be irradiated, which later is directly used as "cathode" in the AMS system. In this context, we did not perform the chemical process referred in [1] to separate the produced ^{26}Al. Instead of that, we filled the cathode with an homogeneous mixture of powder made of a combination of high purity silicon and aluminium powders. The latter was included to normalize the ^{26}Al produced during the irradiation. A series of target/cathodes were mounted (one by one) inside a vacuum chamber to be irradiated for periods of 2 hours with a deuterium beam of an average intensity of 200 nA. A detailed description of this procedure is well described in [6, 7]. In order to monitor the reaction behaviour, two PIP-silicon detectors were placed symmetrically at 150° in both backward sides of the target/cathode. The sketch and a picture of this experimental setup are shown in Fig. 2. A typical RBS spectrum, where scattering on Al/Si can be well identified, is shown in Fig. 3. Here it is worth mentioning that a considerable fraction of the beam intensity is scattered on the copper, according to the spectrum in Fig. 3. Nevertheless, with the spectrum it is possible to estimate the correct amount of beam that could be exploited using a fit program (SIMNRA [8] in this work) along with a numerical calculation. A detailed description of such analysis can be consulted in reference [7].

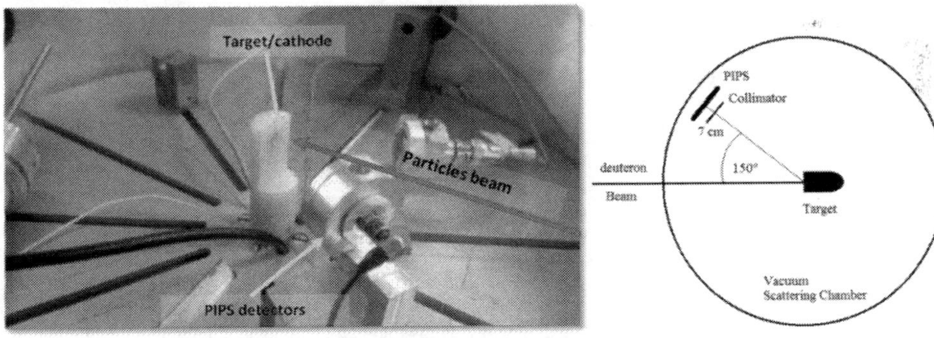

Fig. 2. Experimental setup for the first step of the technique, where a deuterium beam irradiates the target to produce ^{26}Al. Picture (left panel) and sketch (right panel). (see the text for details).

2.2 Counting of ^{26}Al at the AMS system

The Accelerator Mass Spectrometry National Laboratory (LEMA) at IF-UNAM, Mexico City, is a 1 MV tandetron machine, manufactured by HVEE (see right picture at Fig. 1). It is well calibrated to achieve very small concentrations of ^{14}C, ^{10}Be, ^{26}Al, ^{129}I and Pu isotopes [9]. With LEMA is possible to discriminate an ^{26}Al event among 10^{14} ^{27}Al ones.

This characteristic makes LEMA to be an adequate system to test the technique proposed here.

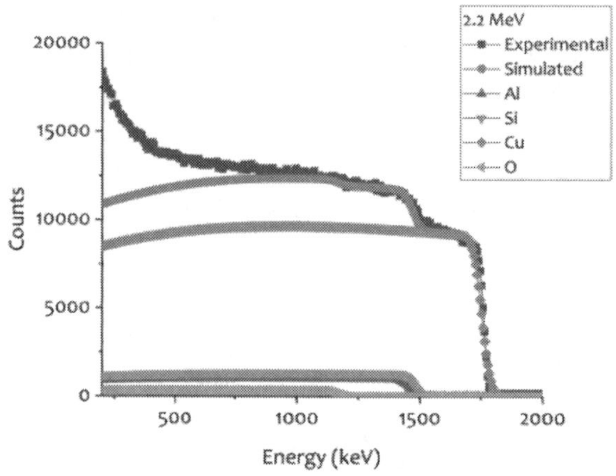

Fig. 3. Typical RBS spectrum (black squares) fitted with SIMNRA (red circles) for a deuterium beam at E_{lab} = 2.2 MeV. Straight and inverted triangles (green and blue lines) account for the contribution of the scattered deuterium in the Al/Si target while diamonds (purple line) account for the scattering on the container capsule, made of copper.

Once the cathodes are irradiated they can be placed in the carrousel for cathodes of LEMA to begin the interaction with a negative Cs ion source in order to produce Al beam. This cathodes are complemented with enriched ^{26}Al cathodes and blanks (both used as standard to calibrate the isotope separator). The number of events related with ^{26}Al are counted by using the ΔE-E discrimination technique, by means of a gas detector placed at the end of the line in the AMS system. The value expected for the ^{26}Al/^{27}Al ratio in the case of the standard was 6.72×10^{-11}, which was well reproduced for the present measurements. In Fig. 4 a typical ΔE-E uncalibrated spectrum is depicted, where the detected events related exclusively with ^{26}Al nuclei for the sample used as standard, can be clearly seen inside the black region of interest.

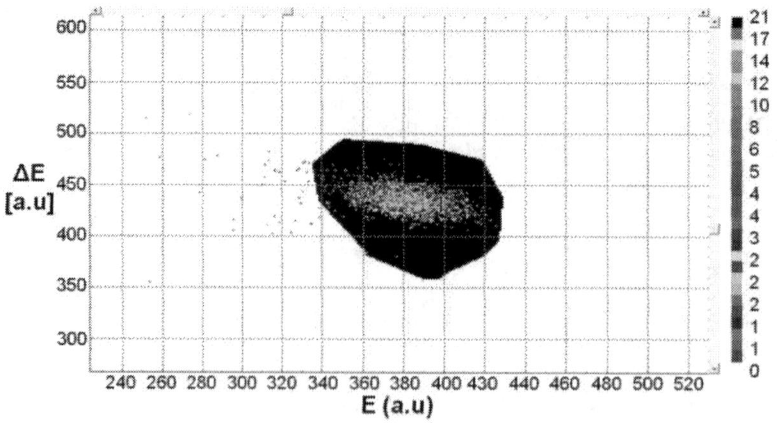

Fig. 4. Typical ΔE-E spectrum for a standard sample used to fine tune the AMS system. The coloured events enclosed by black region are identified as ^{26}Al counts.

3 Preliminary results

Once all possible variables involved in the measurement processes have been taken into account, it is feasible to obtain preliminary values for cross sections related with some specific irradiation energy for ^{26}Al production. The results reported here were obtained for three different energies: 1.1 MeV, 1.5 MeV and 1.8 MeV. The estimated values are 3.9(4) μb, 1.5(2) μb and 1.3(1) μb, respectively. The largest value appears for the lowest energy, showing probably a value closer to a maximum.

The preliminary cross sections estimated could be considered too small for the studied reaction, however to the best of our knowledge they do not exist published values to establish any kind of comparison. It is important to mention that irradiation on the target is in all cases concentrated in the surface of the cathodes, penetrating some tens of Amstrong inside the target material. In this situation, part of the produced radioisotope could get lost at the first interaction with the ion source at the AMS system. Therefore, to validate this methodology, it would be necessary to reach reproducible results repeating the same process, with increasing irradiation period to ensure the radioisotope production, even at the surface level. In order to compare both processes, we will try as well the chemical separation of the radioisotope produced (as it was done in [1] for Mg target), and fill a cathode with the obtained solution. Both goals are nowadays in progress. For the moment we can affirm that we have the capability to address this two-steps method to obtain a value for a determined reaction cross section, particularly for reactions that require of low background systems to be directly measured.

3.1 Perspectives for new measurements involving ^{26}Al reactions

As part of a parallel study, some measurements related with the (p,γ) reaction, previously mentioned, are in progress. Testing this reaction will allow to reproduce some of the values reported in [1,10] by using lower energy AMS system. $^{25}Mg(p,\gamma)^{26}Al$ reaction is the prevailing reaction in the Mg-Al chain, i.e., this is a very important reaction for the nucleosyntesis processes. At temperatures between 0.1 and 1.5 GK in the stellar processes, the ^{26}Al production rate depends on very weak resonances, doing its determination a very difficult task when are used on-line detection techniques to prompt the γ -ray emission.

In spite of those difficulties, quite efforts inside big collaborations have performed to achieve good measurements of several strengths of resonances related with $^{25}Mg(p,\gamma)\,^{26}Al$, combining the on-line and off-line methods [11, 12].

During the last 20 years, most of the interest resonances have been well measured. However, there still remains some discrepancies for specific values as, for instance, 197 keV, 417 keV and 435 keV resonance energies. Considering our present systems, we are able to obtain the strength of resonances around 400 keV, by using a similar methodology that was used in [1, 11].

Since October 2017, a new beam line was mounted and commissioned as part of the Separator Spectrometer in the LEMA laboratory. Before this date, the Spectrometer was entirely dedicated to AMS studies. With this new line, it is possible to produce intense low energy beams of protons: 300 to 2200 keV with 1 μA of stable current. With this conditions it seems feasible to develop the irradiation and later the AMS analysis by using the same accelerator. Considering the very recent implementation of this new beam line, the setting up of such measurements is presently at the very beginning.

4 Conclusions

In this work an alternative method to achieve relevant information regarding particular reactions of difficult access with conventional measured techniques was raised. The proposed methodology combines the use of accelerated low energy beams with accelerator mass spectrometry. The main goal of the used technique is to obtain with the combination mentioned, absolute cross sections of low energy reactions where a selected radioisotope is involved.

To test the proposed technique, the $^{28}Si(d,\alpha)^{26}Al$ reaction was selected. Considering that such reaction involves ^{26}Al nuclei, which plays an important role in astrophysical processes, the scientific interest is justified. Our preliminary results have shown very small values for the cross sections of the three different measured energies (1.1 MeV, 1.5 MeV and 1.8 MeV). However, we could not find experimental values to corroborate the present results. Works are under progress in order to validate the technique at IF-UNAM facilities. In parallel, we are working in the preparation to start some other measurements regarding the (p,γ) reaction, related as well with the study of ^{26}Al production.

Acknowledgments

This research is being partially supported by the grants: CONACYT 51600, 82692,123655 and PAPIIT-DGAPA-UNAM IA101616, IG101016. The authors want to thank to the LCG staff for their help to carry out the measurements here presented. Authors want also thank to the assigned reviewer for the carefully revision of this work with which it was highly improved.

References

1. A. Arazi et. al., Phys. Rev. C. **74**, 025802 (2006).
2. T. Lee, D. a. Papanastassiou, and G. J. Wasserburg, Geophys. Res. Lett. **3**, 1 (1976).
3. J. Knödlseder et. al., Astron. and Astrophys., **344** (1999).
4. B.J. Teegarden J. of Geoph. Res. **72**, 19 (1967).
5. J.Z. Wang et al., Astrophs. J. **564**, 244 (2002).
6. V. Araujo-Escalona et. al., J. of Phys.: Conf. Ser. **730** (2016).
7. V. Araujo-Escalona et. al., Phys. Proc. **90** (2017).
8. M. Mayer, Amer. Inst. of Phys. Conf. Proc. **475** (1999).
9. C. Solís, E. Chávez, M.E. Ortiz, A. Huerta, E. Andrade, E. Barrios. Nucl. Inst. Meth. Phys. B, **331** (2014).
10. D. C. Powell, et al., Nucl. Phys. A **644**, 263 (1998).
11. B. Limata et. al., Phys. Rev. C **82**, 015801 (2010).
12. F. Strieder et. al., Phys. Lett. B **707** (2012).

Nanostructured surfaces for nuclear astrophysics studies in laser-plasmas

Carmen Altana[1], *Vincenzo* Amarù[2], *Giuseppe* Castro[1], *Marco* Frassetto[3, 4], *Gaetano* Lanzalone[1, 5], *Luciana* Malferrari[3], *David* Mascali[1], *Annamaria* Muoio[1], *Fabrizio* Odorici[3] and *Salvo* Tudisco[1]

[1]INFN - LNS, Via S. Sofia 62, 95123 Catania, Italy
[2]Univ. degli Studi di Catania, Via Santa Sofia, 95123 Catania, Italy
[3]INFN- Sez. di Bologna, Viale B. Pichat, 6/2, 40127 Bologna, Italy
[4]Dip. di Fisica e Astronomia, Univ. degli Studi di Bologna, V.le B. Pichat 6/2, 40127 Bologna, Italy
[5]Univ. degli Studi di Enna "Kore", Via delle Olimpiadi, 94100 Enna, Italy

Abstract. The future availability of high-intensity laser facilities capable of delivering tens of petawatts of power (e.g. ELI-NP) into small volumes of matter at high repetition rates will give the unique opportunity to investigate nuclear reactions and fundamental interactions process under extreme plasma conditions [1]. In this context, use of targets with nanostructured surfaces is giving promising indications to reproduce plasma conditions suitable for measurements of thermonuclear reactions rates, in the domain of nanosecond laser pulses.

1 Introduction

Plasma plays a primary role in modern astrophysics since it constitutes the 99% of the matter in the Universe. Therefore, the investigation of the origin of stars and related phenomena deals closely with plasma physics. Direct and indirect measurements of the relevant nuclear cross sections have been performed over the years by using accelerated ion beams [2-5] and innovative experimental apparatus [6, 7].

Although it is impossible to reproduce the extreme properties of stellar matter, keV temperature plasmas can be rather easily generated in laboratory through magnetic confinement, but with densities orders magnitude lower than the stellar ones. An improvement on this direction can be achieved by using Laser-plasmas [8-10]. In the last decades, laser fabrication technology has evolved rapidly and further new physics has emerged from the study of Laser-plasmas [11-15].

In such contest, in order to achieve conditions of astrophysical interest, suitable for nuclear fusion, the application of nano-technologies in this field has also been investigated. Nanostructures on the targets surface, based on metallic nanowires (NWs) embedded in porous alumina [16], can be useful since they can absorb light at very high efficiency. These expectations are confirmed from experiments in different laser pulses regimes (from ns to fs) and various intensities [17, 18].

Moreover, these nanostructures allow to incorporate deuterium into the solid matrix, in order to produce laser deuterated-plasmas. This contribution will focus on the preliminary results of an experiment that aims to produce such plasmas, to study D+D reaction rates in the laboratory, in a condition that is closer to that of the stars.

2 Experimental Set-up

A Q-switched Nd:YAG laser operating at 1064 nm, with a maximum energy of 2 J and a duration of 6 ns impinges on a target at an intensity of about $4\cdot10^{12}$ W/cm^2, with an angle of 30° with respect to the beam. The target is placed inside an interaction chamber, where a pressure of about at 10^{-5} Torr is kept. The chamber, as sketched in Fig. 1a), can be equipped with several diagnostics, in order to characterize the plasma and the nuclear products.

An ICCD for optical fast imaging, placed perpendicularly to the plasma expansion direction, allows investigating the spatio-temporal evolution and velocities of plasmas in the visible domain. A cooled CCD camera, placed at 60° from the target normal axis, sensitive to soft X-rays, is a fundamental diagnostic to characterize the X-ray plasma emission. This camera is equipped with a pinhole array: each pinhole is filtered by using Al-made attenuators of increasing thickness, ranging from 3 to 20 μm, in order to suppress low energy photons and to select X-rays at increasing energies. Another configuration consists of a single Al-layer, 50 μm thick, in order to operate in single photon counting for measuring the X-ray energy spectra and extrapolating plasma temperatures [19]. Ions collectors and SiC detectors [20-23] operating in Time-of-Flight configuration, are placed in the forward direction respect to the plasma propagation, in order to reconstruct the velocity distribution of charged particles escaped from the plasma. A further device is a Thomson Parabola Spectrometer, which allows the identification of ions species and their charge state distribution besides the ability to extract the energy distribution [24-26]. In order to detect and identify neutrons from D+D reactions, the final set-up will include also an array of plastic scintillators with a Silicon Photo-multiplier read-out [27].

In the experiment, we used targets produced at the INFN lab in Bologna [16]. In order to compare performances of nanowires with different composition, identical alumina matrices with the same diameter (20 nm) and thickness (10 μm), were filled with cobalt, iron and nickel. A sample of bulk aluminum was also irradiated, to be used as reference. Since the different nanostructured targets provide similar results in terms of plasma temperature, only the Co-nanowires target analysis is included in this contribution.

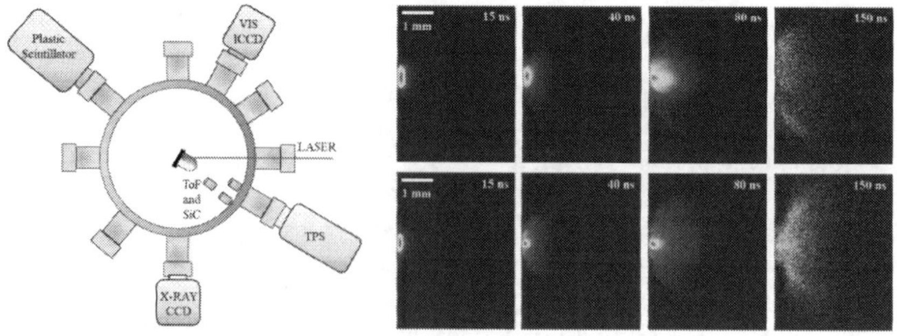

a) b)

Fig. 1. a) Schematic view of the experimental setup; **b)** sequence of ICCD images, showing the spatio-temporal evolution of bulk Al (top) and Co-nanostructured (bottom) plume at various delays.

3 Results

Fig. 1b) shows typical time-resolved ICCD images for bulk Al and Co-nanostructured targets. All images are spectrally integrated in the region 200–700 nm and normalized to their maximum intensity in order to easily compare them. Colours indicate different radiation intensities: higher counts (red) correspond to areas with higher temperature and particle density in the plasma plume. It can be observed that plasma duration of nanostructured target is higher than that of bulk Al: at 150 ns Co-nanostructured plume intensity is about 50% higher than bulk Al. By studying the expansion velocity of the plasma, through the fit of the position-time images of the luminous front, we observe that for bulk Al the plume has a higher velocity than for nanostructured targets, where plasma evolves more slowly and has a longer life. This could be an evidence of plasma stagnation, which implies a hotter and denser plasma.

Furthermore, a previous analysis of the X-ray emission has evidenced an increasing of X-ray flux for nanostructured targets compared to bulk Al and quite similar temperature due to a cold plasma component, extracted by fitting the fluxes with an opportune procedure, as described in [28]. A new set of measurements with a Co-nanowires target, performed by using the X-CCD in Single Photon Counting configuration, has allowed to obtain the X-ray spectra and to provide a more precise estimation of the plasma temperature [19]. Fig. 2a) shows the as-detected X-ray spectrum. The plasma temperature, extracted by fitting the spectrum in the range 4.5-6 keV, after deconvolution for CCD quantum efficiency and Al-filter transmission (see the insert), corresponds to 0.72 ± 0.14 keV.
Moreover, we estimated that deuterium can be incorporated in the alumina matrix (via physical-chemical processes) with a density of about 10^{20} atoms/cm^3, which allow investigating the D+D reaction in a plasma environment. We calculated the expected fusion rates in our experimental conditions, as in [9], using the cross-section from D+D scattering data and assuming that deuterons have a Maxwellian energy distribution, having the same temperature as our plasma. Results are shown in Fig 2b), where the number of fusion events in a single shot is calculated, by assuming plasma temperatures that vary from 100 eV to 1 keV. In particular, the highlighted region represents our experimental condition, with a temperature of 0.72 ± 0.14 keV: in that range, the fusion events in a single shot are close to one. This is a promising result for the next aim to study D+D reactions in plasmas.

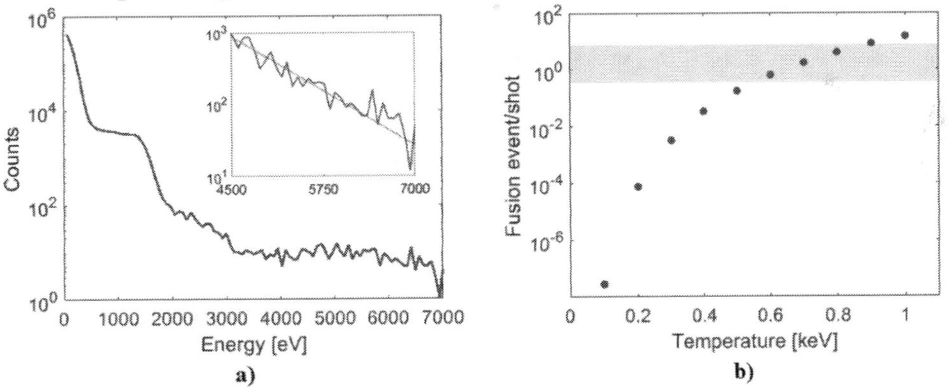

Fig. 2. a) As-detected X-ray spectrum for a Co-nanowires target; the insert shows the fit to the spectrum, after deconvolution for detection efficiencies, in the range 4.5-6 keV; **b)** expected fusion rate in a single shot, as a function of the plasma temperature.

4 Conclusions

A preliminary characterization of plasmas generated during Nd:YAG laser ablation in ns domain from nanostructured targets has been obtained. Different targets, based on Co, Fe and Ni NWs, show similar results in terms of visible and X-ray emission.

By observing the plasma expansion, through time-resolved ICCD images of the visible emission, we find experimental evidence to deem that stagnant plasmas with high temperature and density can be obtained by means of nanostructured materials. Such result is in agreement with the X-ray emission characterization. Although the X-ray energy is similar for bulk Al and NWs targets, a higher X-ray intensity was observed for all nanostructured samples compared to bulk Al: this indicates an increasing in terms of plasma density and/or in terms of plasma stagnation for NWs targets. Moreover, operating in proper conditions, we estimated that a significant number of fusion events could be obtained by means of nanostructured targets containing deuterium. This scenario could be promising since it could allow studying specific nuclear reactions in plasmas for astrophysical applications.

References

1. Negoita, F., et al. Rom. Rep. Phys. **68**, (2016)
2. La Cognata, M., et al. J. Phys. G Nucl. Part. Phys., **35** (2007)
3. Cherubini, S., et al. Eur. Phys. J. A, **20** (2004)
4. La Cognata, M. et al. Phys. Lett. B, **664** (2008)
5. Tumino, A., et al. Nucl. Phys. A, **718** (2003)
6. Del Zoppo, A., et al. Nucl. Instrum. Methods Phys. Res. A, **581** (1996)
7. Cardella, G., et al. Nucl. Instrum. Methods Phys. Res. A, **378** (1996)
8. Mascali, D., et al. Radiat. Eff. Defect. S, **165** (2010)
9. Mascali, D., et al. Proc. Int. Astron. Union, **6** (2010)
10. Ivanovski, S. L. et al., Radiat. Eff. Defect. S, **165** (2010)
11. Mascali, D., et al. Eur. Phys. Lett., **100** (2012)
12. Tudisco, S., et al. Nucl. Instrum. Methods Phys. Res. A, **653** (2011)
13. Gambino, N., et al. Appl. Surf. Sci., **272** (2013)
14. Tudisco, S., et al. Rev Sci Instrum., **87** (2016)
15. Gambino, N., et al. Radiat. Eff. Defect. S, **165** (2010)
16. Malferrari, L., et al. J.Nanotechnol., (2009) 149691
17. Lanzalone, G., et al. Rev. Sci. Instrum., **87** (2016)
18. Cristoforetti, G., et al. Plasma Phys. Contr. F. Journal, **56** (2014)
19. Mascali, D., et al. Rev. Sci. Instrum., **85** (2014)
20. Sciacca, E., et al. IEEE Photonics Technol. Lett., **18** (2006)
21. Mazzillo, M., et al. Sens. Actuators A Phys., **138** (2007)
22. Privitera, S., et al. Sensors, **8** (2008)
23. Muoio, A., et al. EPJ Web of Conferences **117** (2016)
24. Schillaci, F., et al. J. Instrum., **9** (2014)
25. Muoio, A., et al. Nucl. Instrum. Methods Phys. Res. A, **829** (2016)
26. Altana, C., et al. Rev. Sci. Instrum., **87** (2016)
27. Finocchiaro, P., et al. IEE Trans. Nucl. Sci., **52** (2005)
28. Muoio, A., et al. J. Instrum., **12** (2017)

EPJ Web of Conferences **165**, 01003 (2017) DOI: 10.1051/epjconf/201716501003
NPA8 2017

Nuclear physics and astrophysics experiments at ELI-NP: The emerging future

Dimiter L. Balabanski[1,*]

[1]Extreme Light Infrastructure – Nuclear Physics, Horia Hulubei National Institute for R&D in Physics and Nuclear Engineering, Reactorului Str. 30, 077125 Bucharest – Magurele, Romania

Abstract. The status of implementation of the ELI-NP high-power laser system and the high-brilliance gamma beam system is reported. The emerging experimental program at the facility in nuclear physics in astrophysics is discussed, with emphasis of the considered day-one experiments.

1 The ELI-NP Facility

The main research tools at Extreme Light Infrastructure Nuclear Physics (ELI-NP) facility are a high-power laser system (HPLS) and a high-brilliance gamma-beam system (GBS) [1]. They are superior to what is available at present in research laboratories worldwide. The mission of the facility, which will become operational in 2019 as an open access user facility, is cutting-edge research in the field of nuclear photonics. The HPLS has central wavelength of 815 nm, reaching 10 PW peak power in pulses of duration below 22 fs, with a contrast of 10^{13}:1 at 100 ps and 10^{9}:1 at 10 ps ahead of the pulse peak. It has two laser arms, operated by a common front-end. Each laser arm has three optical compressors, at 10 PW, 1 PW and 100 TW. The 10 PW outputs can provide pulses once per minute, the 1 PW outputs operate at 1 Hz and the 100 TW outputs – at 10 Hz. The two arms will be synchronized and experiments with any combination of pulses will be possible [2,3]. The HPLS system and the support target laboratory are placed in clean room areas.

The ELI-NP GBS will provide beams in the range between 200 keV and 19.5 MeV with a bandwidth better than 0.5%, spectral density of about 10^4 photons/(eV·s) and linear polarization higher than 95%. The gamma beam will be produced by inverse Compton scattering of 515 nm green laser light, provided by 100 Hz Yb:YAG lasers, off electron beams accelerated to relativistic energies by a warm linac consisting of S-band photo-injector and C-band acceleration cavities [4]. The electrons will interact with the laser beams at two interaction points, at low and high electron beam energies, producing low (200 keV to 3.5 MeV) and high energy (3 MeV to 19.5 MeV) gamma beams. The gamma beams will be delivered along two beam lines, placed after each interaction point. The electron beam will be bunched with a frequency of 100 Hz, and each bunch will consist of a train of 32 microbunches. Laser re-circulators will be used at the interaction points to

* Corresponding author: dimiter.balabanski@eli-np.ro

 DOI: 10.1051/epjconf/201716501003

ensure the interaction of the laser pulse with each bunch in the train. The gamma-beams will be collimated, cutting the low energy γ-rays, for achieving narrow bandwidth.

The ELI-NP HPLS and GBS are placed on an anti-vibration slab. The latter is separated from the side walls and the roof of the laboratory and it is placed on a system of springs and dampers, mounted on support columns in the basement of the laboratory. The whole structure is designed as a 150,000 ton optical table. It is specified to have the power spectral density of the vibrations at the level of 10^{-10} g^2/Hz in a frequency range between 1 Hz and 200 Hz. With this parameter achieved, the expected pointing fluctuations of the system shall remain below 3 μrad root mean square. For avoiding the long-term drift of the pointing, the temperature in the laboratory building is kept at 22 ºC, stabilized to ± 0.5 ºC, which is achieved by a 5 MW cooling/heating system based on geothermal water. The water is provided from a water layer at 120 m depth through 1000 wells, which are dig within the territory of the lab. After processing, the water is pumped back in the underground water layer. The building was commissioned in September 2016 and the measurements confirmed the stability of the floor and the temperature.

Several types of experiments will be possible at ELI-NP, *e.g.* laser-driven experiments in single or double pulse-shot mode on target, gamma-beam experiments in narrow- or wide-bandwidth mode, and combined laser- and gamma-beam experiments. Thus, the ELI-NP laboratory opens a new dimension for nuclear physics studies with intense electromagnetic probes. The experimental program, which is under preparation at ELI-NP, targets all these experimental modes and a large variety of instruments are under construction for its realization [2,5,6]. In this paper the emerging experimental program in the field of nuclear physics in astrophysics is described, with an emphasis of the considered day-one experiments.

2 Nuclear physics in astrophysics research at the ELI-NP HPLS

The availability of high-power laser pulses with intensities reaching 10^{23} W/cm^2, which corresponds to light pressures of 10^{13} Tor, is expected to open new regimes for laser-driven particle acceleration. Day-one experiments, after commissioning the laser beams, will target proton acceleration up to 200 MeV through laser-matter interaction on ≤ 1 μm CH targets. On a longer term the program will address acceleration of dense ion beams on targets of increasing *Z*, aiming at accelerating actinide beams. It is expected that at such extreme light pressures, the radiation pressure acceleration (RPA) mechanism will become possible. Solid targets, which are foils of nanometre thickness, will be used in experiment. There, as a result of the laser-matter interaction, cold electron compression is achieved in front of the laser pulse, which accelerates the ions via the created dipole electrostatic field. In this way, solid-state density ion bunches, scaling with the Avogadro number, *e.g.* beams of up to 10^{22} ions/pulse are formed [7], many orders of magnitude more dense then the bunches provided by present-day accelerators. Thus, due to the light pressure, the part of the foil which is hit by the laser pulse is accelerated as a whole, like a "light sail", and a good fraction of the laser energy is converted in accelerating ions up to GeV energies.

The nuclear physics program at the ELI-NP HPLS addresses studies of nuclear reactions of astrophysics interest in a controlled plasma environment, which will help understanding the screening effect, *i.e.* the reaction cross section variation due to the presence of the electron cloud around nuclei in hot plasma conditions [8]. One laser beam will be used to create a secondary plasma target with controlled temperature and density, while another laser pulse will accelerate ions from a primary target that will impinge on short time scale, up to 1 ps, on the rapidly expanding plasma target. Plasma environment is expected to produce changes also in other nuclear observables, such as decay modes and half-lives of unstable states. Large number of free electrons, photons and highly charged ions encountered in high-temperature high-density plasmas may induce nuclear excitations and de-excitations at high rate, which become possible through new mechanisms, such as nuclear excitation through electron capture (NEEC) or through electron transition (NEET), schematically shown in Fig. 1. The realization of these processes in hot plasmas might result in changes by many orders of magnitude in half-lives of some states, as have been estimated for ^{93m}Mo and ^{26gs}Al [8], which are considered as candidates for day-one studies.

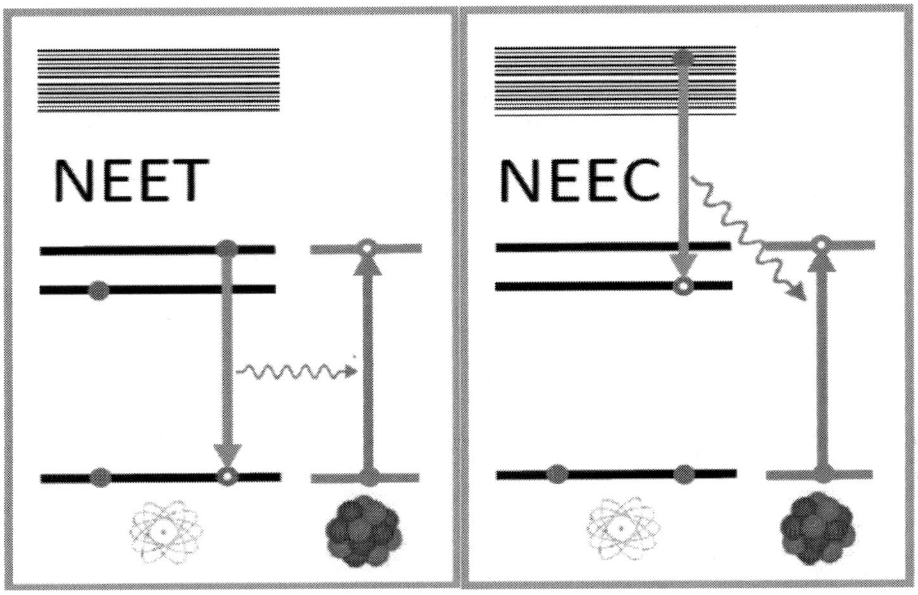

Fig. 1. (colour online) Schematic presentation of the nuclear excitation through electron transition (NEET) process (left) and the nuclear excitation through electron capture (NEEC) process (right).

While indications of the NEET process have been reported in cold targets at 10^{-8} rates in ^{197}Au, ^{189}Os, and ^{183}Ir [9-11], experimental observations of the NEEC process have never been reported.

On a different note, the origin of the screening effect has been disputed a lot over the last 20 years in an attempt to understand the need of exceedingly high values of the screening potential needed to explain the data. Recently it was suggested that the reason for the large screening potential values is in fact due to clusterization effects in nuclear reactions, in particular for reaction involving light nuclei [12]. Experiments at ELI-NP will shed light on this issue.

On a longer term, this program aims at acceleration of dense beams of actinide nuclei at energies above the Coulomb barrier for nuclear reactions, and investigation of the mechanism of the fission-fusion reaction [13]. In the first stage of the reaction, the accelerated actinide nuclei impinge on a secondary target and fission. However, the density of fission fragments is large enough to result in a subsequent fusion. The fusion of light fission fragments would provide a possibility for studies of the structure of nuclei in the vicinity of the $N = 126$ *r*-process waiting point.

3 Nuclear physics in astrophysics research at the ELI-NP GPS

At the ELI-NP GBS the probe for nuclear studies is a photon with well-defined energy and polarization. A number of photonuclear reactions will be possible with these gamma-beams, which are displayed schematically in Fig. 2, namely nuclear resonance fluorescence

(NRF) studies, photo-neutron reactions and photo-charged-particle reactions, studies of large collective motions in nuclei, photo-activation and photo-fission.

The availability of high-brilliance, narrow-bandwidth, pencil-size, polarized ELI-NP beams will result in unique NRF measurements [14]. A most appealing case, which can be done at the low-energy gamma beam line already in the commissioning phase, is the photo-activation measurement of ^{180m}Ta, *i.e.*, a measurement of the subsequent γ decay as a function of the gamma-beam energy. Doorway states in ^{180m}Ta, through which the $I^{\pi} = 9^{-}$, $T_{1/2} > 1.2 \cdot 10^{15}$ year isomer is de-excited, have been studied in bremsstrahlung experiments [15]. This experiment is closely related to nuclear astrophysics research, since the detailed knowledge of the doorway states and the flux which passes through them, provide a sensitive thermometer for studying the star conditions during the nucleosynthesis. Later, on day-one, NRF studies with actinide targets can be performed, which become possible due to the high-brilliance and small spot size of the beam, which require milligram targets to be used.

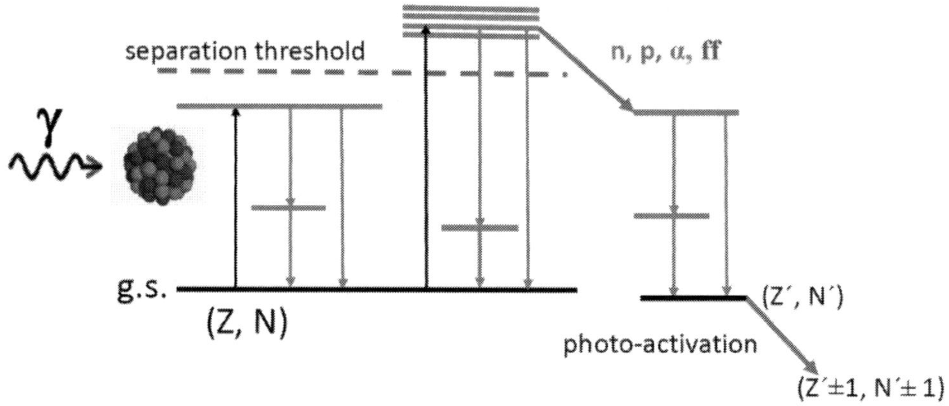

Fig. 2. (colour online) Schematic presentation of the photonuclear processes which can be studied in experiments at the ELI-NP GBS. The decay of excited states below the particle separation threshold (right) is studied with the nuclear resonance fluorescence method. Studies of the decay of excited states above the separation threshold (middle) are related to large-amplitude collective motions, which can be carried out either through neutron emission, or through gamma decay. Studies of charged-particle reactions and photo-fission are also indicated, as well follow-up photo-activation studies in daughter nuclei (right).

Studies of photo-disintegration processes at ELI-NP are related to the Big-Bang nucleosynthesis (BBN), He-burning in stars, and *p*- and *s*- nucleosynthesis processes. The ^{7}Li(γ,t)α reaction addresses two aspects, which need further insight, *e.g.* experimental data below 1.5 MeV disagree with theoretical calculations, and measurements at higher energies can be used for restricting the extrapolation to energies relevant to astrophysics. A first measurement of this reaction has been done recently at the HIγS gamma-beam facility at Duke University, USA [16], the continuation of this study is considered as a day-one experiment at ELI-NP.

In massive star evolution and stellar explosive sites (Type Ia, II, and Ib/c Supernovae), the astrophysical *p*-process is an important way of nucleosynthesis to produce the proton-rich, stable nuclides beyond Fe which cannot be reached by the *s*- and *r*-processes. The common picture is that these nuclides are synthesized by photodisintegration of pre-existing *s*- and *r*-processes nuclei. The dominant nuclear flows of *p*-process go towards the neutron-deficient region through (γ,n) reactions. At present, systematic reaction network calculations are been carried out [17], in order to define the best cases to be explored. For example, astrophysical simulations show that the nucleosynthesis of ^{92,94}Mo and ^{96,98}Ru are systematically under-produced by benchmarking of the predicted abundances of these nuclei to their corresponding abundances observed in solar system composition. Therefore, it is expected that the experimental determination of the astrophysical reaction rates around Mo and Ru would remedy this unsolved problem of *p*-process nucleosynthesis.

The ^{16}O(γ,α)^{12}C reaction, the inverse of ^{12}C(α,γ)^{16}O reaction, is also suggested to be studied at ELI-NP [18]. Modelling the evolution and explosion of massive stars requires a 10% uncertainty on the ^{12}C(α,γ)^{16}O reaction cross section, while a recent analysis of the world data estimated this uncertainty at around 20% [19]. The goal of the measurements at ELI-NP will be to measure detailed cross sections and angular distribution of the reaction products, thus obtaining accurate values of the E2/E1 ratio. One of the advantages of measuring the photo-dissociation of ^{16}O is a gain in cross section due to the detailed balance theorem. The majority of ^{12}C(α,γ)^{16}O experimental and theoretical work over the last 60 years have concentrated in the energy region below 10 MeV. However, measurements at higher energies, all the way to 14 MeV, would allow to include more states in the R-matrix analysis and to reduce the uncertainty of the extrapolation at lower energies [20]. The aim of such measurements at ELI-NP would be to determine angular distributions and cross sections between known resonances with statistical precision of 3% or better. For example, a measurement of the weakest cross section between the 2^+ states at 9.84 and 11.52 MeV will require twelve hours of beam time for each experimental point.

Experiments at and above the neutron separation threshold will aim at detailed studies of E1 strength in atomic nuclei [21], which will provide reliable estimates of nuclear polarizability, an observable, which some theories relate to the symmetry term of the nuclear equation of state (EoS) [22-24]. In addition, detailed studies of the gamma-decay of the GDR both to the ground and excited states will be possible, which will set constraints on different theoretical models. The ELI-NP experimental program will provide high-resolution data of the distribution of the E1 and M1 strength in nuclei using a γ-ray probe, which is complimentary to the results obtained in (p,p′) and (α,α′) measurements. As a first experiment the cross section of ^{9}Be(γ,n) reaction can be studied at γ-beam energies between 1.65 MeV and 3.20 MeV. This cross section has been measured several times. However, there is considerable (> 20%) discrepancy between the recently reported results [25,26] and an independent measurement might resolve this discrepancy in early ELI-NP operation.

Another area of interest is related to neutron-capture rates, which can be determined by inverse photodisintegration and photoexcitation studies [27,28]. Such measurements will be pursued for many of the critical *s*-process branching point nuclei, which will help to determine the neutron flux and the temperature conditions at the *s*-process sites.

Finally, photo-fission experiments will aim at studies of the landscape of the three-dimensional multi-humped fission barrier in the actinide nuclei. This will be done through studies of transmission resonances [29]. Within this approach, a state in the first minimum of the nuclear potential is excited and couples to states in the second or third minima of the potential barrier. This coupling results in resonance states in the fission cross section in the region 5 MeV to 7 MeV [30]. The experiments will aim at the characterization of these transmission resonances through the measurement of angular distributions, and the identification of the fission fragments.

4 Conclusions

A diverse research program, which addresses different topics of nuclear physics in astrophysics research will be carried out at ELI-NP. Cases of flagship experiments were defined, and certain low-hanging fruits, which should be measured first at the start of the facility, have been identified. Experiments at the ELI-NP HPLS will aim at improving our knowledge on nuclear reactions in a plasma environment and at understanding the electron screening effect and the $N = 126$ waiting point of the r-process. Experiments at the ELI-NP GBS will shed light to open problems related to the BBN, the evolution of massive stars and explosive sites, and to a better understanding of the s- and p- processes.

This work is supported by the Extreme Light Infrastructure Nuclear Physics (ELI-NP) Phase II, a project co-financed by the Romanian Government and the European Union through the European Regional Development Fund – the Competitiveness Operational Programme (1/07.07.2016, COP, ID 1334).

References

1. N. V. Zamfir, Nucl. Phys. News **25:3**, 34 (2015).
2. 2. S. Gales *et al.*, Phys. Scr. **91**, 093004 (2016).
3. N. V. Zamfir, Eur. Phys. J. Special Topics **223**, 1221 (2014).
4. O. Adriani *et al.*, arXiv:1407.3669 [physics.acc-ph] (2014).
5. Rom. Rep. Phys. **68** Supplement, *ELI-NP Technical Design Reports* (2016).
6. D. L. Balabsnki *et al.*, Europhys. Lett. **117**, 28001 (2017).
7. A. Macchi, M. Borghesi, M. Passoni, Rev. Mod. Phys. **85**, 751 (2013).
8. F. Negoita *et al.*, Rom. Rep. Phys. **68**, S37 (2016).
9. S. Kishimoto *et al.*, Phys. Rev. Lett. **85**, 1831 (2000).
10. I. Ahmad *et al.*, Phys. Rev. C **61**, 051304 (2000).
11. S. Kashimoto *et al.*, Nucl. Phys. A **748**, 3 (2005).
12. C. Spitaleri *et al.*, Phys. Lett. B **755**, 275 (2016).
13. D. Habs *et al.*, Appl. Phys. B **103**, 471 (2011).
14. C. A. Ur *et al.,* Rom. Rep. Phys. **68**, S483 (2016).
15. D. Belic *et al.*, Phys. Rev. C **65**, 035801 (2002).
16. C. Matei *et al.*, Phys. Rev. C (in preparation).
17. H. Y. Lan *et al.*, Nucl. Inst. Meth. Phys. Res. A (in preparation).
18. O. Tesileanu et al. Rom. Rep. Phys. **68**, S699 (2016).
19. R. J. deBoer *et al*, Rev. Mod. Phys. **89**, 035007 (2017).
20. D.B. Sayre *et al.*, Phys. Rev. Lett. **109**, 142501 (2012).
21. F. Camera *et al.*, Rom. Rep. Phys. **68**, S539 (2016).
22. P.-G. Reinhardt, W. Nazarewicz, Phys. Rev. C **81**, 051303 (2010).
23. J. Piekarewicz, Phys. Rev. **83**, 034319 (2011).
24. X. Roca-Maza *et al.*, Phys. Rev. Lett. **106**, 252501 (2011).
25. H. Unsunomiya *et al.*, Phys. Rev. C **92**, 064323 (2015).
26. C. W. Arnold *et al.*, Phys. Rev. C **85**, 044605 (2012).
27. R. Raut *et al.*, Phys. Rev. Lett. **111**, 2501 (2013).
28. A. Sauerwein *et al.*, Phys. Rev. C **89**, 035803 (2014).
29. D. L. Balabanski *et al.*, Rom. Rep. Phys. **68**, S621 (2016).
30. A. Krasznahorkay, in: *Handbook of Nuclear Chemistry* (Springer Verlag, Dordrecht, 2011), p. 281.

EPJ Web of Conferences **165**, 01004 (2017) DOI: 10.1051/epjconf/201716501004
NPA8 2017

Asymptotic normalization coefficients and halo radii of ^{12}B in the excited states

T.L. Belyaeva[1], *S.A.* Goncharov[2], *A.S.* Demyanova[3], *A.A.* Ogloblin[3], *A.N.* Danilov[3], *V.A.* Maslov[4], *Yu.A.* Sobolev[4], *W.* Trzaska[5], *S.V.* Khlebnikov[6], *G.P.* Tyurin[6], *N.* Burtebaev[7], *D.* Janseitov[7], and *E.* Mukhamejanov[8]

[1] *Universidad Autónoma del Estado de México, C. P. 50000, Toluca, México*
[2] *Lomonosov Moscow State University, GSP-1, Leninskie Gory, Moscow 119991, Russia*
[3] *NRC Kurchatov Institute, Moscow RU-123182, Russia*
[4] *Flerov Laboratory for Nuclear Research, JINR, 141980, Dubna, Moscow region, Russia*
[5] *Department of Physics, University of Jyväskylä, FIN-40014 Jyväskylä, P.O. Box 35, Finland*
[6] *V. G. Khlopin Radium Institute, 194021, St. Petersburg, Russia*
[7] *Institute of Nuclear Physics, National Nuclear Center of Republic of Kazakhstan, Almaty, 050032, Republic of Kazakhstan*
[8] *al-Farabi Kazakh National university, Almaty, 050040, Republic of Kazakhstan*

Abstract. We present the results of measurements and analysis of the differential cross sections of the $^{11}B(d,p)^{12}B$ reaction leading to formation of the 1^+ ground state and the 0.953-MeV 2^+, 1.674-MeV 2^-, 2.621-MeV 1^-, 2.723-MeV 0^+, 3.389-MeV 3^- excited states of ^{12}B at E_d = 21.5 MeV. The analysis of the data was carried out within the coupled-reaction-channels method for the direct neutron transfer and the Hauser-Feshbach formalism of the statistical compound-nucleus model. We deduced the spectroscopic factors, asymptotic normalization coefficients, and rms radii of the last neutron in all states studied. The existence of the neutron halos in the 1.674-MeV 2^- and 2.621-MeV 1^- states was found in consistence with the earlier published data. New information about the enlarged rms radii (6.5 fm) of the last neutron in the unbound 3.389-MeV 3^- states of ^{12}B was obtained, which may indicate the evidence of the neutron halo with the orbital momentum of the last neutron equal to two.

1 Introduction

The neutron- and proton-transfer reactions with stable and radioactive beams in the traditional and inverse kinematics are justifiably regarded as an important source of spectroscopic and astrophysical information. An important application of the nucleon-transfer reactions is related to the determination of the last neutron and proton radii, including radii of nuclei in the short-lived excited states [1, 2]. The evidence of enlarged radii of nuclei in the excited states found by different methods [3–5] indicates the existence of neutron halos not only in the ground states of exotic nuclei, but also in the excites states of "normal" nuclei.

2 Results

The differential cross sections of d+ ^{11}B elastic scattering and the $^{11}B(d, p)^{12}B$ (g.s., 0.95, 1.67, 2.62, 2.72, 3.39 MeV) reaction were measured at incident deuteron energy E_{lab} = 21.5 MeV in the angular range (~ 5° – 85°) at Jyvaskyla University cyclotron using the Large Scattering Chamber (LSC).

The experimental elastic-scattering angular distribution was fitted with the optical potentials of the standard Woods-Saxon form, which included the real, spin-orbital, and imaginary (surface) components. The parameters were chosen on the base of the global parameterization. Calculations were carried out with code FRESCO [6].

The compound-nucleus (CN) analysis of the $^{11}B(d, p)^{12}B$ differential cross sections was carried out within the statistical Hauser-Feshbach formalism by using the computer code CNCOR [7]. It was found that the CN mechanism provides less than 0.1% of the cross sections at forward angles and about 1-3% at medium angles 60-80°.

The coupled-reaction-channels (CRC) analysis included the finite-range neutron transfer mechanism. The neutron single-particle (sp) overlap wave function in the deuteron with orbital angular momentum l = 0 was chosen from Ref. [8]. The normalized sp overlap ^{11}B+n wave functions were generated by the ^{11}B+n interaction potential for each state of ^{12}B.

Figure 1 shows the results of the CRC (dashed lines) and CN calculations (dotted lines) in comparison with the measured differential cross sections of the $^{11}B(d, p)^{12}B$ reaction populating the 2.723-MeV 0^+ and 3.389-MeV 3^- states. The solid curves represent the incoherent sum of the CN model and direct transfer calculations.

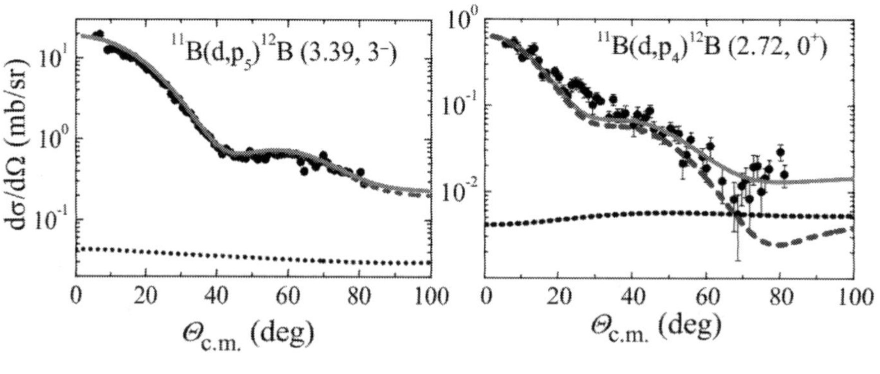

Figure 1. Differential cross sections of the $^{11}B(d, p)^{12}B$ reaction leading to the fourth excited 2.723-MeV 0^+ state and the fifth excited 3.389-MeV 3^- state of ^{12}B measured in the present work (points) in comparison with the CRC (dashed line), CN (dotted line) calculations, and their sum (solid line).

Table I shows the deduced spectroscopic factors (SFs), the neutron ANCs and the rms neutron radii in comparison with the results obtained by the DWBA analysis of this reaction in Refs. [1, 9, 10]. We found that the rms radii of the last neutron in all excited states studied are greater than that in the g.s. Thus for the 2^- state, the excess is a factor of 1.55, and for the 1^- state, it is a factor of 2.05, with respect to the rms radius of the g.s. The rms radius of the last neutron wave function of the fifth 3.389-MeV 3^- excited state was found R_{rms} = 6.5 fm. This value is a factor of 1.8 larger than that

 DOI: 10.1051/epjconf/201716501004

of the g.s. and exceeds considerably the rms radii of the neutron wave function in the 1.674-MeV 2^- state of ^{12}B.

Table 1. Summary of the neutron spectroscopic factors, asymptotic normalization coefficients, rms radii, D_1 and D_2 coefficients for the states of ^{12}B.

E_x (MeV), J_f^π	n_2l_2	$S_{n_2l_2}^{\rm expt}$	$\left(C_{^{11}{\rm B}n}^{\rm expt}\right)^2$ (fm^{-1})	$R_{\rm rms}$ (fm)	D_1 %	D_2 %	Ref.
0.0, 1^+	1 1		1.35 ± 0.23	3.16 ± 0.32	19.9	70.2	[1]
		0.69					[9, 10]
		0.67	1.33 ± 0.07	3.6 ± 0.2	11	58	this work
1.674, 2^-	2 0		1.80 ± 0.43	4.01 ± 0.61	53.6	91.9	[1]
		0.57					[9, 10]
		0.36	1.57 ± 0.15	5.58 ± 0.26	48.5	91	this work
2.621, 1^-	2 0		0.88 ± 0.15	5.64 ± 0.90	66.8	96.3	[1]
		0.75					[9, 10]
		0.625	1.11 ± 0.10	7.40 ± 0.35	70	97	this work
3.389, 3^-	2 2	0.5					[9, 10]
		0.251		6.5 ± 0.3	47	94	this work

Coefficients $D_1(R_N)$ determining the weight of the asymptotic part of the wave function and D_2 estimating the contribution of the asymptotic part of the wave function to the rms radius define a probability of the halo-nucleon to be outside the range of the core potential, which is expected to be more than 50% for halo states.

It is evident that the 2.621-MeV 1^- state satisfies all criteria of the halo state with the enormous rms neutron radius, $R(1^-) = 7.4 \pm 0.35$ fm, and D_1 and D_2 coefficients equal to 70% and 97%, respectively. The 1.674-MeV 2^- state apparently also can be considered as the halo state with the last neutron spending about 50% of its time outside the range of the core potential. The neutron rms radius in the 3.389-MeV 3^- excited state was found equal to 6.5 fm that is 1.16 larger than that for the 1.674-MeV 2^- state. The D_1 and D_2 are almost the same as for the 2^- state. Thus, we can suggest that ^{12}B in the 3.389-MeV 3^- excited state also possesses the neutron halo. Note that this observation reveals the first halo excited state with a non-zero $l_2 = 2$ orbital momentum of the last neutron.

References

[1] Z. H. Liu *et al.*, Phis. Rev C **64**, 034312 (2001).
[2] T. L. Belyaeva *et al.*, Phys. Rev. C **90**, 064610 (2014).
[3] T. Otsuka, N. Fukunishi, and H. Sagawa, Phys. Rev. Lett. **70**, 1385 (1993).
[4] A. A. Ogloblin *et al.*, Phys. Rev. C **84**, 054601 (2011).
[5] A. A. Ogloblin *et al.*, Nuclear size isomers: The excited states of light nuclei with cluster structure and nonstandard sizes, in *Nuclear Particle Correlations and Cluster Physics*. (World Scientific, Pekin, 2017) pp. 311-338.
[6] I. J. Thompson, FRESCO user's manual and code, available from the author.
[7] T. L. Belyaeva, N. S. Zelenskaya, and N. V. Odintsov, Comput. Phys. Commun. **70**, 161 (1992).
[8] H. Esbensen, G. F. Bertsch, and K. A. Snover, Phys. Rev. Lett. **94**, 042502 (2005).
[9] J. E. Monhan, H. T. Fortune, C. M. Vincent, R. E. Segel, Phys. Rev. C **3**, 2192 (1971).
[10] H. Y. Lee *et al.*, Phys. Rev. C **81**,015802 (2010).

EPJ Web of Conferences **165**, 01005 (2017)
NPA8 2017

DOI: 10.1051/epjconf/201716501005

Studies of X-ray burst reactions with radioactive ion beams from RESOLUT

J. C. Blackmon[1,⋆], *I.* Wiedenhöver[2], *J.* Belarge[2], *S. A.* Kuvin[2], *M.* Anastasiou[2], *L. T.* Baby[2], *J.* Baker[2], *K.* Colbert[3], *C. M.* Deibel[1], *H. E.* Gardiner[1], *D. L.* Gay[3], *E.* Good[1], *P.* Höflich[2], *A. A. D.* Hood[1], *N.* Keely[4], *J.* Lai[1], *A.* Laminack[1], *L. E.* Linhardt[1], *J.* Lighthall[1,2], *K. T.* Macon[1], *E.* Need[1], *N.* Quails[3], *B. C.* Rasco[1], *N.* Rijal[2], and *A.* Volya[2]

[1]*Louisiana State University, Baton Rouge, LA 70803 USA*
[2]*Florida State University, Tallahassee, FL 32306 USA*
[3]*University of North Florida, Jacksonville, FL 32224 USA*
[4]*National Center for Nuclear Research, Otwock, Poland*

Abstract. Reactions on certain proton-rich, radioactive nuclei have been shown to have a significant influence on X-ray bursts. We provide an overview of two recent measurements of important X-ray burst reactions using in-flight radioactive ion beams from the RESOLUT facility at the J. D. Fox Superconducting Accelerator Laboratory at Florida State University. The ^{17}F(d,n)^{18}Ne reaction was measured, and Asymptotic Normalization Coefficients were extracted for bound states in ^{18}Ne that determine the direct-capture cross section dominating the ^{17}F(p,γ)^{18}Ne reaction rate for $T \lessapprox 0.45$ GK. Unbound resonant states were also studied, and the single-particle strength for the 4.523-MeV (3^+) state was found to be consistent with previous results. The ^{19}Ne(d,n)^{20}Na proton transfer reaction was used to study resonances in the ^{19}Ne(p,γ)^{20}Na reaction. The most important 2.65-MeV state in ^{20}Na was observed to decay by proton emission to both the ground and first-excited states in ^{19}Ne, providing strong evidence for a 3^+ spin assignment and indicating that proton capture on the thermally-populated first-excited state in ^{19}Ne is an important contributor to the ^{19}Ne(p,γ)^{20}Na reaction rate.

1 Introduction

X-ray bursts occur in binary systems when hydrogen-rich matter from a main-sequence star accretes onto a neutron star and ignites under degenerate conditions. These are the most common stellar explosions, with over 100 systems known to exhibit bursts recurring with timescales typically from hours to days. Comparisons of observations of many bursts (e.g. see [1]) to improved computational simulations using a variety of astrophysical models (e.g. see [2]) are providing insights into the composition, dynamics, and evolution of these systems. Simulations also show that certain nuclear reactions involving proton-rich, radioactive nuclei have a direct impact on energy generation, nucleosynthesis, and the resulting light curve [3, 4]. The rates of many of these reactions are purely theoretical or have large uncertainties due to experimental challenges in studying short-lived nuclei that hinder our understanding of X-ray bursts.

⋆e-mail: blackmon@lsu.edu

Some of the important reactions in X-ray bursts are those that involve breakout of the hot-CNO cycle, either through the ^{17}F(p,γ)^{18}Ne(α,p)^{21}Na or ^{15}O(α,γ)^{19}Ne(p,γ)^{20}Na reaction sequences. We measured the ^{17}F(d,n)^{18}Ne and ^{19}Ne(d,n)^{20}Na reactions to improve our understanding of the ^{17}F(p,γ)^{18}Ne and ^{19}Ne(p,γ)^{20}Na capture reactions, respectively, that are important in breakout from the hot-CNO cycle. Both measurements were conducted at the J. D. Fox Superconducting Accelerator Laboratory at Florida State University using "in-flight" radioactive ion beams. Stable beams of ^{16}O and ^{19}F bombarded a cryogenic gas cell to produce ^{17}F and ^{19}Ne from the ^{16}O(d,n)^{17}F and ^{19}F(p,n)^{19}Ne reactions, respectively. The radioactive products were collected and separated by the RESOLUT facility [5]. The secondary radioactive beams bombarded a deuterated polyethylene target, and a suite of detector systems was used to detect neutrons, gamma rays, light charged particles and heavy ions.

2 ^{17}F(p,γ)^{18}Ne

The ^{17}F(p,γ)^{18}Ne reaction rate is determined mainly by contributions from direct capture and a single 3^+ resonance at E_{cm} = 600 keV. The 600-keV resonance strength was directly measured at the Holifield Radioactive Ion Beam Facility [6], but the direct-capture contribution that dominates the reaction rate at $T \lessapprox 0.45$ GK remains uncertain. A measurement of the ^{17}F(p,γ)^{18}Ne direct-capture cross section at $E_{cm} < 600$ keV would require a ^{17}F beam intensity of about 10^9s^{-1}. Such intensities are beyond the reach of current facilities, and indirect techniques must be applied to better constrain the direct-capture cross section.

We measured the ^{17}F(d,n)^{18}Ne proton-transfer reaction to study states in ^{18}Ne that are important for the ^{17}F(p,γ)^{18}Ne reaction rate, including bound states that determine the direct-capture cross section [7]. A 95.5-MeV beam of ^{17}F from RESOLUT with an intensity of about 3×10^4 s^{-1} bombarded a 0.52 mg/cm^2 CD_2 target. Neutrons from the (d,n) reaction were detected at $\theta_{\text{lab}} = 145° - 165°$ using the RESONEUT array of P-terphenyl scintillators that provided good $n - \gamma$ discrimination for energies greater than about 50 keV$_{\text{ee}}$ [8]. All heavy ions (beam particles and heavy reaction products) were detected using a fast-counting, position-sensitive gas ionization detector covering $\theta_{\text{lab}} < 6°$ that determined the atomic number through relative energy loss and measured the position (in 2 dimensions perpendicular to the beam axis) with a resolution of better than 3 mm [9]. Gamma rays emitted from bound, excited states were detected with a total efficiency of 12% using an array of 20 NaI(Tl) scintillators [10] arranged in a barrel configuration subtending $\theta_{\text{lab}} = 35° - 100°$. The triple coincidence of neutrons, γ rays, and heavy ions identified by atomic number provided clean selection of the population of bound states in ^{18}Ne except for the ground state, which is expected to be weakly populated.

The relative timing between the neutrons and the accelerator *rf* signal was used to determine the E_x of states in ^{18}Ne by time-of-flight with a corresponding energy resolution of about $\Delta E_{cm} \approx$ 150 keV. For a triple coincidence of $n - \gamma -{}^{18}$Ne, we observe two peaks in the time-of-flight spectrum corresponding to the 1.887-MeV (2^+) state and to the $4^+ - 0^+ - 2^+$ triplet of states at about 3.5 MeV. While the triplet is not resolved, population of the 3.57-MeV (0^+) state is expected to be very weak. Assuming the 0^+ to be negligible, we determined the cross sections for population of the 1.887-MeV (2^+), 3.376-MeV (4^+) and 3.616-MeV (2^+) states from a combined fit to the neutron time-of-flight spectrum (see Fig. 13 in [7]). These cross sections (corresponding to very forward angles for θ_{cm}) determine Asymptotic Normalization Coefficients for the states.

Direct capture is expected to be dominated by $\ell = 0$ capture to the 2^+ states in ^{18}Ne below the proton threshold. We are unable to constrain the relative population of $s_{1/2}$ and $d_{5/2}$ for the 2^+ states from our data due to the limited statistics and angular range of the detected neutrons. Assuming the same relative $s_{1/2}$ and $d_{5/2}$ mixing as in the mirror states in ^{18}O [11], we find $C^2_{s_{1/2}} = 16\pm8$ fm^{-1} for the

 DOI: 10.1051/epjconf/201716501005

1.887-MeV state and $C^2_{s_{1/2}} = 150 \pm 60$ fm^{-1} for the 3.616-MeV state, assuming the reaction to occur at an average energy of 5.53 MeV/u corresponding to the center of the target. Using these values, we calculated the contribution to the ^{17}F(p,γ)^{18}Ne reaction rate from direct capture. We find a 30% larger contribution from direct capture than estimated from properties of the mirror [12], though the results are consistent within uncertainties, which are dominated in our result by a statistical uncertainty of 30% in the cross section for populating the 3.616-MeV state.

We were also able to study resonant states above the proton-separation energy, even though they predominantly decay by proton emission, providing no γ-ray signature and giving the same species of recoiling heavy ion as the beam (^{17}F). We cleanly selected the events of interest in this case by identifying the emitted protons in an annular telescope array of silicon-strip detectors subtending $\theta_{lab} = 8° - 21°$. Correlating the measured protons with the ^{17}F ions in the gas ionization chamber allowed proton-unbound states in ^{18}Ne to be accurately reconstructed from an invariant mass analysis using only the position and energy of the protons and ^{17}F ions, thus inferring information about the outgoing neutron without using data from the RESONEUT neutron detector array. We do find consistent results when requiring coincidence neutron detection with RESONEUT, though the efficiency reduces the statistics by a factor of about 30. The reconstructed excitation spectrum in ^{18}Ne from ^{17}F+p coincidences (see Fig. 8 in [7]) is dominated by one strong resonance corresponding to the 3^+ state in ^{18}Ne at 4.523 MeV that known to be the strongest resonance in the ^{17}F(p,γ)^{18}Ne reaction [6]. The higher statistics achieved without neutron detection allowed the distribution of events in angle/energy to be compared to a FRESCO coupled-channels calculation. We find dominant population of the state through $\ell = 0$ transfer, confirming the 3^+ spin assignment, with a single-particle spectroscopic factor of $C^2S = 0.78 \pm 0.06$ corresponding to a proton partial width of $\Gamma_p = (14.2 \pm 1.1)$ keV, where the uncertainty is dominated by the systematic uncertainties in the analysis. The extracted partial width is about 20% lower than determined from ^{17}F+p elastic scattering [13]. It should be noted that the triple coincidence of neutrons, protons, and ^{17}F ions showed strong population of this resonant state in the neutron time-of-flight spectrum, serving as a check on the time-of-flight analysis for the bound states.

3 ^{19}Ne(p,γ)^{20}Na

The ^{19}Ne(p,γ)^{20}Na reaction rate is dominated by contributions from resonances corresponding to states just above the proton-separation energy ($S_p = 2.19$ MeV) in ^{20}Na. While the structure of ^{20}Na is relatively well established for bound states, properties of proton-unbound levels are uncertain. The level scheme is shown in Figure 1 compared to the isospin-mirror nucleus ^{20}F, where we have included spin-parity (J^π) information only when unambiguous. There are 6 observed states in ^{20}Na within 1 MeV above the proton-separation energy. Excitation energies are relatively well determined from charge-exchange reactions, but these studies have placed only weak constraints on J^π values due to challenges from unresolved states and increased background above the separation energy, problems exacerbated by difficulties with neon targets [14–16]. Only the 3.00-MeV (1^+) and 3.09-MeV (0^+) states have clear assignments based upon ^{19}Ne+p elastic scattering [17] and their population in allowed β-decay transitions from ^{20}Mg [18, 19]. These states have been identified as mirror levels to the 3.49-MeV and 3.53-MeV states in ^{20}F, with a large Coulomb shift resulting from a substantial $2s_{1/2}$ component in the wavefunction as expected from *sd* shell model calculations.

The other levels at 2.65, 2.85, 2.98 and 3.07 MeV in ^{20}Na have generally been assumed to be mirrors to the only 4 states observed in ^{20}F in a comparable energy range [20]. Only the 2.966-MeV state in ^{20}F has majority *sd* single-particle character, being strongly populated in the ^{19}F(d,p)^{20}F and ^{18}O(^{3}He,p)^{20}F reactions with angular distributions giving a 3^+ assignment [21]. The other 3 states in ^{20}F have more complex configurations. The 2.864-MeV and 3.171-MeV states in ^{20}F are weakly populated in transfer reactions. Angular distributions indicate negative parity for the 2.864-MeV state, with arguments favoring a 3^- assignment [22]. The 3.171-MeV state is populated predominantly by $\ell = 2$ transfer in both ^{19}F(d,p)^{20}F and ^{18}O(^{3}He,p)^{20}F, indicating positive-parity and a likely 1^+ assignment, but composed primarily of $6p - 2h$ core excitations. Finally, the 2.968-MeV state is a high-spin state, likely $J^\pi \geq 4^-$ [23].

Most important is to determine the properties of the first state above the separation energy in ^{20}Na at 2.65 MeV that could dominate the ^{19}Ne(p,γ)^{20}Na reaction rate given the low (E_{cm} = 457 keV) resonance energy. Some arguments favor a 1^+ assignment for this state but with a small single-particle component (see [24] for example). This is also supported an upper limit on the proton-capture resonance strength ($\omega\gamma$ < 16 meV) set from the a ^{19}Ne(p,γ)^{20}Na reaction measurement [25]. However, some information argues otherwise. The lack of observation of the 2.65-MeV state in ^{20}Mg beta-decay strongly favors a forbidden transition [19], and other charge-exchange measurements have argued for a 3^+ assignment [26, 27]. These charge-exchange studies also indicated a tentative 3^- assignment of the 2.85-MeV state in ^{20}Na as a potential mirror to the 2.86-MeV level in ^{20}F, though these states are not well resolved from neighboring states and are weakly populated.

We studied the properties of unbound states in ^{20}Na using the ^{19}Ne(d,n)^{20}Na reaction [28]. The experimental configuration and approach were similar to that in the study of unbound levels in ^{18}Ne. A 86-MeV beam of ^{19}Ne from RESOLUT bombarded a 0.52 mg/cm^2 CD_2 target. Protons were detected and identified in an annular telescope of silicon strip detectors subtending $\theta_{\text{lab}} = 8° - 21°$ while heavy ions were detected using the position-sensitive gas ionization detector covering $\theta_{\text{lab}} < 6°$. The ^{19}Ne beam intensity ($< 2000\ s^{-1}$) was more than an order of magnitude less intense than in the ^{17}F(d,n)^{18}Ne study described previously, which prohibited the coincident detection of neutrons due to the reduced efficiency. As with the ^{18}Ne, however, we are able to reconstruct the unbound states of interest from an invariant mass analysis using the kinematics of the protons and ^{19}Ne ions (see Fig. 1 in [28]) with a center-of-mass energy resolution of about 200 keV (FWHM)

One prominent feature in our reconstructed spectrum is a well-resolved resonance at E_{cm} = 220 keV. There is no known level in ^{20}Na near E_x = 2.41 MeV that would correspond to proton emission to the ground state of ^{19}Ne for this resonance. However, the E_{cm} matches proton decay from the important 2.65-MeV state in ^{20}Na to the first-excited state ($\frac{5}{2}^+$) in ^{19}Ne. Emission of a 220-keV proton strongly argues for $\ell = 0$ angular momentum due to the prohibitive penetrability for larger ℓ values. An $\ell = 2$ proton-decay branch at this energy would be about an order of magnitude weaker than the expected gamma-decay branch and would not be observed. This supports the 3^+ assignment for the 2.65-MeV resonance as indicated from the weak population in beta decay. We also extract an angle/energy distribution for the emitted ^{20}Na (and inferred neutron) and model the ^{19}Ne(d,n)^{20}Na cross section following the procedure of [29]. A fit to the angular distribution using $\ell = 2$ proton-transfer onto the ^{19}Ne ground state (required to populate a 3^+ resonant state) results in $\chi^2/\nu = 1.8$, while a corresponding fit using $\ell = 0$ proton transfer (likely for a 1^+ resonance) results in $\chi^2/\nu = 4.3$. Both the angular distribution and the strength by which the first-excited state is populated in proton emission provides strong evidence supporting the 3^+ assignment for the 2.65-MeV level.

The second prominent feature in our E_{cm} spectrum is a strong resonance at 660 keV, corresponding to proton emission from the 2.85-MeV state in ^{20}Na to the ground state in ^{19}Ne. This resonance has a significant low-energy shoulder best described by two additional weaker resonances at E_{cm} = 440 keV and 540 keV. The 440-keV resonance corresponds to proton emission from the 2.65-MeV state in ^{20}Na to the ^{19}Ne ground state. We find the branching ratio for proton decay from the 2.65-MeV state to the ^{19}Ne ground state to be approximately equal to that for decay to the first-excited state. These equal decay branches imply that the wavefunction for the 2.65-MeV state has a $s_{1/2}$ proton component (coupled to the ^{19}Ne first-excited state) with C^2S = 0.5, about 7 times greater than the $d_{5/2}$ component, which is consistent with the 3^+ mirror state in ^{20}F at 2.966 MeV.

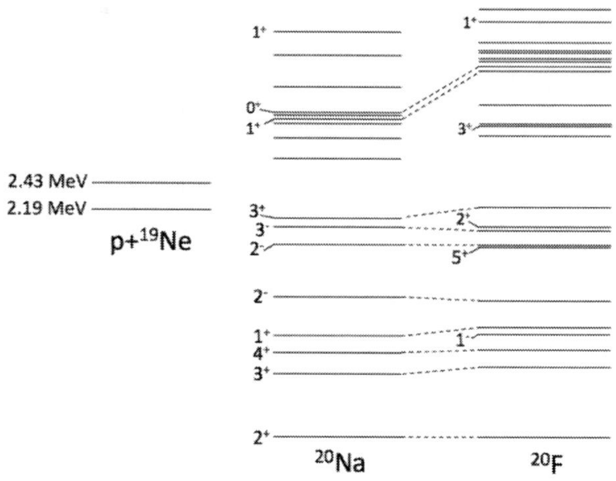

Figure 1. Level scheme of ^{20}Na compared to the mirror nucleus ^{20}F. Figure is adapted from the evaluation of [31]. The ^{19}Ne ground state ($J^{\pi} = 1/2^{+}$) couples to an $\ell = 0$ proton with channel spin $S = 0, 1$ corresponding to $E_x = 2.19$ MeV in ^{20}Na, while the first-excited state in ^{19}Ne ($J^{\pi} = 5/2^{+}$) couples to an $\ell = 0$ proton with channel spin $S = 2, 3$ corresponding to $E_x = 2.43$ MeV in ^{20}Na.

The strong coupling of the 2.65-MeV state to the first-exited state in ^{19}Ne significantly increases the ^{19}Ne(p,γ)^{20}Na reaction rate. The first-excited state in ^{19}Ne is sufficiently low in excitation energy that significant thermal population is possible (for example, a few percent at $T = 1GK$). The lower energy required for proton capture onto the thermally-excited ^{19}Ne $\frac{5}{2}^{+}$ state results in a contribution to the reaction rate that is comparable to that of capture on the ground state. Including this effect, we find a ^{19}Ne(p,γ)^{20}Na reaction rate that is significantly greater than previously estimated.

The strong population of the 2.85-MeV state in our measurement is somewhat harder to interpret. Analyzing angular distributions for the ^{19}Ne(d,n)^{20}Na reaction results in $\chi^2/\nu = 3.5$ for $\ell = 0$ proton transfer and $\chi^2/\nu = 5.1$ for $\ell = 2$ proton transfer, with the $\ell = 0$ fit giving a $s_{1/2}$ proton spectroscopic factor that is consistent only with the the 3.49-MeV (1^{+}) mirror state in ^{20}F ($C^2S = 0.40$) [30]. However, population of the 2.85-MeV state is not observed in ^{20}Mg beta decay, and the decay branch must be on the order of 0.1% [19]. This is very unlikely for an allowed Gamow-Teller transition, and difficult to reconcile with the strong population of the 2.85-MeV state that we observe in ^{19}Ne(d,n)^{20}Na. Further measurements are required to resolve the nature of the 2.85-MeV state, but it is important to note that even with a 1^{+} assignment, the 2.85-MeV state contributes no more than a few percent to the overall ^{19}Ne(p,γ)^{20}Na reaction rate.

This work was partially supported by the National Science Foundation, under Grants No. PHY-1401574, No. PHY-1064819, No. PHY-1126345 and partially supported by the U.S. Department of Energy, Office of Science, Office of Nuclear Physics under Grants No. DE-FG02-02ER41220 and No. DE-FG02-96ER40978.

References

[1] D.K. Galloway, M.P. Muno, J.M. Hartman, D. Psaltis, D. Chakrabarty, The Astrophysical Journal Supplement Series **179**, 360 (2008)
[2] N. Lampe, A. Heger, D.K. Galloway, The Astrophysical Journal **819**, 46 (2016)
[3] A. Parikh, J. José, F. Moreno, C. Iliadis, The Astrophysical Journal Supplement Series **178**, 110 (2008)
[4] R.H. Cyburt, A.M. Amthor, A. Heger, E. Johnson, L. Keek, Z. Meisel, H. Schatz, K. Smith, The Astrophysical Journal **830**, 55 (2016)

[5] I. Wiedenhöver et al., *Studies of Exotic Nuclei at the Resolut Facility of Florida State University*, in *Proceedings, 5th International Conference on Fission and Properties of Neutron Rich Nuclei (ICFN5): Sanibel Island, Florida, USA, November 4-10, 2012* (2013), pp. 144–151
[6] K.A. Chipps, D.W. Bardayan, J.C. Blackmon, K.Y. Chae, U. Greife, R. Hatarik, R.L. Kozub, C. Matei, B.H. Moazen, C.D. Nesaraja et al., Phys. Rev. Lett. **102**, 152502 (2009)
[7] S. Kuvin et al., submitted to Phys. Rev. C (under review)
[8] L. Baby et al., submitted to Nucl. Instrum. Methods Phys. Res. A (in press)
[9] J. Lai et al., submitted to Nucl. Instrum. Methods Phys. Res. A (under review)
[10] N. Kaloskamis, K. Chan, A. Chishti, J. Greenberg, C. Lister, S. Freedman, M. Wolanski, J. Last, B. Utts, Nuclear Instruments and Methods in Physics Research Section A: Accelerators, Spectrometers, Detectors and Associated Equipment **330**, 447 (1993)
[11] T.K. Li, D. Dehnhard, R.E. Brown, P.J. Ellis, Phys. Rev. C **13**, 55 (1976)
[12] T. Al-Abdullah, F. Carstoiu, X. Chen, H.L. Clark, C.A. Gagliardi, Y.W. Lui, A. Mukhamedzhanov, G. Tabacaru, Y. Tokimoto, L. Trache et al., Phys. Rev. C **89**, 025809 (2014)
[13] D.W. Bardayan, J.C. Blackmon, C.R. Brune, A.E. Champagne, A.A. Chen, J.M. Cox, T. Davinson, V.Y. Hansper, M.A. Hofstee, B.A. Johnson et al., Phys. Rev. Lett. **83**, 45 (1999)
[14] S. Kubono, H. Orihara, S. Kato, T. Kajino, Astrophys. J. **344**, 460 (1989)
[15] N.M. Clarke, P.R. Hayes, M.B. Becha, C.N. Pinder, S. Roman, Journal of Physics G: Nuclear and Particle Physics **16**, 1547 (1990)
[16] J. Görres, M. Wiescher, Phys. Rev. C **52**, 412 (1995)
[17] R. Coszach, T. Delbar, W. Galster, P. Leleux, I. Licot, E. Liénard, P. Lipnik, C. Michotte, A. Ninane, J. Vervier et al., Phys. Rev. C **50**, 1695 (1994)
[18] L.J. Sun, X.X. Xu, D.Q. Fang, C.J. Lin, J.S. Wang, Z.H. Li, Y.T. Wang, J. Li, L. Yang, N.R. Ma et al., Phys. Rev. C **95**, 014314 (2017)
[19] J. Wallace, P. Woods, G. Lotay, A. Alharbi, A. Banu, H. David, T. Davinson, M. McCleskey, B. Roeder, E. Simmons et al., Physics Letters B **712**, 59 (2012)
[20] J.P. Wallace, P.J. Woods, Phys. Rev. C **86**, 068801 (2012)
[21] D.J. Crozier, H.T. Fortune, Phys. Rev. C **10**, 1697 (1974)
[22] H.T. Fortune, R.R. Betts, Phys. Rev. C **10**, 1292 (1974)
[23] J.C. Legg, D.J. Crozier, G.G. Seaman, H.T. Fortune, Phys. Rev. C **18**, 2202 (1978)
[24] M. Smith, P. Magnus, K. Hahn, A. Howard, P. Parker, A. Champagne, Z. Mao, Nuclear Physics A **536**, 333 (1992)
[25] M. Couder, C. Angulo, E. Casarejos, P. Demaret, P. Leleux, F. Vanderbist, Phys. Rev. C **69**, 022801 (2004)
[26] N.M. Clarke, S. Roman, C.N. Pinder, P.R. Hayes, Journal of Physics G: Nuclear and Particle Physics **19**, 1411 (1993)
[27] B.D. Anderson, B. Wetmore, A.R. Baldwin, L.A.C. Garcia, D.M. Manley, R. Madey, J.W. Watson, W.M. Zhang, B.A. Brown, C.C. Foster et al., Phys. Rev. C **52**, 2210 (1995)
[28] J. Belarge, S.A. Kuvin, L.T. Baby, J. Baker, I. Wiedenhöver, P. Höflich, A. Volya, J.C. Blackmon, C.M. Deibel, H.E. Gardiner et al., Phys. Rev. Lett. **117**, 182701 (2016)
[29] N. Keeley, N. Alamanos, V. Lapoux, Phys. Rev. C **69**, 064604 (2004)
[30] H.T. Fortune, G.C. Morrison, R.C. Bearse, J.L. Yntema, B.H. Wildenthal, Phys. Rev. C **6**, 21 (1972)
[31] D. Tilley, C. Cheves, J. Kelley, S. Raman, H. Weller, Nuclear Physics A **636**, 249 (1998)

 DOI: 10.1051/epjconf/201716501006

^{23}Na(p,γ)^{24}Mg Cross Section Measurements

Axel Boeltzig[1,2,⋆], *Andreas* Best[3], *Richard J.* deBoer[1], *Antonino* DiLeva[3], *Gianluca* Imbriani[3], *Matthias* Junker[4], and *Michael* Wiescher[1]

[1] *University of Notre Dame, Notre Dame, USA and Joint Institute for Nuclear Astrophysics (JINA)*
[2] *Gran Sasso Science Institute (GSSI), L'Aquila, Italy*
[3] *Università degli Studi di Napoli "Federico II" and INFN Sezione di Napoli, Naples, Italy*
[4] *INFN Laboratori Nazionali del Gran Sasso, Assergi, Italy*

Abstract. As a link between the NeNa and MgAl cycles in stellar burning, the reaction ^{23}Na(p,γ)^{24}Mg is of interest for various astrophysical scenarios, such as AGB stars. A combined effort at the Laboratory for Underground Nuclear Astrophysics (LUNA) and the Nuclear Science Laboratory (NSL) at the University of Notre Dame aims at a cross section determination for the this reaction, to constrain the astrophysical reaction rate by improving the knowledge of the resonance strengths and the non-resonant component. Experiments at LUNA benefit from the underground location at the Gran Sasso National Laboratory which allows for the measurement of resonances at low energies with high sensitivity in a low background environment. Measurements at the University of Notre Dame pursue a determination of the non-resonant cross section at higher energies. We present the two experiments and the status of the data analysis.

1 Introduction

The reaction ^{23}Na(p,γ)^{24}Mg links the NeNa and MgAl cycles in stellar hydrogen burning. For temperatures up to approximately 100 MK, typical for Red Giant Branch (RGB) and low and intermediate mass Asymptotic Giant Branch (AGB) stars, the rate of this reaction is predominantly determined by the non-resonant component of the cross section and possibly by a narrow resonance at $E_{\mathrm{c.m.}}$ = 138 keV [1]. At slightly higher temperatures a narrow resonance at $E_{\mathrm{c.m.}}$ = 240 keV starts to become more influential. An upper limit for the strength of the 138 keV resonance has been established in [2]. The non-resonant cross section of ^{23}Na(p,γ)^{24}Mg has not been observed in a direct experiment yet (cf. [3]). The uncertainty of these two contributions to the cross section yields large uncertainties in the astrophysical reaction rate in the temperature range given above.

2 Measurements Underground at LUNA

We used two complementary detector setups to study the ^{23}Na(p,γ)^{24}Mg reaction underground at LUNA: one setup employed a segmented BGO summing-detector [4] with high efficiency but modest energy resolution (e. g. about 4-5% energy resolution and 60% full energy detection efficiency for

⋆e-mail: axel.boeltzig@nd.edu

a single 12 MeV gamma ray), the other setup used a HPGe detector (ORTEC GEM120225-P-ST, located at 55° relative to the beam) that offered much better energy resolution but lower efficiency and angular coverage than the first setup (efficiency and solid angle depending on precise location of the detector relative to the target, but typically about two orders of magnitude lower compared to the BGO setup). The two setups are shown in the contribution of Pantaleo et al. in this issue.

Thin solid targets (e. g. with an energy loss of in the main target layer on the order of 10 keV for a a 300 keV proton beam) were produced by evaporating sodium compounds on 0.4 mm thick tantalum backings. Most of the measurements used targets that were produced with sodium tungstate (Na_2WO_4); other materials that were tested included Na_2SiO_3 and NaCl, but sodium tungstate appeared to provide the best compromise of uniformity and reproducibility of the production and target stability during the irradiation. We scanned the narrow E_{lab} = 309 keV resonance in regular intervals to monitor the target degradation during bombardment. Targets were exchanged when degradation appeared to become significant (e. g. the thickness of the active target layer was significantly reduced). Ex-situ elastic recoil-detection analyses (ERDA) were performed on selected targets to obtain depth profiles of the chemical composition on various spots on the target surface.

The main purpose of the BGO detector setup was the search for the weak resonance at 138 keV. The HPGe detector was employed to study the properties of narrow resonances at higher energies, and to search for a signal from non-resonant capture or the tail of broader resonances at higher energies.

The primary signature of ^{23}Na$(p,\gamma)^{24}$Mg in the BGO summing detector is a signal at $E_{\mathrm{c.m.}} + Q$ in the spectrum of the total energy deposition (i. e. in all six detector segments combined). As the environmental backgrounds in this energy region (Q = 11.693 MeV for ^{23}Na$(p,\gamma)^{24}$Mg [5]) are very small, due to the underground location, the limiting factor for the sensitivity to this signature was beam-induced background. The reactions ^{11}B(p,γ) and ^{7}Li(p,γ) can both contribute to background in the ^{23}Na region of interest, as their Q-values are higher than the studied reaction on ^{23}Na. After tests with raw materials from various suppliers the level of beam-induced background could be reduced substantially (in particular the lithium contamination) by careful selection of the raw materials. Nonetheless, beam-induced backgrounds were still non-negligible at these beam energies.

For the measurement of an off-resonance cross section, the presence of weaker narrow resonances between the 309 keV resonance and the 400 keV proton energy limit at LUNA (as well as the comparatively strong 309 keV resonance itself, $\omega\gamma$ = (105 ± 19) meV [6]) posed a problem. Resonant reactions occurring at depths beyond the main deposited layer of target material in the target (i. e. at lower proton energies) act as a background for the capture to excited states in ^{24}Mg that occurs at higher proton energies in the main layer of the target material. The observation of sodium in deeper layers of the target, caused either by diffusion or by a pre-existing contamination of sodium in the backing material, therefore limited the sensitivity to the signal of non-resonant capture to excited states.

Preliminary results from the measurements at LUNA show promise in achieving a better sensitivity for the search of the 138 keV resonance, as well as a reduced uncertainty of the strength of a narrow resonance at 240 keV and a first observation of the non-resonant capture to the ground state of ^{24}Mg.

3 Measurements at the University of Notre Dame

Whilst gamma rays from off-resonance reactions have been observed at the highest proton energies that are available at LUNA, additional information is required to constrain the direct-capture component of the cross section. This information includes the capture to excited states in ^{24}Mg, and the proper accounting for the tails of non-narrow resonances at higher energies.

DOI: 10.1051/epjconf/201716501006

To obtain this information, an experiment was conducted at the Nuclear Science Laboratory at the University of Notre Dame. A proton beam, provided by the 5U accelerator, was used to bombard a solid sodium tungstate target of the same type as previously described for the experiment at LUNA. Typical beam currents were on the order of $10\,\mu$A. We used an HPGe detector at an angle of 55° to acquire gamma ray energy spectra.

The studied range of beam energies includes resonances at $E_{\text{lab}} = 309\,\text{keV}$ and $512\,\text{keV}$, both with known resonance strengths [6], and extends up to proton energies slightly above 1 MeV. The measured range thus covers several resonances in ^{23}Na$(p,\gamma)^{24}$Mg [6, 7], and data was acquired on these resonance as well as at proton energies in between them.

The analysis of this measurement is ongoing. Area and shape of the observed primary gamma lines are analyzed to obtain information on the reaction cross section and disentangle the direct capture and resonant contributions.

4 Summary

The cross section of ^{23}Na$(p,\gamma)^{24}$Mg has been measured in the astrophysical energy region of interest underground at LUNA, and these measurements are expected to yield more precise information on the narrow resonances affecting the reaction rate for stellar temperatures below 100 MK. With hints of the non-resonant cross sections observed at LUNA, additional measurements were performed at the University of Notre Dame, with the aim to obtain better constraints on the non-resonant contribution to the cross section.

The analysis of both experiments is underway, and a combination of the results of both measurements is expected to reduce the uncertainty of reaction rate calculations for temperatures below 100 MK.

References

[1] Boeltzig et al., Eur. Phys. J. A **52**: 75 (2016)
[2] Cesaratto et al., Phys. Rev. C **88**, 065806 (2013)
[3] Hale et al., Phys. Rev. C **70**, 045802 (2004)
[4] Casella et al., Nucl. Instr. Meth. Phys. Res. A **489**, 160 (2002)
[5] Meng Wang et al., Chin. Phys. C **41**, 030003 (2017)
[6] Endt, Nucl. Phys. A **521**, 1 (1990)
[7] Switkowski et al., Aust. J. Phys. **28**, 141-53 (1975)

EPJ Web of Conferences **165**, 01007 (2017) DOI: 10.1051/epjconf/201716501007
NPA8 2017

The astrophysical S-factor of the direct ^{18}O(p, γ)^{19}F capture by the ANC method

V. Burjan[1,*], *Z. Hons*[1], *V. Kroha*[1], *J. Mrázek*[1], Š. Piskoř[1], *A. M. Mukhamedzhanov*[2], *L. Trache*[2], R. E. Tribble[2], *M. La Cognata*[3], *L. Lamia*[3], *G. R. Pizzone*[3], *S. Romano*[3], *C. Spitaleri*[3], and A. Tumino[3]

[1]Nuclear Physics Institute of Czech Academy of Sciences, 250 68 Řež, Czech Republic
[2]Cyclotron Institute, Texas A&M University, College Station, TX 77843
[3]Università di Catania and INFN Laboratori Nazionali del Sud, Catania, Italy

Abstract. We attempted to determine the astrophysical S-factor of the direct part of the ^{18}O(p, γ)^{19}F capture by the indirect method of asymptotic normalization coefficients (ANC). We measured the differential cross section of the transfer reaction ^{18}O(^{3}He, d)^{19}F at a ^{3}He energy of 24.6 MeV. The measurement was realized on the cyclotron of the NPI in Řež, Czech Republic, with the gas target consisting of the high purity ^{18}O (99.9 %). The reaction products were measured by eight ΔE-E telescopes composed from thin and thick silicon surface-barrier detectors. The parameters of the optical model for the input channel were deduced by means of the code ECIS and the analysis of transfer reactions to 12 levels of the ^{19}F nucleus up to 8.014 MeV was made by the code FRESCO. The deduced ANCs were then used to specify the direct contribution to the ^{18}O(p, γ)^{19}F capture process and were compared with the mutually different results of two works.

1 Introduction

From the astrophysical point of view the radiative capture reaction ^{18}O(p, γ)^{19}F is important for the study of processes in AGB stars. The hydrogen burning of ^{18}O via the (p, γ) capture in AGB stars, i. e., in stars several times heavier than our Sun, is competing with the ^{18}O(p, α)^{15}N reaction [1]. The (p, α) reaction brings ^{15}N back into the CNO cycles while the (p, γ) capture, which rate is at solar energies about 3 orders lower, causes the escape to other cycles. The ratio ^{18}O/^{16}O represents an important parameter which can be used for the comparison of processes in the Sun and presolar nucleosynthesis with different stages of the evolution of AGB stars. The direct part of the (p, γ) process constitutes the substantial part of the whole cross section of the (p, γ) capture at astrophysical energies where only several channels for the decay of the compound nucleus are open.

A very thorough measurement of the ^{18}O(p, γ)^{19}F in the energy range E_p = 0.08-2.2 MeV was performed by Wiescher et al. [2] where the direct part of the (p, γ) capture was determined experimentally and also calculated theoretically. Later, Buckner et al. [3], when investigating reaction rates of ^{18}O(p, γ)^{19}F, measured the cross section and determined the direct part of this capture using different capture models. However, there is a substantial difference between the astrophysical S-factor of

*e-mail: burjan@ujf.cas.cz

DOI: 10.1051/epjconf/201716501007

Wiescher et al. and of Buckner et al. at low energies. The tendency of values of Wiescher et al. is decreasing while that of Buckner et al. is increasing. Also their absolute values differ considerably. We therefore decided to determine the astrophysical S-factor of the direct capture ^{18}O(p, γ)^{19}F by the indirect method of astrophysical normalization coefficients (ANC). The method of ANC [4] was developed to determine the astrophysical S-factor of direct capture reactions at very low energies where the cross section is either too small to be measured directly or the nuclei participating in the reactions are radioactive. We compared our results of the direct capture obtained by the ANC method with the two above mentioned direct measurements and calculations. We measured the differential cross section of the ^{18}O(^{3}He, d)^{19}F transfer reaction and from it we deduced ANC values.

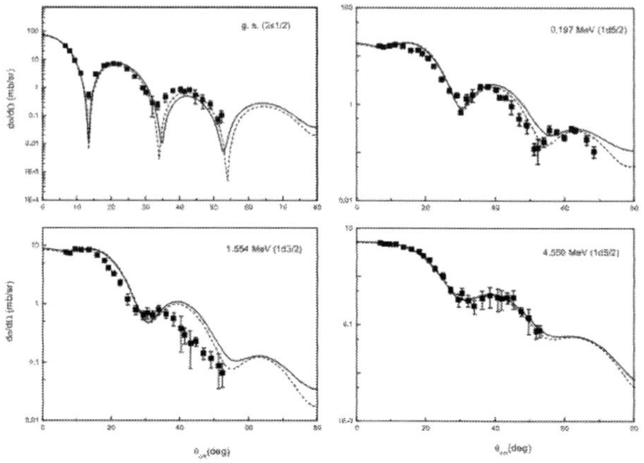

Figure 1. The angular distribution of the ^{18}O(^{3}He, d)^{19}F reaction for the transitions to the ground state and to the 0.197, 1.554 and 4.550 MeV states, respectively. Sets of optical model parameters are taken from our fits of elastic scattering with volume and surface absorbtion (solid and dashed line respectively) and for the output channel were taken from [5].

2 Analysis and results

The measurement of the cross section of the ^{18}O(^{3}He, d)^{19}F reaction on an oxygen gas target was carried out using a momentum analyzed 24.6 MeV ^{3}He beam from the isochronous cyclotron U-120M at the Nuclear Physics Institute of the Czech Academy of Sciences. The gas target chamber contained the high purity ^{18}O isotope (99.9%). The entrance window of the gas target was covered with a 2 μm thick Ti foil of 4 mm in diameter. The output window was covered with a 3 μm thick havar foil at the angular range from $-40°$ to $65°$. The gas pressure was kept at the level of 150 mbar and was monitored by a gas control system. Reaction products were registered by eight ΔE-E telescopes consisting of about 250-μm and 5-mm thick Si(Li) surface barrier detectors. All telescopes were equipped with a pair of collimating slits of dimensions 1 x 3 mm^2 and 1 x 4 mm^2. The front sides of the near 1 x 4 mm^2 and the far 1 x 3 mm^2 slits were located 105 mm and 185 mm, respectively, from the center of the gas target chamber. The effective target thickness, different at different reaction angles, was determined for the given geometry by calculation. Three telescopes were fixed on an upper plate at the angles of 17°, 27° and 37°, respectively, as monitors and the remaining five telescopes on the lower plate separated by 10° angular steps were movable in the angular range between 6.0° and 62°.

We have seen and analyzed 12 deuteron peaks corresponding to transitions to the bound states of the ^{19}F nucleus and to the first unbound state 8.014 MeV (proton emission threshold 7.9936 MeV). Other deuteron levels were populated weakly and were not considered in further analysis. For the analysis of the measured angular distributions the phenomenological optical potential was used with the sets of optical model parameters for the input channel obtained by the fit of the angular distribution of elastically scattered ^{3}He particles on the ^{18}O nucleus measured at the same energy as the transfer

 DOI: 10.1051/epjconf/201716501007

reaction itself. The optical model parameters for the outgoing deuterons were taken from the global formula of Perey and Perey [5]. The calculations of the theoretical angular distributions of the transfer reaction $^{18}O(^3He, d)^{19}F$ were performed using the computer code FRESCO by Thompson [6]. The example of the $^{18}O(^3He, d)^{19}F$ experimental angular distributions compared with calculated angular distributions are given in Fig. 1. From these comparisons we can deduce ANCs and also spectroscopic factors which are given in Table 1. The transitions with the largest ANCs in descending order are to the ground state and to the levels 0.197, 8.014 and 1.554 MeV of ^{19}F. ANC for the 8.014 MeV state was calculated assuming the proton binding energy 100 keV. The ANCs are less model dependent than spectroscopic factors and were therefore used in further calculations.

Table 1. ANCs and spectroscopic factors S_{lj} from the $^{18}O(^3He, d)^{19}F$ reaction. Experimental and model dependent uncertainties of ANCs are 20%.

^{19}F state $[MeV]$	j_{tr}	b_{lj} (see [4]) $[fm^{-1/2}]$	ANC(C^2_{Aplj}) $[fm^{-1}]$	S_{lj} this work	S_{lj} [7] Green	S_{lj} [8] Schmidt
0 $(1/2^+)$	2s1/2	16.364	78.3	0.292	0.45	0.32
0.197 $(5/2^+)$	1d5/2	5.1638	16.2	0.609	0.63	0.61
1.554 $(3/2^+)$	1d3/2	3.5842	4.14	0.322	0.44	0.38
4.550 $(5/2^+)$	1d5/2	1.9767	0.347	0.089	—	—
5.535 $(5/2^+)$	1d5/2	1.7195	7.55E-2	0.026	—	—
6.088 $(3/2^-)$	2p3/2	5.0714	8.96E-1	0.035	—	—
6.497 $(3/2^+)$	1d3/2	1.1438	2.42E-2	0.019	0.05	0.05
6.787 $(3/2^-)$	2p3/2	4.48	0.422	0.021	0.11	0.032
6.927 $(7/2^-)$	1f7/2	0.2625	6.82E-3	0.099	—	0.072
7.113 $(5/2^+)$	1d5/2	1.066	3.45E-2	0.030	0.13	0.022
7.540 $(5/2^+, T = 3/2)$	1d5/2	1.0907	0.464	0.39	0.41	0.332
8.014 $(5/2^+)$	1d5/2	7.3883	4.23	0.077	—	0.065

The direct capture cross section $\sigma(E)_{cap}$ was calculated by the computer code FRESCO [6] using derived ANCs from Table 1 under the assumption that the transfer reaction was of the surface character. The surface character was proved, e. g., by testing the dependence of the differential cross section on the cut-off radius. As the interaction we assumed the E1 multipole electromagnetic operator. By FRESCO we calculated the bound state wave function of the given level of the target nucleus on which a proton is captured and the wave function of the incoming proton in the field of the target nucleus. We calculated the bound state wave functions using the Woods-Saxon potential [2]. The scattered wave function of the proton can be also calculated by the Woods-Saxon potential or for instance by the optical model potential. We did calculations with three kinds of potentials for incoming protons: with the complex optical potential of Perey and Perey [5] and with the zero nuclear potential keeping only the Coulomb part to stress the fact that at low bombarding energies of protons the Coulomb potential could play a dominant role. For comparison we also simulated the hard sphere potential by putting V = -300 MeV as the repulsive potential into the code FRESCO. Resulting curves of the total S-factor from all three potentials are given in Fig. 2. Dominant contributions to the total S-factor come from the capture to the ground and 0.197 MeV states (60% together). We compared these curves with the low-energy calculations of $^{18}O(p, \gamma)^{19}F$ by Buckner et al. [3] and also with the direct measurement by Wiescher et al. [2]. While the S(E)-factor generated by the hard-sphere potential is almost constant at the value of 2 keV.b, our other two curves based on the optical and Coulomb potentials exhibit a rising tendency in the energy interval 0 to 2 MeV which is in correspondence with the calculation of Buckner et al. ($S(E) = 7.06 + 2.98 \times 10^{-3}E - 2.6 \times 10^{-7}E^2$ [keV.b]) but in contradiction with the direct measurement of Wiescher et al. ($S(E) = 15.7 - 0.34 \times 10^{-3}E - 1.21 \times 10^{-6}E^2$ [keV.b]). Our calculation of S-factor with the Coulomb potential when normalized to the measured direct capture cross section at $E_{c.m.} = 1751.9$ keV by Wiescher et al. [2] is even almost identical with the curve of Buckner et al.

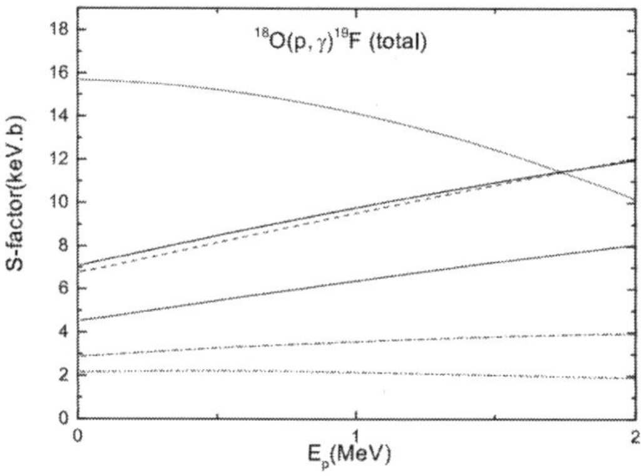

Figure 2. S-factor of the total direct proton capture $^{18}O(p,\gamma)^{19}F$ determined from the transfer reaction. Our blue solid, dash-dotted and dash-dot-dotted lines correspond respectively to the calculations with the zero nuclear potential, the global potential of Perey and Perey [5] and with the V = -300 MeV potential simulating the hard sphere potential. The red solid decreasing curve is based on the measurement of the direct contribution (p, γ) by Wiescher et al. [2] and the black solid increasing curve is the calculation by Buckner et al. [3]. The curve with the zero nuclear potential normalized to the measured point at $E_{c.m.} = 1751.9$ keV by Wiescher et al. is depicted by the blue dashed line.

3 Conclusion

The deduced values of ANCs were used for the determination of the direct part of the total astrophysical S-factor of the radiative capture $^{18}O(p,\gamma)^{19}F$. Our values of the total astrophysical S-factor calculated with three different interaction potentials show in two cases similar tendency as was determined by Buckner et al. [3] and for the Coulomb interaction potential, after normalization, reproduce these data. The calculations with our interaction potentials without normalization are lower than the values of Buckner et al. at least about 35%.

Acknowledgements: This work was partially supported by Grants No. LH11001, No. M10480902 and No. RBFR082838 (FIRB2008) and by Project Spiral 2-CZ.

References

[1] Lorentz-Wirzba H., Schmalbrock P., Trautvetter H. P., Wiescher M. and Rolfs C., Nucl. Phys. **A313**, 346 (1979)

[2] Wiescher M., Becker H. W., Görres J., Kettner K.-U., Trautvetter H. P., Kieser W. E., Rolfs C., Azume R. E., Jackson K. P. and Hammer J. W., Nucl.Phys. **A349**, 165 (1980)

[3] Buckner M. Q., Iliadis C., Cesaratto J. M., Howard C., Clegg T. B., Champagne A. E. and Daigle S., Phys. Rev. **C86**, 065804 (2012)

[4] Xu H. M., Gagliardi C. A., Mukhamedzhanov A. M., Timofeyuk N. K., Tribble R. E., Phys. Rev. Lett. **73**, 2027 (1994)

[5] Perey C. M. and Perey F. G., At. Data Nucl. Data Tables **17**, 1-101 (1976)

[6] Thompson I. J., Comp. Phys. Rep. **7**, 167 (1988)

[7] Green L. L., Lennon C. O. and Naqib I. M., Nucl. Phys. **A142**, 137 (1970)

[8] Schmidt C. and Duhm H. H., Nucl. Phys. **A155**, 644 (1970)

EPJ Web of Conferences **165**, 01008 (2017) DOI: 10.1051/epjconf/201716501008
NPA8 2017

High-precision mass measurements for the *rp*-process at JYFLTRAP

Laetitia Canete[1*], *Tommi* Eronen[1], *Ari* Jokinen[1], *Anu* Kankainen[1], *Ian D.* Moore[1], *Dimitry* Nesterenko[1], and *Sami* Rinta-Antila[1]

[1]University of Jyväskylä, P.O. Box 35 (YFL) FI-40014 University of Jyväskylä, Finland

Abstract. The double Penning trap JYFLTRAP at the University of Jyväskylä has been successfully used to achieve high-precision mass measurements of nuclei involved in the rapid proton-capture (rp) process. A precise mass measurement of ^{31}Cl is essential to estimate the waiting point condition of ^{30}S in the rp-process occurring in type I x-ray bursts (XRBs). The mass-excess of ^{31}Cl measured at JYFLTRAP, -7034.7(3.4) keV, is 15 more precise than the value given in the Atomic Mass Evaluation 2012. The proton separation energy S_p determined from the new mass-excess value confirmed that ^{30}S is a waiting point, with a lower-temperature limit of 0.44 GK. The mass of ^{52}Co effects both ^{51}Fe$(p,\gamma)^{52}$Co and ^{52}Co$(p,\gamma)^{53}$Ni reactions. The mass-excess value measured, -34 331.6(6.6) keV is 30 times more precise than the value given in AME2012. The Q values for the ^{51}Fe$(p,\gamma)^{52}$Co and ^{52}Co$(p,\gamma)^{53}$Ni reactions are now known with a high precision, 1418(11) keV and 2588(26) keV respectively. The results show that ^{52}Co is more proton bound and ^{53}Ni less proton bound than what was expected from the extrapolated value.

1 JYFLTRAP: a double Penning trap for mass measurements

Exotic nuclei are produced at the University of Jyväskylä by using a stable ion beam delivered by the K130 cyclotron to the Ion Guide Isotope Separator On-Line (IGISOL) facility. The produced ions created either by fission or fusion reactions on a thin target are extracted out from the ion guide gas cell and accelerated to 30 keV. They are then mass separated with a 55° dipole magnet and cooled in a radiofrequency cooler and buncher (RFQ). The RFQ periodically releases a bunch of ions to the double Penning trap setup JYFLTRAP [1] where a static quadrupolar electric field and an homogenous magnetic field is applied.

In a Penning trap, an ion has three eigenmotions: the axial (ν_z), the magnetron (ν_-) and the reduced cyclotron motion (ν_+). The cyclotron frequency is connected to the mass of the ion and the magnetic field of the trap by:

$$\nu_c = \nu_+ + \nu_- = qB/2\pi m \quad (1)$$

* Corresponding author: lacanete@student.jyu.fi

The ions of interest are highly purified in the first trap (the purification trap) and sent to the second trap (the precision trap) where their cyclotron frequency is determined using the time-of-flight ion cyclotron resonance (TOF-ICR) technique [2].

2 Mass measurements of ^{31}Cl

2.1 Astrophysical implication

In an XRB, a neutron star accretes matter from a companion star via Roche lobe overflow. The combination of high temperature and hydrogen/helium rich environment induce a nucleosynthesis flow through the rp-process. In the XRBs rp-process, the ^{30}S acts as a waiting point [3]: the half-life of ^{30}S is long (1.178(5) s) and its proton-capture Q value is low. At typical temperature range of XRBs, the proton-capture rate on ^{30}S is dominated by the resonant proton-capture to the two lowest excited states in ^{31}Cl. However, the proton-capture reaction of ^{30}S is counterbalanced by the photodisintegration ($\lambda_{\gamma,p}$) of ^{31}Cl. To estimate the ratio of $\lambda_{\gamma,p}$ to the proton-capture Q value on ^{30}S, a precise value of the S_p of ^{31}Cl has to be determined.

2.2 Results

A 40 MeV proton beam has been used on a 1.8 mg/cm^2-thick ZnS target to produce ^{31}Cl$^+$ ions. The mass-excess obtained, -7034.7(3.4) keV [4], is 15 times more precise than the value given in the AME2012 (−7070(50) keV [5]). From the measured mass-excess value, the S_p of ^{31}Cl has been determined as 264.6(3.4) keV. The new S_p value shows that ^{31}Cl is less proton bound than previously expected. It indicates also that for temperatures higher than 0.44 GK, the ^{30}S is confirmed to be a waiting point: at least 20% of the rp-process flow must wait for the β^+ decay of ^{30}S. The update ratio of $\lambda_{\gamma,p}$ to the proton-capture Q value on ^{30}S (Fig.1) shows that the photodisintegration of ^{31}Cl dominate in the rp-process of XRBs at lower temperature than expected based on AME2012.

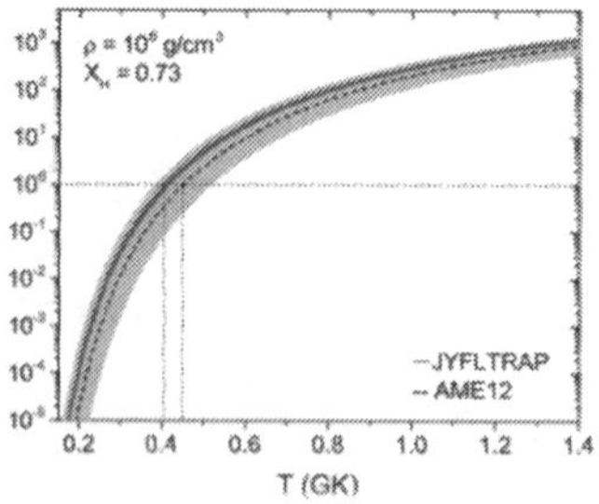

Fig. 1. Ratio of the photodisintegration to the proton-capture rates for typical XRB conditions. The uncertainties related to the JYFLTRAP Q value are shown by the blue lines and to the AME12 value by the gray-shaded area.

3 Mass measurement of ^{52}Co

In the XRBs, the rp-process flow toward heavier elements is affected by the proton-capture Q value of ^{51}Fe and ^{52}Co. The proton-capture reactions, ^{51}Fe(p,γ)^{52}Co and ^{52}Co(p,γ)^{53}Ni are in competition with their inverse photodisintegration reactions, ^{52}Co(γ,p)^{51}Fe and ^{53}Ni(γ,p)^{52}Co. To estimate the ratio of the inverse photodisintegration to the total proton-capture rate, a precise Q value of these reactions has to be determined. The ^{52}Co^{+} has been produced using a 50 MeV proton beam on a 1.8 mg/cm^2 thick ^{54}Fe target. The TOF-ICR technique has been applied with a 100 ms long quadrupolar radiofrequency (RF) excitation. A new mass-excess value for ^{52}Co was measured: -34 331.6(6.6) keV [6]. The proton-capture Q values for ^{51}Fe(p,γ)^{52}Co and ^{52}Co(p,γ)^{53}Ni are 1418(11) keV and 2588(26) keV, respectively. The precision were highly improved in comparison to the extrapolated Q values given in AME2012 (1077(196)# keV and 2930(197)# keV respectively [5]). ^{52}Co is around 340 keV more proton-bound than expected based on the AME2012 value: the photodisintegration of ^{52}Co is less probable to occur in XRBs (Fig.2a). ^{53}Ni is less proton bound, and thus, the photodisintegration of ^{53}Ni is more dominant in the XRBs rp-process than previously expected (Fig.2b).

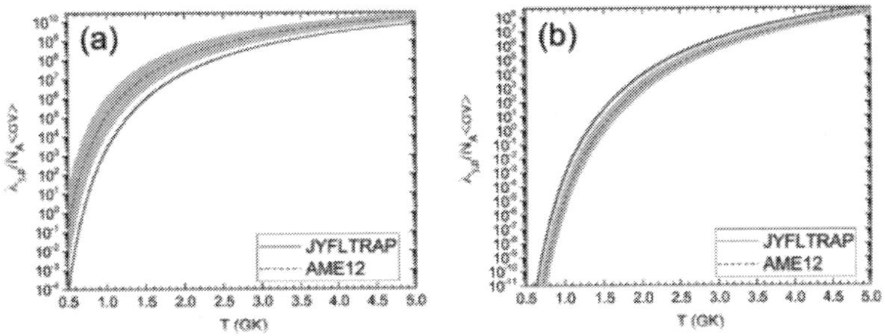

Fig.2. Ratio of the photodisintegration to the proton-capture rates for (a) ^{51}Fe(p,γ)^{52}Co–^{52}Co(γ, p)^{51}Fe and (b) ^{52}Co(p,γ)^{53}Ni –^{53}Ni(γ, p)^{52}Co reactions.

This work has been supported by the Academy of Finland under projects No. 275389 and 284516 as well as under the Finnish Centre of Excellence Programme 2012-2017 (Nuclear and Accelerator Based Physics Research at JYFL)

References

1. T. Eronen, V.S. Kolhinen, V.-V. Elomaa, D. Gorelov, U. Hager, J. Hakala, A. Jokinen, A. Kankainen, P. Karvonen, S. Kopecky, I.D. Moore, H. Penttilä, S. Rahaman, S. Rinta-Antila, J. Rissanen, A. Saastamoinen, J. Szerypo, C. Weber, J. Äystö, Eur. Phys. J. A **48,** 46 (2012)
2. M. König, G. Bollen, H.-J. Kluge, T. Otto, J. Szerypo, Int. J. Mass Spec. Ion Proc. **142**, 95 (1995)
3. J. L. Fisker, F.-K. Thielemann, M. Wiescher, Astrophys. J. **608**, L61 (2004)
4. A. Kankainen, L. Canete T. Eronen, J. Hakala, A. Jokinen, J. Koponen, I. D. Moore, D. Nesterenko, J. Reinikainen, S. Rinta-Antila, A. Voss, J. Äystö, Phys. Rev. C **93**, 041304 (R) (2016)
5. M. Wang, G. Audi, A. Wapstra, F. Kondev, M. MacCormick, X. Xu, B. Pfeiffer, Chin. Phys. C **36**, 1603 (2012)
6. D.A. Nesterenko, A. Kankainen, L. Canete, M. Block, D. Cox, T. Eronen, C. Fahlander, U. Forsberg, J. Gerl, P. Golubev, J. Hakala, A. Jokinen, V.S. Kolhinen, J. Koponen, N. Lalović, C. Lorenz, I. D. Moore, P. Papadakis, J. Reinikainen, S.

DOI: 10.1051/epjconf/201716501008

Rinta-Antila, D. Rudolph, L. G. Sarmiento, A. Voss, J. Äystö, J. Phys. G: Nucl. Part. Phys. **44**, 065103 (2017)

EPJ Web of Conferences **165**, 01009 (2017) DOI: 10.1051/epjconf/201716501009
NPA8 2017

A new method for the determination of very small Γ_γ partial widths

Giuseppe Cardella[1], *Luis* Acosta[1,7], *Lucrezia* Auditore[1,4], *Alberto* Camaiani[8], *Enrico* De Filippo[1], *Saverio* De Luca[1,4], *Nicla* Gelli[8], *Brunilde* Gnoffo[1,3], *Francisco* Favela[1], *Bogdan* Fornal[9], *Gaetano* Lanzalone[2,5], *Silvia* Leoni[6], *Concetta* Maiolino[2], *Nunzia Simona* Martorana[2,3], *Adriana* Nannini[8], *Sebastianella* Norella[1,4], *Angelo* Pagano[1], *Emanuele Vincenzo* Pagano[2,3], *Massimo* Papa[1], *Sara* Pirrone[1], *Giuseppe* Politi[1,3], *Francesco* Porto[2], *Lucia* Quattrocchi[4], *Francesca* Rizzo[2,3], *Paolo* Russotto[2,3], *Domenico* Santonocito[2], *Antonio* Trifirò[1,4] and *Marina* Trimarchi[1,4]

[1] INFN-Sez.Catania, Catania, Italy

[2] LNS-INFN, Catania, Italy

[3] Dipartimento di Fisica e Astronomia Università di Catania, Catania, Italy

[4] Dip. di Scienze Mat. e Inf., Scienze Fisiche e Scienze della Terra Università di Messina, Italy

[5] Facoltà di Ingegneria e Architettura, Universit Kore, Enna, Italy

[6] INFN sezione di Milano and Dip. di Fisica Università di Milano, Italy

[7] Instituto de Fsica, Universidad Nacional Autnoma de Mexico, Mexico

[8] INFN sezione di Firenze and Dip. di Fisica Università di Firenze,Italy

[9] Institute of Nuclear Physics, Polish Academy of Sciences, Krakow, Poland

Abstract. We present a new method for the measurement of very small Γ_γ partial width that is important for the synthesis of elements in astrophysics. The method is based on the simultaneous detection of scattered beam, residual nucleus and decay γ rays. This method is optimized for the use of the CHIMERA detector at LNS. Experimental details are described.

1 Introduction

Gamma ray decay partial widths of nuclear levels, at excitation energy larger than the particle decay energy threshold, are generally quite small. Due to large background, it is difficult to measure them with enough accuracy. However, in some cases they are rather important in order to explain the synthesis of the elements in stellar environments. For instance, only after a γ-decay a ^{12}C nucleus is produced in the triple alpha reaction, passing through the Hoyle state, see for instance [1] and references therein. Therefore, the exact knowledge of its decay gamma width is fundamental in order to explain the ^{12}C abundance in the universe. Many experiments were performed in the 70s of last century, in order to get reliable values for the γ-rays decay widths of high energy levels of ^{12}C. Two main methods were used: the direct measurements of coincidence γ-rays (two γ-rays are emitted in the case of the decay of the 0^+ Hoyle state through the 4.44 MeV 2^+ level) [2]; or the detection of ^{12}C residual nuclei in coincidence with inelastically scattered projectile particles [3-5]. This last method was more used due to the small detection efficiency of γ-ray detectors available at that time. Another measurement has been recently performed mainly to improve the knowledge of the 9.64 MeV gamma decay width [6]. The measurement relied on the very precise detection of residual ^{12}C nucleus with the Grand Rayden magnetic spectrometer. In order to decrease background beam has been used (therefore excluding ^{12}C contaminants in the target, the limiting factor of ref. [7]) and a solid hydrogen target has been developed.

At LNS Catania, we recently proposed a new method to measure such kind of partial gamma widths by using a multiple fold coincidence technique with the CHIMERA 4π detector [8]. The first test experiment will be performed at the beginning of November 2017. The partial gamma width of the ^{12}C Hoyle state [9] will be measured in order to confirm the validity of the method. At the same time, we expect also to measure or, at least, to improve the known upper limit of the γ–decay width of the 9.64 MeV 3^- level [7]. The population of this level seems in fact important for the understanding of the ^{12}C production in very hot astrophysical environments [6]. We will also investigate on the possible population and decay of the 2^+ level, expected in the same energy region that should belong to the rotational band based on the Hoyle state [10]. This technique can be extended to radioactive nuclei, produced by the LNS fragmentation beam line [11], using inverse kinematic reactions on proton targets [12]. Details of the technique and preliminary results will be shown in the following.

2 The experimental method

At LNS Catania, we proposed the following new approach, taking advantage of the opportunities offered by the CHIMERA 4π detector. Such array is composed by 1192 telescopes, with a first stage of 300 μm thick silicon detectors and a second stage of CsI(Tl) scintillators, with photodiode readout, with length from 3 to 12 cm [8]. The new GET full digital electronics [13] is used for the readout of the CsI(Tl) stage, allowing the storage of the preamplifier signals. The shape analysis of the signal is a simple way in order to discriminate particles from γ-rays and isotopically identify charged particles up to Lithium as described in [14,15]. An example of the difference in shape of the signals collected for different particles is given in fig.1. With CHIMERA, we can detect both the residual ^{12}C (stopped in the 300 μm first stage silicon detectors) and γ-rays (in the thick, second stage CsI(Tl) scintillator), therefore in this case the high expected random background will be reduced using the cleaning effect of up to four fold time coincidence techniques. The cut of the background relies not only on the pure coincidence time, (quite good, of the order of ns, for silicon detectors, reasonable, of the order of hundreds of ns, for the CsI(Tl) signals) but also on the particle identification quality of the detector, on kinematic properties, and on energy conservation rules. The simple coincidence between ^{12}C and the scattered projectile produces a reduction of random background of the order of 10^{5}, (assuming a rate of the order of 1 kHz) due to the relatively good typical time resolution of few ns of silicon detectors. A further background reduction, by at least another order of magnitude, is obtained fulfilling the constraints due to energy conservation. The request of kinematical coincidence, due to the opening angle of our detectors, reduces the random coincidences by a factor of the order of 30 [12]; particle identifications, with time of flight, ΔE-E, and pulse shape analysis of particles stopped in silicon detectors [16], produce another order of magnitude reduction.

DOI: 10.1051/epjconf/201716501009

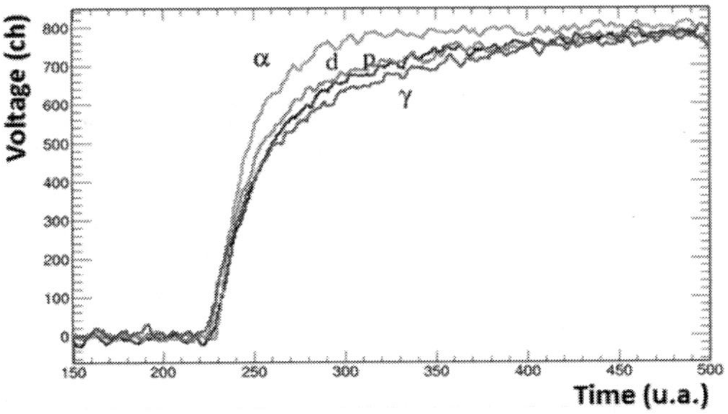

Fig. 1. Examples of signals collected with GET electronics for different particles and γ-rays.

In total, two-particle coincidences in silicon detectors produce a background suppression of the order of 10^8 (see table 1).This is not enough (see ref. [7]) to clean from ^{12}C background in the target. However, a further suppression factor can be attained detecting also γ-rays. Looking to the 9.64 level, because it has a spin 3 and negative parity, we can have both E3 direct decay to the ground state and E1 decay to the 4.44 MeV 2^+ level, with a subsequent E2 decay to ground state. Looking to the decay strengths the emission of 2 γ-rays of 5.2 and 4.44 MeV seem the most probable. The coincidence measurement of each γ-ray, still assuming a rate of the order of one kHz in the detector, but with the worse time resolution of the CsI(Tl), produces a background suppression factor of at least 3 order of magnitude. Further constraints are given by γ-rays detected energy, and by particle-gamma discrimination properties in CsI(Tl). Therefore, the 4-fold coincidence measurement proposed results in a suppression factor of the background of the order of 10^{15} at least.

In ref. [12] we have already shown the quality of kinematical coincidences measurements performed with CHIMERA detector. Gamma-ray detection efficiency and the possibility to measure γ-ray angular correlations with CHIMERA was shown in ref. [17]. Angular correlation measurements will be very important to identify the possible decay of a 2+ level for which evidences near 10 MeV exists [10]. This level is very interesting because it is expected to belong to the presumable rotational band based on the Hoyle state so could be another important brick of the ^{12}C production in the universe.

TABLE 1 summary of suppression factors for background with 2,3 and 4 fold coincidences

	10 ns coincidence time (si-det)	kinematical coincidence	energy conservation	particle identification	total 2 particle coincidence	1 μs coincidence time(CsI-det)	energy conservation	particle identification	total 3 fold coincidence	total 4 fold coincidence
suppression factor	1,0E+05	3,0E+01	1,0E+01	3,0E+00	9,0E+07	1,0E+03	1,0E+01	3,0E+00	2,7E+12	8,1E+15

The main trigger of the experiment will be the coincidence between projectile and target in silicon detectors. In order to reduce the contribution of the elastic channel we plan to trigger only with detectors belonging to the spherical region of the CHIMERA detector (θ_{lab}>30°). In this region, we expect to have a relative contribution of elastic and inelastic channels similar to what we have already measured with proton beam, at 24 MeV, shown in fig.2. In this figure, we report the Q-value spectrum showing the elastic channel and excited levels. We note that the cross section to populate the resonance at 9.63 MeV is of the same order of magnitude of the 4.44 MeV level, and a factor about 20 larger than the probability to populate the Hoyle state at 7.65 MeV. This will help to get enough statistics also for the 9.63 MeV notwithstanding the smaller expected γ-decay probability.

In summary, we are planning to perform a new measurement of the γ-decay probability of the Hoyle state and of the group of levels around 9.63 MeV, including a possible 2+ excited around 10 MeV, which should belong to a rotational band based on the Hoyle state. We will use a 4-fold coincidence method decreasing the background of at least 15 orders of magnitude. We expect to collect at least 5x10^{6} event for the decay of the Hoyle state, i.e. about 2000 coincidences with ^{12}C, and around 100 4-fold coincidences with 2 gamma rays. We will get some10^{5} events for the 9.63 MeV level. From which we expect to have up to 100 coincidences α-^{12}C and up to 20 3-fold and few 4-fold coincidences with gamma rays.

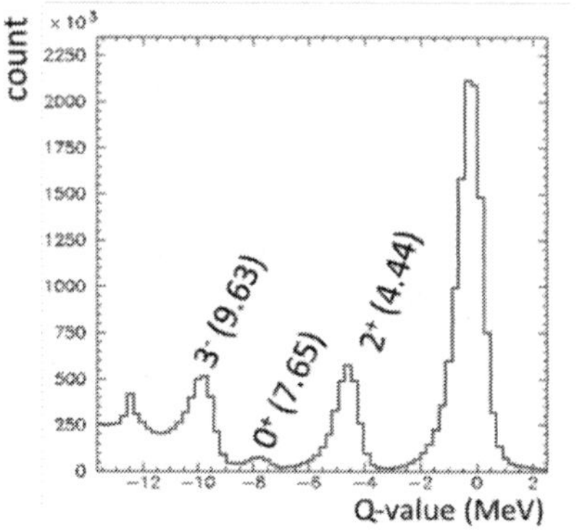

Fig. 2. Q-value spectrum collected at angles θ>30° in the reaction p+^{12}C at 24 MeV.

References

1. F.Herwig S.M. Austin and J.C. Lattanzio Phys. Rev. C **73** (2006) 025802.
2. A. W. Obstt and W. J. Braithwaite phys.Rev.C **13** (1976) 2033
3. D.Chamberlin et al., Phys.Rev.C **9** (1974) 69
4. N. Davids et al Phys Rev C **11** (1975) 2063
5. H.-B. Mak et al Phys.Rev.C **12** (1975) 1158
6. M.Tsumura et al., Journal of Physics: Conference Series **569** (2014) 012051.
7. D. Chamberlin et al., Phys.Rev.C **10** (1974) 909
8. A.Pagano et al., Nucl.Phys. A **734** (2004) 504
9. R.G.Markham et al., Nucl.Phys. A**270** (1976) 489
10. M.Freer and H.O.U.Fynbo, prog.part.and nucl. phys. 78(2014)1
11. https://www.lns.infn.it/it/acceleratori/fribs-lns.html
12. L.Acosta et al., NIM A **715** (2013) 56
13. E. Pollacco et al., submitted to NIM
14. M.Alderighi et al NIM A **489** (2002) 257
15. F.Amorini et al IEEE trans. on Nucl. Sci. **1** (2012) 1772
16. R. Bassini et al., IEEE Trans. on Nucl. Sci. **1** (2006) 507
17. G.Cardella et al NIM A **799** (2015) 64

EPJ Web of Conferences **165**, 01010 (2017) DOI: 10.1051/epjconf/201716501010
NPA8 2017

Three-body radiative capture reactions

J. Casal[1,2,⋆], *M. Rodríguez-Gallardo*[2], *J. M. Arias*[2], and *J. Gómez-Camacho*[2,3]

[1] *European Centre for Theoretical Studies in Nuclear Physics and Related Areas (ECT*) and Fondazione Bruno Kessler, Villa Tambosi, Strada delle Tabarelle 286, I-38123 Villazzano (TN), Italy*
[2] *Departamento de Física Atómica, Molecular y Nuclear, Facultad de Física, Universidad de Sevilla, Apartado 1065, E-41080 Sevilla, Spain*
[3] *Centro Nacional de Aceleradores, Universidad de Sevilla, Junta de Andalucía, CSIC, Tomas Alva Edison 7, E-41092 Sevilla, Spain*

Abstract. Radiative capture reaction rates for ^{6}He, ^{9}Be and ^{17}Ne formation at astrophysical conditions are studied within a three-body model using the analytical transformed harmonic oscillator method to calculate their states. An alternative procedure to estimate these rates from experimental data on low-energy breakup is also discussed.

1 Introduction

Radiative capture reactions are crucial for the stellar models aiming to describe the evolution in composition, energy production and temperature structure of different astrophysical environments [1]. For two-body reactions involving stable nuclei, the direct experimental measurement of the relevant cross section gives directly the astrophysical quantity. In many cases, however, reaction cross sections may not be measured directly. This may occur if the initial nucleus is exotic [2], or when the capture process is a three-body reaction [3]. Three-body radiative capture processes are essential in overcoming the $A = 5, 8$ instability gaps, and traditionally they have been described as two-step sequential reactions [4]. However, it has been shown that the direct three-body capture may play an important role at low temperatures, where the particles have no energy to populate intermediate resonances [5]. Radiative capture reaction rates can be obtained from the inverse photodissociation cross section, which can be calculated within a proper three-body model for the compound nucleus [6]. In this work, we summarize some recent results on three-body capture reactions and discuss future developments.

2 Theoretical framework and applications

The energy-averaged reaction rate for three-body $a + b + c \rightarrow A + \gamma$ radiative capture processes can be calculated from the inverse photodissociation reaction as a function of the temperature T by the expression [6, 7]

$$\langle R_3(\varepsilon)\rangle(T) = \frac{C_3}{(k_B T)^3} e^{\frac{|\varepsilon_B|}{k_B T}} \int_{|\varepsilon_B|}^{\infty} d\varepsilon_\gamma \, \varepsilon_\gamma^2 \sigma_\gamma(\varepsilon_\gamma) e^{\frac{-\varepsilon_\gamma}{k_B T}}, \quad (1)$$

⋆e-mail: jcasal@us.es

 DOI: 10.1051/epjconf/201716501010

where $\varepsilon = \varepsilon_\gamma + \varepsilon_B$ is the initial three-body kinetic energy, ε_γ is the energy of the photon emitted, and ε_B is the ground-state energy of the compound nucleus A. Here, C_3 is a constant depending on the masses and spins of the initial particles. The photodissociation cross-section $\sigma_\gamma(\varepsilon_\gamma)$ of the nucleus A can be expanded into electric and magnetic multipoles, λ, as [8]

$$\sigma_\gamma^{(\lambda)}(\varepsilon_\gamma) = \frac{(2\pi)^3(\lambda+1)}{\lambda[(2\lambda+1)!!]^2}\left(\frac{\varepsilon_\gamma}{\hbar c}\right)^{2\lambda-1}\frac{dB(O\lambda)}{d\varepsilon}, \tag{2}$$

where $B(O\lambda)$ is the order λ of electric or magnetic transition probability ($O = E, M$). The previous expressions make no assumption about the reaction mechanism, thus including sequential and direct, resonant and non-resonant contributions on an equal footing. In computing the reaction rate given by Eq. (1), only the corresponding transition probability distributions of the compound nucleus A are required. To this purpose, we recently applied the analytical transformed harmonic oscillator (THO) method to study the Borromean nuclei ^{6}He [7], ^{9}Be [9, 10] and ^{17}Ne [11] and their formation at astrophysical conditions.

2.1 Formation of ^{6}He, ^{9}Be and ^{17}Ne

It has been proposed that the $\alpha(2n,\gamma)^6$He reaction may play a role in the evolution of neutron star mergers [12], one of the possible scenarios for the r-process nucleosynthesis. The reaction rate of the $\alpha(2n,\gamma)^6$He process is governed by $E1$ transitions between the 0^+ ground state and 1^- continuum states [6]. In Fig. 1a, we show our computed reaction rate for ^{6}He formation, using the analytical THO method to generate ^{6}He states. The details about these structure calculations can be found in Ref. [7]. We compare our results with a previous sequential estimation assuming a dineutron formation [12], showing important differences between them.

The two-proton capture reaction ^{15}O$(2p,\gamma)^{17}$Ne may play a role for the rp-process in Type I X-ray bursts [13]. As in the preceding example, its formation is governed by $E1$ transitions. In this case, it is due to the presence of a low-energy $1/2^+$ resonance in ^{17}Ne. Details can be found in Ref. [11]. Our computed reaction rate, shown in Fig. 1b, is several orders of magnitude larger than the previous calculations in Ref. [14]. The possible implications of this difference should be investigated.

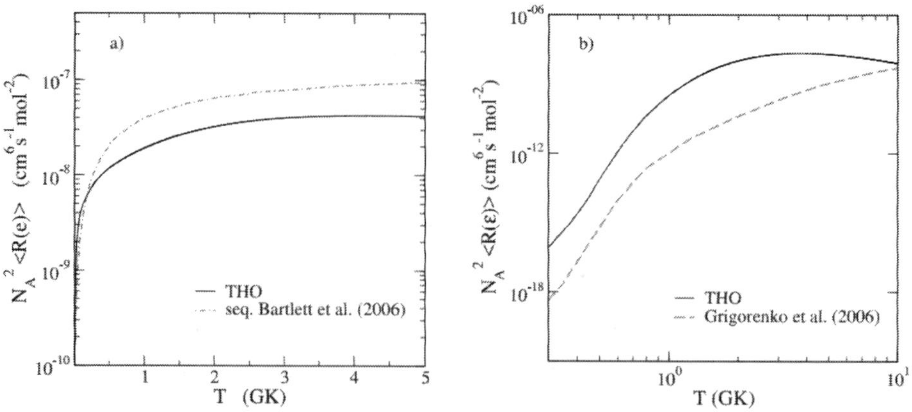

Figure 1. Reaction rate of (a) $\alpha(2n,\gamma)^6$He and (b) ^{15}O$(2p,\gamma)^{17}$Ne using the THO method to generate ^{6}He and ^{17}Ne states. Our results are compared with previous estimations from Refs. [12, 14].

DOI: 10.1051/epjconf/201716501010

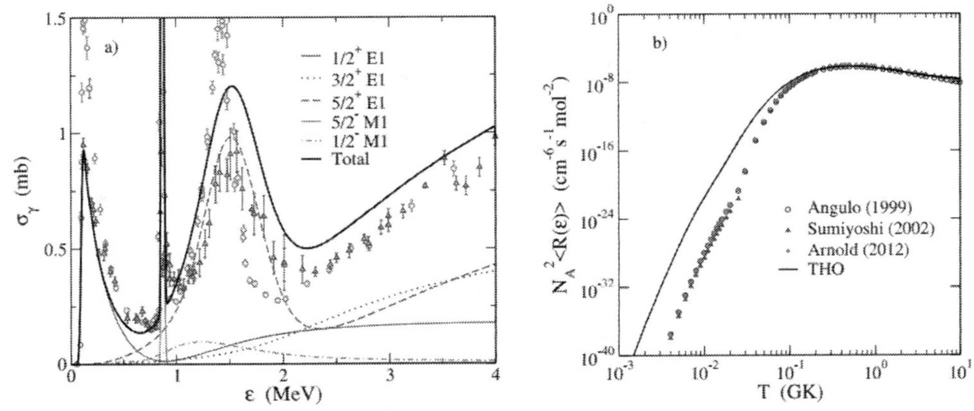

Figure 2. (a) Photodissociation cross section of ^{9}Be using the THO method compared with the data from Refs. [3] (triangles) and [15] (circles). (b) Reaction rate of $\alpha(\alpha n,\gamma)^9$Be compared with sequential estimations from Refs. [3, 4, 15] .

The formation of ^{9}Be has been also linked to the r-process, in this case in type II supernovae [3, 15]. The reaction rate of $\alpha(\alpha n,\gamma)^9$Be is governed by low-energy resonances in ^{9}Be [6], especially the near-threshold $1/2^+$ state. In Fig. 2, we compare our calculations with the experimental data on the photodissociation cross section for ^{9}Be, including $E1$ and $M1$ transitions, and we show the corresponding reaction rates. Details can be found in Ref. [9]. Our calculations agree with sequential estimations at high temperature, but show an enhancement at low temperature coming from the three-body direct $E1$ capture to low-energy $1/2^+$ states. This effect cannot be described using the definition of the rate in Refs. [3, 15], where a sequential process is assumed.

3 Reaction rates from inclusive breakup data

Radiative capture rates computed within different theoretical frameworks may exhibit important differences, as shown in Fig. 1. We recently proposed an alternative method to estimate these reaction rates from inclusive low-energy breakup data on the compound nucleus [16]. When the process is $E1$-dominated, such as in the previous examples, the reaction rate can be obtained as

$$\langle R_3(\varepsilon)\rangle(T) = C_3' t^3 e^{|\varepsilon_B|t} \frac{d^2}{dt^2}\left(\frac{1}{t^2}P_r(t)\right), \tag{3}$$

where P_r is the reduced breakup probability [17] and $t = \frac{a_0}{\hbar v}\left(\pi + \frac{2}{\sin(\theta/2)}\right)$ is the collision time, associated with the temperature through $t = 1/k_B T$. Here, a_0 is half the classical closest approach distance, and v is the relative projectile-target velocity. Equation (3) is valid only at small scattering angles, where the process is Coulomb dominated, and allows us to explore different astrophysical temperatures by measuring the breakup at different kinematic conditions. This is shown in Fig. 3 for ^{6}He + ^{208}Pb, which can provide information on the $\alpha(2n,\gamma)^6$He reaction. In Ref. [16], the method was applied to the case of ^{11}Li, for which suitable experimental data were available [17]. For ^{6}He, however, the existing data do not cover an optimal angular region for the method to be valid. Therefore, a dedicated experiment is required to make a reliable estimation of the rate and to assess the validity of the different theoretical approaches. Such data have to be restricted to the low-energy and small-angle sector of Fig. 3, providing the reaction rates in the region between 1-3 GK.

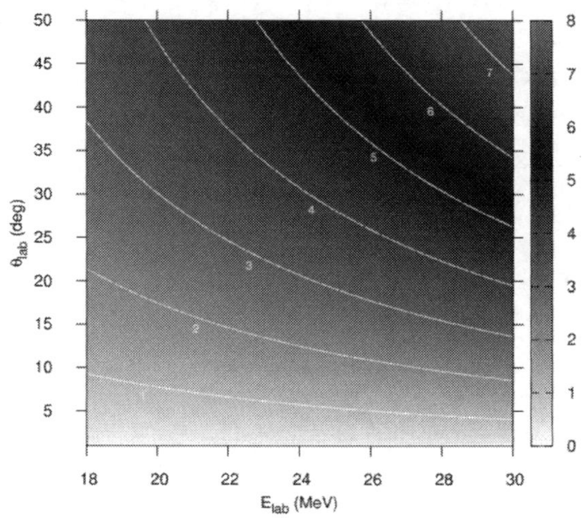

Figure 3. Temperatures, in a color scale, at which the reaction rates for the $\alpha(2n,\gamma)^6$He process can be obtained, as a function of the scattering angle and bombarding energy in the ^{6}He + ^{208}Pb reaction. This is obtained from the definition of the collision time and the relation $t = 1/k_BT$. White lines represent fixed temperatures in GK.

Acknowledgements

This work has been partially supported by the Spanish Ministerio de Economía y Competitividad under Project No. FPA2016-77689-C2-1-R, No. FIS2014-53448-c2-1-P, and No. FIS2014-51941-P, by Junta de Andalucía under Group No. FQM-160 and Project No. P11-FQM-7632, and by the European Union's Horizon 2020 research and innovation program under grant agreement No. 654002. M. R.-G. acknowledges postdoctoral support from the Universidad de Sevilla under the V Plan Propio de Investigación, Contract No. USE-11206-M.

References

[1] K. Langanke, M. Wiescher, Rep. Prog. Phys. **64**, 1657 (2001)
[2] M. Arnould, K. Takahashi, Rep. Prog. Phys. **62**, 395 (1999)
[3] K. Sumiyoshi, H. Utsunomiya, S. Goko, T. Kajino, Nucl. Phys. A **709**, 467 (2002)
[4] C. Angulo *et al.*, Nucl. Phys. A **656**, 3 (1999)
[5] E. Garrido, R. de Diego, D.V. Fedorov, A.S. Jensen, Eur. Phys. J. A **47**, 102 (2011)
[6] R. de Diego, E. Garrido, D.V. Fedorov, A.S. Jensen, Eur. Phys. Lett. **90**, 52001 (2010)
[7] J. Casal, M. Rodríguez-Gallardo, J.M. Arias, Phys. Rev. C **88**, 014327 (2013)
[8] C. Forssén, N.B. Shul'gina, M.V. Zhukov, Phys. Rev. C **67**, 045801 (2003)
[9] J. Casal, M. Rodríguez-Gallardo, J.M. Arias, I.J. Thompson, Phys. Rev. C **90**, 044304 (2014)
[10] J. Casal, M. Rodríguez-Gallardo, J.M. Arias, Phys. Rev. C **92**, 054611 (2015)
[11] J. Casal et al., Phys. Rev. C **94**, 054622 (2016)
[12] A. Bartlett et al., Phys. Rev. C **74**, 015802 (2006)
[13] J. Görres, M. Wiescher, F. Thielemann, Phys. Rev. C **51**, 392 (1995)
[14] L.V. Grigorenko, K. Langanke, N. Shul'gina, M. Zhukov, Phys. Lett. B **641**, 254 (2006)
[15] C.W. Arnold et al., Phys. Rev. C **85**, 044605 (2012)
[16] J. Casal et al., Phys. Rev. C **93**, 041602(R) (2016)
[17] J.P. Fernández-García *et al.*, Phys. Rev. Lett. **110**, 142701 (2013)

EPJ Web of Conferences **165**, 01011 (2017) DOI: 10.1051/epjconf/201716501011
NPA8 2017

Characterization of X3 Silicon Detectors for the ELISSA Array at ELI-NP

S. Chesnevskaya[1,*], *D. L.* Balabanski[1], *D.* Choudhury[1], *M. La* Cognata[2], *P.* Constantin[1], *D. M. Filipescu*[1], *D. G. Ghita*[1], *G. L.* Guardo[1,2], *D.* Lattuada[1,2], *C. Matei*[1], *A.* Rotaru[1], *C.* Spitaleri.[2], *A.* State[1], and *Y.* Xu[1]

[1] *ELI-NP/IFIN-HH, 30 Reactorului Street, 077125, Magurele, Romania*
[2] *INFN, Laboratori Nazionali del Sud, Via Santa Sofia 62, 95123 Catania, Italy*

Abstract. Position-sensitive silicon strip detectors represent one of the best solutions for the detection of charged particles as they provide good energy and position resolution over a large range of energies. A silicon array coupled with the gamma beams at the ELI-NP facility would allow measuring photodissociation reactions of interest for Big Bang Nucleosynthesis and on heavy nuclei intervening in the p-process. Forty X3 detectors for our ELISSA (ELI-NP Silicon Strip Detectors Array) project have been recently purchased and tested. We investigated several specifications, such as leakage currents, depletion voltage, and detector stability under vacuum. The energy and position resolution, and ballistic deficit were measured and analyzed. This paper presents the main results of our extensive testing. The measured energy resolution for the X3 detectors is better than results published for similar arrays (ANASEN or ORRUBA).

1 Introduction

A 4π silicon-strip detector array with removable solid target, ELISSA, has been designed in collaboration with INFN-LNS, Catania [1–3]. The ELISSA array consisting of three rings of twelve single-sided X3 detectors, and two end-caps made up of four double-sided QQQ3 detectors from Micron Semiconductor [4] is under construction at the Extreme Light Infrastructure-Nuclear Physics (ELI-NP) facility in Magurele, Romania. This type of array has already been successfully applied for nuclear astrophysics studies, e g. ORRUBA (Oak Ridge Rutgers University Barrel Array) or ANASEN (Array for Nuclear Astrophysics Studies with Exotic Nuclei) [5, 6].

2 X3 detector testing: results and discussion

The X3-type detectors are 1000 μm thick and have an active area of 7.5 cm × 4 cm. A resistive layer of doped silicon divided into four 1-cm wide strips is the front junction side. The rear ohmic side is non-resistive. The experimental test setup consists of the X3 detector and α-source assembly located inside a large cubicle vacuum chamber, also designed to accommodate the signal read-out chain. The X3 detector was attached to a holder fixed to a heavy aluminum plate. A standard two-peak open

*e-mail: svetlana.chesnevskaya@eli-np.ro

 DOI: 10.1051/epjconf/201716501011

α-source, ^{241}Am/^{239}Pu, with an activity of 1.035 kBq, was placed in a cylindrical canister closed with a lid of selectable collimator hole. The hole diameter of 2 mm and the distance between the α-source and the surface of the detector of 25 cm are chosen so as to illuminate two strips simultaneously. The α-source can be moved vertically and horizontally with respect to the fixed X3 detector using two stepper motors controlled via LabView Software. The motors are driving threaded rods with a thread pitch of 1 mm.

The energy resolution was measured for all X3 detectors using the read-out chain based on the preamplifiers MPR-16 [7]. Typical energy spectra measured at the front and rear side of the X3 detector are shown in Figure 1.

The peaks are well separated, but finer structure of the peaks cannot be resolved on either the front or the rear side.

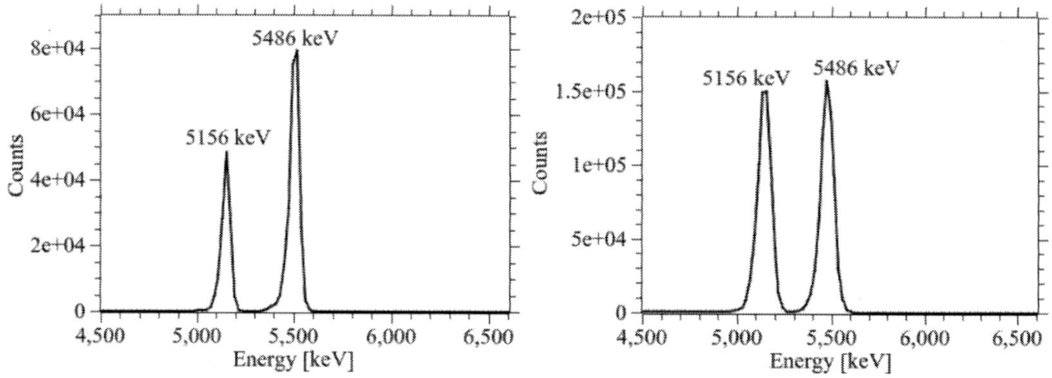

Figure 1. Energy spectra on the front and on the rear side of the X3 detector (SN:3199-14) measured with an ^{241}Am/^{239}Pu α-source using read-out chain with preamplifiers MPR-16.

The minimum and maximum energy resolution (FWHM) measured on the front side is of 37 keV–57 keV, depending on the specific detector. This is better than the previously published results of about 60 keV with an α-source for similar arrays using X3-type detectors and read-out chain with preamplifiers [6, 8]. The results obtained on the rear side are worse, i.e. they are between 63 keV –88 keV in our tests and 54 keV–81 keV in the data reported by Micron Semiconductor Ltd. with ^{241}Am α-source. The search for improving the resolution on the rear side of the X3 detectors continues, one solution is more segmentation. The best X3 detector was tested with multiplexers MUX-16. Measurements using the read-out chain with multiplexers [7] found an energy resolution which is about 50 % worse than the preamplifier chain.

The position resolution of the X3 detector was also investigated. A mask of 3 mm thick was placed at a distance of 5 mm from the front side of the detector. Slits 1 mm wide are located 1 mm apart. The α-source was set at 100 mm from the X3 detector to illuminate its entire front surface. The Figures 2 show the results of the measurement. The measured position resolution using the α-source confirms the values published for the X3-type detectors [8, 9] and is about 1.5 mm. The observed position resolution was not an intrinsic resolution, since the source had a finite size, and there was a distance between the mask and the X3 detector. According to our estimates, the intrinsic resolution of the X3 detector is less than 1 mm.

Monitoring leakage currents during measurements showed them to be higher than specified by the manufacturer. In several cases the X3 detector displayed high leakage currents (above 700 nA) even at

 DOI: 10.1051/epjconf/201716501011

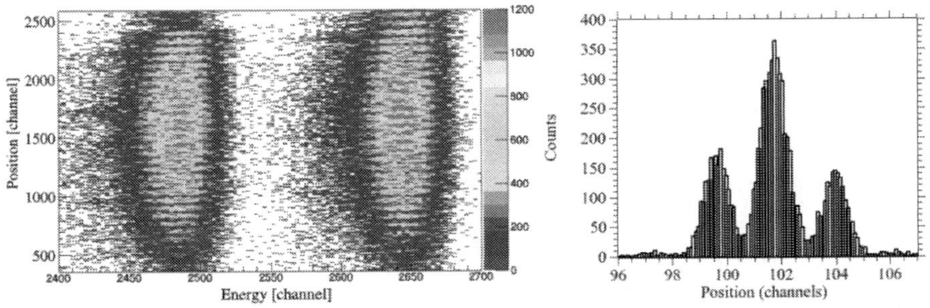

Figure 2. Energy vs position. Narrow slits of 1 mm are separated (on the left side). Calculation of the position resolution in the case with three illuminated slits: taking the average value of FWHM for the peaks 1, 3 and dividing it by the number of channels/mm (on the right side).

lower voltages. We also found an increase of leakage currents by 30-50% with the room temperature rises by 3–4°C.

The ballistic deficit manifested in the dependence of pulse amplitude of the amplifier output on the charge collection time in the detector was investigated at four shaping times of 0.25, 0.5, 1.0 and 2 μs. The effect decreased with increasing shaping time to 1 μs, but it again increased at 2 μs due to a decrease in the 'signal-to-noise' ratio. The same behavior was found for energy resolutions. They worsened by 20-30 % at the shortest and longest shaping times. The effect is most pronounced at the shaping time of 0.25 μs and is 30 keV.

3 Conclusion

At ELI-NP, a batch of forty silicon strip detectors X3 was tested, and the results were presented at the NPA8 conference. Measurement of the energy resolution on the front side showed better results than those published for similar arrays using X3 type detectors and read-out chain with preamplifiers. The leakage currents are 50 % higher, depending on the detector X3 than the values measured by the manufacturer, but no correlation was found between their magnitude and energy resolution. Two read-out chains were compared: the MUX-16 multiplexers and the MPR-16 preamplefiers. The read-out chain with preamplifiers is selected for the ELISSA project, because the energy resolutions measured with the MUX-16 multiplexers are 50 % worse.

References

[1] O. Tesileanu et al., Rom. Rep. in Phys., **68**, 699 (2016).
[2] C. Matei et al., Il Nuovo Cimento C, **39**, 360 (2016).
[3] M. La Cognata et al., JINST, **12** C03079, (2017).
[4] Micron Semiconductor Ltd, Silicon detector catalogue http://www.micronsemiconductor.co.uk/silicon-detector-catalogue/.
[5] S. D. Pain et al., Nucl. Instr. Meth. Phys. Res. B, **261**, 1122 (2007).
[6] M. Matos et al., PoS (NIC XI), 226 (2010).
[7] Mesytec GmbH & Co. KG - Detector Readout Systems, http://www.mesytec.com/.
[8] L. E. Pratt, PhD Thesis, LSU, (2014).
[9] E. Koshchiy, G.V. Rogachev et al., Nucl. Instr. Meth. Phys. Res A, **870**, (2017).

EPJ Web of Conferences **165**, 01012 (2017)
NPA8 2017
DOI: 10.1051/epjconf/201716501012

Target characterizations for direct measurement of the $^{13}C(\alpha, n)^{16}O$ reaction at LUNA 400

G.F. Ciani[1,2,★], *A.* Best[3], *L.* Csedreki[2], *Gy.* Gyürky[4], and *I.* Kochanek[2]

[1] *Gran Sasso Science Institute, Viale Francesco Crispi 7, 67100 L' Aquila, Italy*
[2] *INFN, Laboratori Nazionali del Gran Sasso (LNGS), Via G. Acilelli, 67100 Assergi, Italy*
[3] *Università di Napoli Federico II and INFN, Sezione di Napoli, Strada Comunale Cintia, 80126 Napoli, Italy*
[4] *Institute for Nuclear Research (MTA Atomki), PO Box 51, HU-4001 Debrecen, Hungary*

Abstract. The $^{13}C(\alpha, n)^{16}O$ reaction is the main neutron source for the s-process in low mass AGB stars. Although several direct measurements have been performed, no dataset reaches the Gamow window due to the low cross section. Moreover a dominant component of systematic uncertainty comes from target behaviour under beam irradiation.
This work presents a detailed characterization of enriched ^{13}C targets in order to test reproducibility and uniformity of the evaporation method, stability and purity of targets under a high intensity proton beam (100-200 uA) for the study of this reaction at the Laboratory for Underground Nuclear Astrophysics (LUNA) facility installed at Laboratori Nazionali del Gran Sasso (LNGS, INFN). A procedure to evaluate the target thickness and to monitor its degradation is discussed and preliminary results are presented.

1 Status of the art and next LUNA measurement

The $^{13}C(\alpha, n)^{16}O$ reaction is the major neutron source for the main component of the *s*-process in low mass ($1 - 3M_{\odot}$) Asymptotic Giant Branch (AGB) stars, whose temperature of interest is around $1 - 2 \cdot 10^8$ K. This corresponds to a Gamow window between 120 and 250 keV. As shown in Figure 1, altough several direct measurements have been performed [1–4], no dataset covering this energy range is available yet. This causes differences of one order of magnitude in the extrapolations towards low energies. In addition, the low energy cross section is affected by the presence of the 6.356 MeV near threshold resonance in ^{17}O.

The LUNA collaboration plans the measurement of the cross section of this reaction going down in energy approaching the Gamow window with an overall accuracy of 10%. This will be performed at LNGS, combining a neutron background reduction of 3 orders of magnitude compared with the surface [5] with an intense alpha beam delivered by the LUNA 400 accelerator [6].

★e-mail: giovanni.ciani@gssi.it

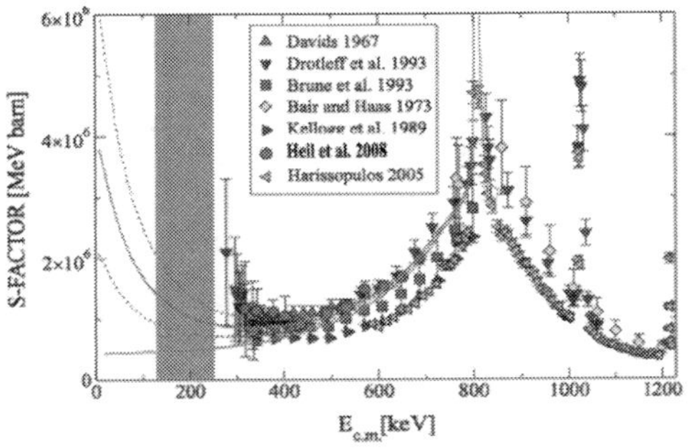

Figure 1. State of the art of the direct measurement for the $^{13}C(\alpha, n)^{16}O$ reaction adapted from Heil et al. [1]. The violet band represents the Gamow window, the red solid line is the extrapolation of the astrophysical S-factor down to this region and its error bands are indicated by red dashed lines. The green line is an extrapolation in which the near threshold resonance is omitted.

2 Characterization of ^{13}C enriched targets

Targets under investigation have been produced in MTA Atomki (Debrecen, HU) evaporating 99% enriched ^{13}C powder on tantalum backing using the electron gun technique. A total number of 10 targets were produced.
An accurate determination of the target thickness can usually be obtained by means of the Nuclear Resonant Reaction Analysis (NRRA), scanning a well known narrow resonance [7].
Due to the fact that in the LUNA 400 energy range (50-400 keV) neither $^{13}C(\alpha, n)^{16}O$ nor the $^{13}C(p, \gamma)^{14}N$ reactions can excite any resonance state, an alternative approach needs to be tested. The idea is to use the gamma signature in a HPGe detector using the off resonance cross section of $^{13}C(p, \gamma)^{14}N$ reaction (Q = 7550.56 keV) of the broad resonance at E_p = 550 keV (Γ = 23 keV), so its low energy tail could reach energies in the LUNA 400 accelerator range. This technique has already been successfully used at LUNA[8]. The feasibility study was divided in two phases, described in the next sections.

2.1 Measurements at Atomki

After the evaporation, all the enriched targets have been exposed to a proton beam at the 2MV Tandetron built at MTA Atomki, in order to scan the narrow resonance at E_p = 1748 keV (Γ = 122 eV) through the well known NRRA method and estimate their thickness to use as reference value at LUNA.
The current during this measurement was limited to 500 nA to prevent damage to the targets. γ-rays were detected using a HPGe detector at 0° with respect to the beam direction.
A dedicated set up permitted to rotate an off axis target holder and thus irradiate targets in different points, evaluating the uniformity of the evaporation process (Figure 2(a)). Figure 2(b) shows that there are variations in the width of the profile at a level of < 5%: the difference of height in the plateau is due to a different relative position between target and detector that changes slightly the efficiency; the displacement of the leading edge (~100 eV) was not fully understood. This means that in the error budget a 2% of systematic uncertainty will be added. Further investigation on this will be done in the future.

DOI: 10.1051/epjconf/201716501012

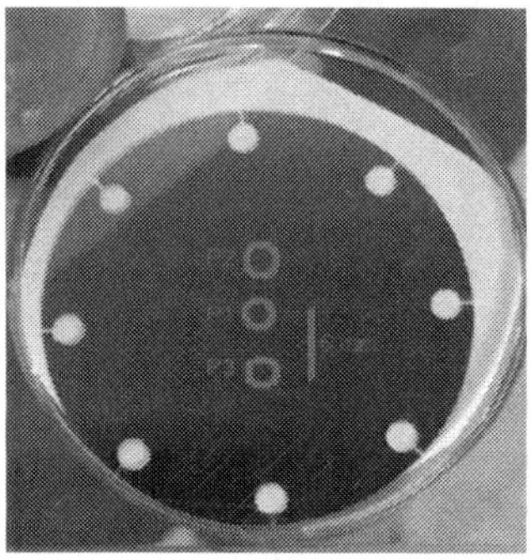

(a)

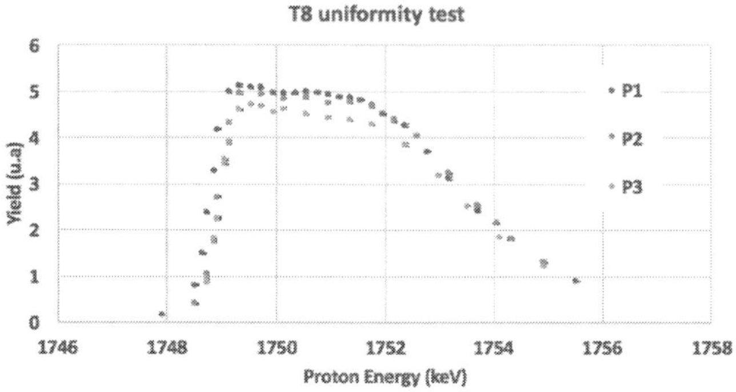

(b)

Figure 2. (a) An evaporated target; the red circles indicate the points where the target was irradiated to evaluate the uniformity of evaporation. (b) Resonance scans at positions P1, P2 and P3. For each point enough counts were accumulated to reach the 1% of statistical uncertainty in the plateau and 3% far from the resonance energy. A 3% of systematic uncertainty should be added from the charge measurement.

2.2 Measurements at LUNA 400

After the test in Atomki, targets were transported to LNGS and two of them were irradiated at the LUNA-400 facility. Here the stability of the targets under proton beam with higher current (120 uA) was tested. After an exposition of 33 C, the yield in the peak of direct capture at $E_p = 310$ keV is reduced by $\sim$ 17% (Figure 3(a)). Spectra have been acquired with a HPGe at 55° with a relative efficiency of 120%.
The tail of the broad resonance at 550 keV increases the cross section of the $^{13}C(p,\gamma)^{14}N$ reaction. The shape of the primary peak comes from the behaviour of the cross section that changes with the decreasing of the beam energy in the targets. Moreover the shape is also influenced by the energy loss of the projectiles in carbon, since the stopping power is energy dependent.
The thickness of the target was evaluated by the fit of the primary full energy peak from a transition directly to the ground state (Figure 3(b)).

3 Results and conclusions

The thickness of the evaporated targets measured in two independent ways at Atomki and LUNA cannot be compared directly; it is necessary to normalise the two with the stopping power, different for the two beam energies, taken from the software package SRIM [9] and a term that takes into account the different angle between the beam line and the target (90° at Atomki, 55° at LUNA). Thickness comparison is shown in Table 1. The surface densities evaluated using the above normalisations agree within the uncertainties even if the discrepancy between the two values is at level of 10%. The systematic uncertainties (in percentage) of the method is expected to be reduced at 5% thanks to an improvement in the peak fitting procedure of the Gamma Shape Method.
As a conclusion the targets characterization confirmed the reproducibility and homogeneity of the ^{13}C enriched evaporated targets prepared with electron gun evaporation method. Moreover, under a proton beam, targets show a high stability with a deterioration of 17% after an accumulated charge of 33 C. Further investigation is needed to understand targets' behaviour after the alpha beam irradiation,

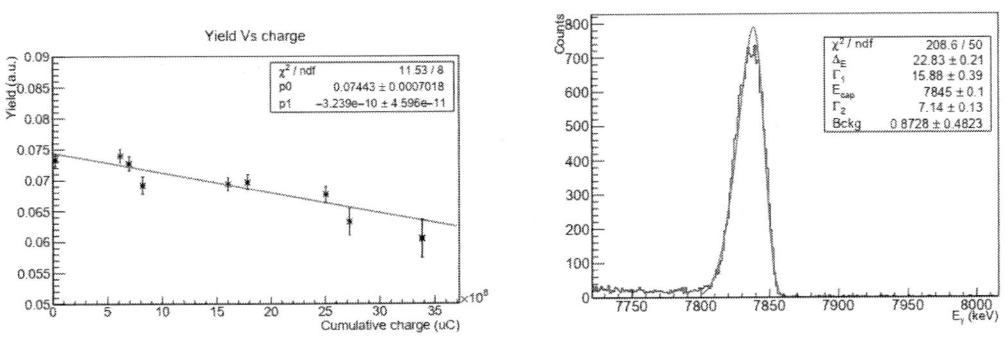

(a) (b)

Figure 3. (a)Yield as a function of the cumulative charge on the target at the reference energy $E_p = 310$ keV. The solid line is a linear best fit to the data. (b) The direct capture peak of the transition of the $^{13}C(p,\gamma)^{14}N$ at $E_p = 310$ keV to the ground state. The red curve represents the peak fit and the values of the parameters are inside the textbox.

Table 1. Thickness comparison with the two methods

	Atomki	LUNA
Thickness from fit (keV)	4.61±0.02	22.83±0.2
Stopping power (keV/10^{18}atoms/cm^2)	3.1±0.2	9.7 ± 0.6
Surface density (10^{18}atoms/cm^2)	1.49±0.09	1.35± 0.08

because higher mass of projectiles could cause a faster degradation.
The gamma-shape analysis showed good agreement with the target thickness determination using the resonance scan method and confirm its applicability for direct measurement of $^{13}C(\alpha,n)^{16}O$ reaction.

References

[1] Heil et al., Phys. Rev. C 78, 025803 (2008)
[2] H.W. Drotleff et al., ApJ, 414, 735-739, (1993)
[3] C.R. Brune et al., Phys. Rev. C 48, 6 (1993)
[4] S. Harissopulos et al., Phys. Rev. C 72 062801(R) (2005)
[5] Best et al., Nucl. Instr. Meth. Phys. Res A 812 (2016) 1
[6] A. Formicola et al., Nucl. Instrum. Methods Phys. Res., Sect. A, 507, 609 (2003)
[7] C.E. Rolfs, and W.S. Rodney, Cauldrons in the Cosmos (The University of Chicago Press, 1988)
[8] Imbriani et al., Eur. Phys. J. A 25, 455-466 (2005)
[9] J. F. Ziegler and J. P. Biersack. SRIM 2013. http://srim.org/

Informing neutron capture nucleosynthesis on short-lived nuclei with (d,p) reactions

Jolie A. Cizewski[1*], *Andrew* Ratkiewicz[1,2], *Jutta E.* Escher[2], *Alexandre* Lepailleur[1], *Steven D.* Pain[3], and *Gregory* Potel[4] for the STARLiTeR and GODDESS collaborations

[1]Department of Physics and Astronomy, Rutgers University, New Brunswick, NJ 08901 USA
[2]Nuclear & Chemical Sciences Division, Lawrence Livermore National Laboratory, CA, 94550 USA
[3]Physics Division, Oak Ridge National Laboratory, Oak Ridge, TN 37831 USA
[4]Department of Physics & Astronomy, Michigan State University and the Facility for Rare Isotope Beams, East Lansing, MI 48824 USA

Abstract. Neutron capture on unstable nuclei is important in understanding abundances in r-process nucleosynthesis. Previously, the non-elastic breakup of the deuteron in the (d,p) reaction has been shown to provide a neutron that can be captured by the nucleus and the gamma-ray decay of the subsequent compound nucleus can be modelled to predict the gamma-ray decay of the compound nucleus in the (n,γ) reaction. Preliminary results from the ^{95}Mo(d,pγ) reaction in normal kinematics support the (d,pγ) reaction as a valid surrogate for neutron capture. The techniques to measure the (d,pγ) reaction in inverse kinematics have been developed.

1 Introduction

Neutron capture reactions are responsible for the synthesis in stars of essentially all of the elements heavier than iron through either the *s* or *r* process. While the *s* process proceeds near stable nuclei, the *r*-process waiting points are short-lived and far from stability. Recent studies [1] have demonstrated that unknown (n,γ) rates on nuclei near the *r*-process path, and in particular near closed neutron shells, could have significant impact on predicting abundances with *r*-process network calculations. Constraining (n,γ) rates could also serve to inform our knowledge of the site of the *r* process.

Neutron capture near closed shells can proceed by two processes. Direct (including semi-direct) capture can be deduced if the spectroscopic factors of low-spin states have been measured, for example, with (d,p) reactions with radioactive ^{126,128,130,132}Sn beams [2]. For open neutron shell nuclei, neutron capture is expected to proceed predominantly through the population of a compound nucleus with gamma decay that proceeds by many paths. While the population of the compound nucleus can be calculated with optical models, the decay depends upon the level density and γ-ray strength function, properties that cannot be accurately extrapolated to weakly bound nuclei, far from stability. Because

* Corresponding author : cizewski@physics.rutgers.edu

of the importance of (n,γ) cross sections on short-lived neutron-rich nuclei in understanding r-process nucleosynthesis, and also in applications of nuclear physics, it is important to validate a surrogate for neutron capture and in particular to inform the decay of compound nuclei [3].

The (d,pγ) reaction is a promising surrogate for (n,γ) and for measurements with radioactive ion beams (RIBs). Potel and collaborators have recently developed the framework [4] in which the non-elastic breakup of the deuteron provides a neutron that can be captured by the nucleus. This framework also provides an angular momentum decomposition as a function of excitation energy (effective neutron energy) as well as the fraction of the (d,p) reaction cross section in the elastic breakup channel. Escher and colleagues [5] have developed the techniques to fit surrogate reaction gamma-ray decay probabilities as a function of excitation energy to extract the Hauser-Feshbach (HF) decay of the compound nucleus. The decay of the compound nucleus is corrected for the differences in entry angular momentum between the (d,p) reaction (predicted within the Potel framework) and neutron capture. This corrected decay of the compound nucleus is combined with the cross section for the target + neutron system calculated with optical models to predict (n,γ) cross sections as a function of neutron energy.

The (d,p) reaction can also be used to deduce the direct-semi-direct component of neutron capture as a function of neutron energy important in nuclei with low neutron separation energies (S_n), for example near closed neutron shells. Finally the (d,p) reaction is well suited for measurements with RIBs in inverse kinematics, where events measured at angles >90° in the laboratory are dominated by reaction protons, with other channels, such as elastic and inelastic scattering, observed at angles forward of 90°.

2 Validating (d,pγ) as (n,γ) surrogate

The ^{95}Mo+neutron system [6] was chosen to validate the (d,pγ) reaction as a neutron capture surrogate because the (n,γ) cross section has been measured up to 200 keV [7,8]. Also, the final nucleus ^{96}Mo is even-even, with a relatively simple decay scheme at low excitation energies, and the level scheme is well-established [8]. The primary validation experiment was conducted at the Texas A&M Cyclotron Institute with ≈12.5-MeV deuteron beams interacting with an enriched ^{95}Mo target. The reaction charged particles and gamma rays were measured in singles and coincidence with the STAR-LiTeR system of annular silicon strip and Compton-suppressed high-purity clover Ge detectors [9]. A preliminary gamma-ray energy as a function of excitation energy (measured with protons) matrix is displayed in Figure 1. Below the neutron separation energy S_n, all of the gamma rays are in ^{96}Mo. Above S_n, surrogate neutron capture (transitions in ^{96}Mo) and surrogate (n,n') transitions in ^{95}Mo are observed. By gating on excitation energy, the gamma-ray emission probability as a function of excitation energy (or effective neutron energy) can be deduced, as displayed in Figure 2. The 778-keV $2^+\rightarrow0^+$ transition in ^{96}Mo collects essentially all of the gamma-ray intensity with very little intensity bypassing the 2^+ state, e.g., via the $2_2^+\rightarrow0^+$ transition. Also the 778-keV transition is about a factor of five stronger than the 850-keV $4^+\rightarrow2^+$ yrast transition, with the 812-keV $6^+\rightarrow4^+$ yrast transition a factor of 20 lower. This supports a relatively low angular momentum transfer even at high excitation energies approaching S_n.

Preliminary results for fitting the measured gamma-ray intensity probabilities are displayed in Figure 2. These account for the angular momentum transferred in the surrogate reaction. The decay model resulting from the fit was combined with optical model predictions for the formation of the ^{95}Mo+n compound nucleus to deduce the surrogate

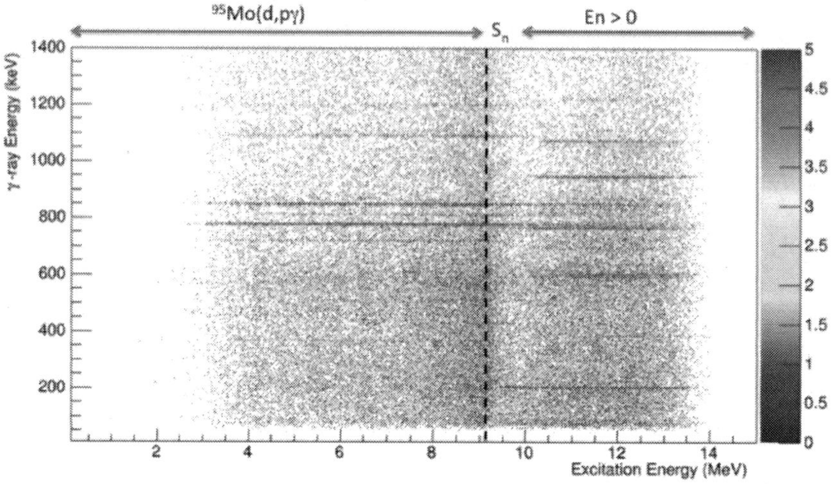

Figure 1. Preliminary gamma-ray energy as a function of excitation energy (measured with protons) matrix for the ^{95}Mo(d,pγ) reaction. The vertical line represents neutron separation energy S_n that is also the zero for effective neutron energies.

(n,γ) cross sections. Preliminary results indicate good agreement between measured, evaluated and surrogate-reaction deduced (n,γ) cross sections. Next steps include incorporating uncertainties in the modelling of the ^{95}Mo+d system.

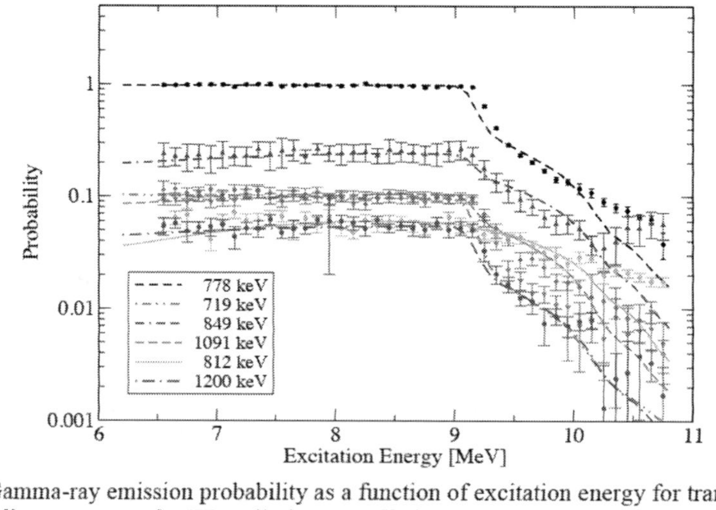

Figure 2. Gamma-ray emission probability as a function of excitation energy for transitions in ^{96}Mo. The dashed lines represent the HF preliminary predictions of the γ-ray emission probabilities.

3 Developing techniques to measure (d,pγ) reactions with RIBs

We have developed the capabilities to measure the (d,pγ) reaction in inverse kinematics on the path to realizing such measurements with RIBs interacting with CD_2 targets. Light reaction products are detected in the Oak Ridge Rutgers University Barrel Array (ORRUBA) of position-sensitive silicon-strip detectors augmented by annular arrays of graduated silicon-strip detectors at back and forward angles in the laboratory. Reaction gamma rays are detected in the Gammasphere array of Compton-suppressed Ge detectors to realize the GODDESS system [10]: Gammasphere ORRUBA Dual Detectors for

Experimental Structure Studies. As a result of analyzing the particle-gamma coincidences, new states at $E_x \approx 2$ MeV have been identified as candidates for $2f_{7/2}$ and $3p_{3/2}$ excitations above the N=82 shell gap. Above the neutron separation energy, transitions in the (n,n') surrogate ^{134}Xe nucleus are observed. Further analysis of the (d,pγ) reactions with ^{134}Xe and ^{95}Mo beams (to demonstrate a surrogate in inverse kinematics) are in progress.

4 Summary and conclusions

In summary, we are demonstrating that the (d,pγ) reaction is a valid surrogate for the (n,γ) reaction important in understanding observed abundances in *r*-process nucleosynthesis. The Potel theoretical framework has shown how non-elastic breakup of the deuteron in the (d,p) reaction provides a neutron to be captured. Level density and gamma-ray strength functions from (d,pγ) reactions can be deduced and used in calculating (n,γ) cross sections. The techniques to measure (d,pγ) reactions with RIBs have been developed. First experiments with N≈82 nuclei near the *r*-process path of nucleosynthesis are being planned.

This work is supported in part by the U.S. Department of Energy National Nuclear Security Administration under the Stewardship Science Academic Alliances program DE-FG52-09NA29467 and DE-NA0000979, LLNL DE-AC52-07NA27344 and LDRD 16-ERD-022 and the Office of Nuclear Physics ORNL DE-AC05-00OR22725 and the National Science Foundation.

References

1. M. Mumpower *et al.*, Prog. Nucl. Part. Phys. **86**, 86 (2016)
2. R. L. Kozub *et al.*, Phys. Rev. Lett. **109**, 172501 (2012) and B. Manning *et al.*, (submitted 2017) and references therein.
3. Almudena Arcones *et al.*, Prog. Nucl. Part. Phys. **94**, 1 (2017)
4. G. Potel, F. M. Nunes, and I. J. Thompson, Phys. Rev. C **92**, 034611 (2015)
5. J. E. Escher *et al.*, Rev. Mod. Phys. **84**, 353 (2012) and J. E. Escher *et al.*, EPJ Web of Conferences **122**, 12001 (2016)
6. A. Ratkiewicz *et al.*, EPJ Web of Conferences **93**, 02012 (2015)
7. A. De L. Musgrove *et al.*, Nucl. Phys. A **270**, 108 (1976)
8. ENDF and ENSDF databases http://www.nndc.bnl.gov
9. S. Lesher *et al.*, Nucl. Inst. & Meth. in Phys. Res. A **621**, 286 (2010) and R. J. Casperson *et al.*, Phys. Rev. C **90**, 034601 (2014)
10. S.D. Pain, *et al.*, Physics Procedia (in press, 2017)

EPJ Web of Conferences **165**, 01014 (2017) DOI: 10.1051/epjconf/201716501014
NPA8 2017

The Nuclear Astrophysics program at n_TOF (CERN)

N. Colonna[1,*], O. Aberle[8], J. Andrzejewski[7], L. Audouin[16], M. Bacak[11,8,12], J. Balibrea[6], M. Barbagallo[1], F. Bečvář[19], E. Berthoumieux[12], J. Billowes[17], D. Bosnar[18], A. Brown[20], M. Caamaño[4], F. Calviño[21], M. Calviani[8], D. Cano-Ott[6], R. Cardella[8], A. Casanovas[21], F. Cerutti[8], Y. H. Chen[16], E. Chiaveri[8], G. Clai[15,30], G. Cortés[21], M. A. Cortés-Giraldo[2], Cosentino[22], S. Cristallo[43], L. A. Damone[1,23], M. Diakaki[12], C. Domingo-Pardo[3], L. R. Dressler[14], E. Dupont[12], I. Durán[4], B. Fernández-Domínguez[4], A. Ferrari[8], P. Ferreira[10], P. Finocchiaro[22], V. Furman[24], K. Göbel[5], A. R. García[6], A. Gawlik[7], S. Gilardoni[8], T. Glodariu[9], I. F. Gonçalves[10], E. González[6], E. Griesmayer[11], C. Guerrero[2], F. Gunsing[12,8], H. Harada[25], S. Heinitz[14], A. Hernandez-Prieto[21], J. Heyse[26], D. G. Jenkins[20], E. Jericha[11], F. Käppeler[27], Y. Kadi[8], A. Kalamara[28], P. Kavrigin[11], A. Kimura[25], N. Kivel[14], M. Kokkoris[28], J. Kroll[19], M. Krtička[19], D. Kurtulgil[5], C. Langer[5], E. Leal-Cidoncha[4], C. Lederer[29],J. Lerendegui-Marco[2], H. Leeb[11], S. Lo Meo[15,30], S. J. Lonsdale[29], S. Losito[8], D. Macina[8], A. Mallik[41], J. Marganiec[7,13], T. Martínez[6], A. Masi[8], C. Massimi[30,31], P. Mastinu[32], M. Mastromarco[1], E. A. Maugeri[14], A. Mazzone[1,33], E. Mendoza[6], A. Mengoni[15], F. Mingrone[8], P. M. Milazzo[34], M. Mirea[9], A. Musumarra[22,35], A. Negret[9], R. Nolte[13], A. Oprea[9], N. Patronis[36], A. Pavlik[37], J. Perkowski[7], L. Piersanti[43], I. Porras[38], J. Praena[38], J. M. Quesada[2], D. Radeck[13], T. Rauscher[39,40], R. Reifarth[5], A. Riego[21], M.S. Robles[4], C. Rubbia[8], J. A. Ryan[17]M. Sabaté-Gilarte[8,2], A. Saxena[41], P. Schillebeeckx[26], D. Schumann[14], P. Sedyshev[24], A. G. Smith[17], N. V. Sosnin[17], A. Stamatopoulos[28], G. Tagliente[1], J. L. Tain[3], A. Tarifeño-Saldivia[21]L. Tassan-Got[16], A.Tsinganis[8], S. Valenta[19], G. Vannini[30,31], V. Variale[1], P. Vaz[10], A. Ventura[30], V. Vlachoudis[8], R. Vlastou[28], A. Wallner[42], S. Warren[17], C. Weiss[11], P. J. Woods[29]T. Wright[17], P. Žugec[18,8]
The n_TOF Collaboration

[1]*Istituto Nazionale di Fisica Nucleare, Sezione di Bari, Italy;* [2]*Universidad de Sevilla, Spain;* [3]*Instituto de Física Corpuscular, Universidad de Valencia, Spain;* [4]*University of Santiago de Compostela, Spain;* [5]*Goethe University Frankfurt, Germany;* [6]*Centro de Investigaciones Energéticas Medioambientales y Tecnológicas (CIEMAT), Spain;* [7]*University of Lodz, Poland;* [8]*European Organization for Nuclear Research (CERN), Switzerland;* [9]*Horia Hulubei National Institute of Physics and Nuclear Engineering, Romania;* [10]*Instituto Superior Técnico, Lisbon, Portugal;* [11]*Technische Universität Wien, Austria;* [12]*CEA Saclay, Irfu, Gif-sur-Yvette, France;* [13]*Physikalisch-Technische Bundesanstalt (PTB), Braunschweig, Germany;* [14]*Paul Scherrer Institut (PSI), Villingen, Switzerland;* [15]*Agenzia nazionale per le nuove tecnologie (ENEA), Bologna, Italy;* [16]*Institut de*

[*] Corresponding author: nicola.colonna@ba.infn.it

DOI: 10.1051/epjconf/201716501014

Physique Nucléaire, CNRS-IN2P3, Univ. Paris-Sud, Université Paris-Saclay, France; [17]University of Manchester, United Kingdom; [18]University of Zagreb, Croatia; [19]Charles University, Prague, Czech Republic; [20]University of York, United Kingdom; [21]Universitat Politècnica de Catalunya, Spain; [22]INFN Laboratori Nazionali del Sud, Catania, Italy; [23]Dipartimento di Fisica, Università degli Studi di Bari, Italy; [24]Joint Institute for Nuclear Research (JINR), Dubna, Russia; [25]Japan Atomic Energy Agency (JAEA), Tokai-mura, Japan; [26]European Commission, Joint Research Centre, Geel, Belgium; [27]Karlsruhe Institute of Technology, IKP, Karlsruhe, Germany; [28]National Technical University of Athens, Greece; [29]School of Physics and Astronomy, University of Edinburgh, United Kingdom; [30]Istituto Nazionale di Fisica Nucleare, Sezione di Bologna, Italy; [31]Dipartimento di Fisica e Astronomia, Università di Bologna, Italy; [32]Istituto Nazionale di Fisica Nucleare, Sezione di Legnaro, Italy; [33]Consiglio Nazionale delle Ricerche, Bari, Italy; [34]Istituto Nazionale di Fisica Nucleare, Sezione di Trieste, Italy; [35]Dipartimento di Fisica e Astronomia, Università di Catania, Italy; [36]University of Ioannina, Greece; [37]University of Vienna, Faculty of Physics, Vienna, Austria; [38]University of Granada, Spain; [39]Department of Physics, University of Basel, Switzerland; [40]Centre for Astrophysics Research, University of Hertfordshire, United Kingdom: [41]Bhabha Atomic Research Centre (BARC), India; [42]Australian National University, Canberra, [43] INAF – Osservatorio Astronomico di Collurania, Teramo, Italy

Abstract. An important experimental program on Nuclear Astrophysics is being carried out at the n_TOF since several years, in order to address the still open issues in stellar and primordial nucleosynthesis. Several neutron capture reactions relevant to s-process nucleosynthesis have been measured so far, some of which on important branching point radioisotopes. Furthermore, the construction of a second experimental area has recently opened the way to challenging measurements of (n, charged particle) reactions on isotopes of short half-life. The Nuclear Astrophysics program of the n_TOF Collaboration is here described, with emphasis on recent results relevant for stellar nucleosynthesis, stellar neutron sources and primordial nucleosynthesis.

1 Introduction

Neutron-induced cross sections play a fundamental role in Nuclear Astrophysics, being a key nuclear physics ingredient in stellar nucleosynthesis of heavy elements [1], as well as in light element production in Big Bang Nucleosynthesis [2]. In stars, neutron capture reactions are responsible for the production of the majority of elements heavier than Fe, with two processes contributing more or less equally to the overall abundance pattern: the s- and the r-process. In the first one, which involves low neutron densities and mostly stable isotopes, progressively heavier elements are produced starting from the Fe seed nuclei, by neutron captures and subsequent β-decays. An interesting case occurs when an unstable nucleus with a relatively long half-life is produced in the process, as it can either decay or undergo a further neutron capture reaction, thus generating a branching in the s-process path. The competition between neutron capture and β-decay depends on the capture cross section, on the stellar half-life of the branching isotope and, most importantly, on the stellar thermodynamic conditions, in particular the neutron density and stellar temperature. As a consequence, the study of the branching points can shed some light on the stellar environment in which the s-process takes place, provided that accurate nuclear data on capture cross section and decay half-life are available.

Apart from heavy elements, neutron capture cross sections are needed for some light elements acting as neutron poison, or linked to stellar neutron sources. In this last case, neutron spectroscopy may provide information that are not easily accessible with other experimental techniques. Finally, neutron cross sections are needed to clarify one of the most intriguing and long-lasting problems in Big Bang Nucleosynthesis (BBN), i.e. the gross overestimate in BBN models of the primordial abundance of Lithium.

Although neutron-induced reactions of relevance for Nuclear Astrophysics are being studied since many decades at neutron facilities worldwide, some open issues in stellar and primordial nucleosynthesis still remain to be addressed in order to reach a more comprehensive understanding of the elemental abundance distribution and of the galactic chemical evolution. To this end, an intense experimental program is undergoing at the neutron facility n_TOF (CERN) since almost two decades, in order to reduce the uncertainty on neutron capture and (n, charged particle) cross sections for some key isotopes, and ultimately improve the reliability of astrophysical models. Thanks to the innovative characteristics of the neutron beam, to the development of high-performance detection and data acquisition systems, and to the use of refined analysis and simulation techniques, a number of valuable results of relevance for the field of Nuclear Astrophysics have been produced so far. After a brief description of the facility, the most important results recently obtained at n_TOF are here discussed.

2 The n_TOF facility at CERN

The n_TOF facility was proposed and built at CERN with the aim of addressing the needs of new data of interest for Nuclear Astrophysics, for basic Nuclear Physics and for technological applications [3]. Neutrons are produced by spallation of a pulsed proton beam of 20 GeV/c momentum from the Proton Synchroton (PS) impinging on a monolithic water-cooled Pb target. A further layer of demineralized or borated water in the forward direction ensures additional moderation of the neutron energy, so to produce an almost isolethargic spectrum. For the first thirteen years of operation, only one neutron beam line, in the horizontal direction, was available, with the experimental area (EAR1) located at 185 m distance from the spallation target. In this area, the neutron beam is characterized by an instantaneous intensity of 10^{6} neutrons/bunch, an energy spectrum extending over almost eleven orders of magnitude, from thermal to 1 GeV, and a high energy resolution ($\Delta E/E<10^{-3}$ in most of the energy range). More details on this area can be found in Ref. [4].

A second beam line in the vertical direction was recently completed at n_TOF. The corresponding experimental area (EAR2) is located at 20 m distance from the spallation target [5]. Compared with EAR1, the neutron beam in the new area has a much higher intensity ($>10^{7}$ neutrons/bunch), at the expenses of a worse energy resolution and slightly smaller energy range. A schematic drawing of the facility is shown in Figure 1 (left panel), together with the neutron flux distribution in the two areas (right panel).

The two experimental areas are somewhat complementary. The characteristics of the neutron beam in EAR1 are ideal for high resolution measurements of neutron capture reactions on stable or long-lived radioactive isotopes ($t_{1/2}>100$ y), while the much higher flux in EAR2 allows one to perform measurements on radioisotopes with short half-life, on samples available in small (sub-mg) mass, on reactions with low cross section, or on all of these at the same time. In fact, for radioactive isotopes the combination of the higher flux and shorter time-of-flight results in an increase of the signal-to-background ratio of more than two orders of magnitude, relative to EAR1, when considering the background related to the radioactivity of the sample. As a consequence, in EAR2 it is possible to perform challenging measurements on isotopes with half-life as short as a few months.

DOI: 10.1051/epjconf/201716501014

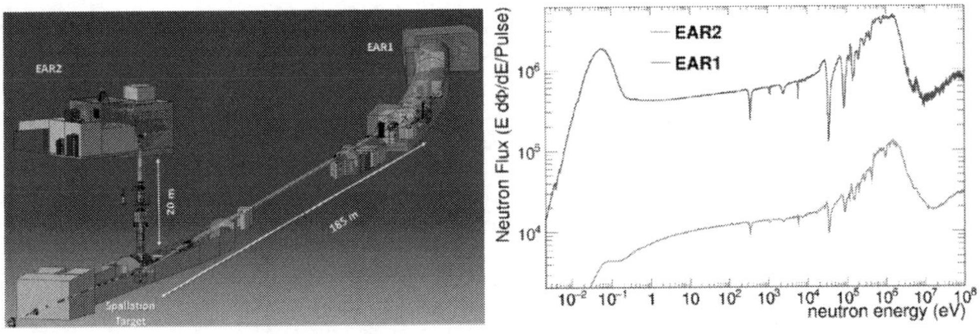

Fig. 1. Schematic drawing of the n_TOF facility, showing the horizontal and vertical beam lines and corresponding experimental areas (left panel) and, number of neutrons per bunch for the beams in the two areas (right panels).

3 Recent results in Nuclear Astrophysics

The experimental program of the n_TOF Collaboration in the field of Nuclear Astrophysics has, up to recently, mostly regarded neutron capture cross sections. Since the start of operation, in 2001, a large number of capture reactions have been measured, in EAR1, on a variety of subjects, from neutron magic nuclei, acting as bottleneck for the reaction flow of the s-process, to branching point isotopes, from light nuclei acting as neutron poison, to end-point nuclei, and, finally, to isotopes of special interest, such as those involved in the Os/Re nuclear cosmochronometer (see [6] for the full list of n_TOF results, or visit the site www.cern.ch/ntof). In all cases, a considerable reduction of the uncertainty on the capture cross section has been obtained, thanks to the peculiar features of the n_TOF neutron beam, in particular the high instantaneous neutron intensity and the high resolution. With the construction of the second experimental area, new lines of research have become accessible, in particular on reactions leading to light charged particle emission. In the following, a few indicative results recently obtained on some of the above-mentioned classes of reactions are described more in details.

3.1 Neutron capture cross section of branching point isotopes

Branching points isotopes are unstable nuclei whose half-life is such that a competition exists between neutron capture and β-decay. An accurate knowledge of the capture cross section of these isotopes can help characterize the stellar conditions in which s-process occurs. A comprehensive list of the most important branching point isotopes can be found in [1].

At present, the neutron capture cross section of many of these isotopes is poorly known, mainly due to the difficulties in producing and handling samples of sufficient mass and adequate purity. Furthermore, the natural radioactivity of the sample is a source of background that might heavily affect the capture yield. For these reasons, few measurements have been performed on these isotopes so far, at a time-of-flight facility. At n_TOF, two important measurements, on ^{151}Sm [7] and ^{63}Ni [8], were successfully completed in the first experimental area. The relatively long half-life and large sample masses resulted in a high signal-to-background ratio, and allowed the n_TOF Collaboration to determine the Maxwellian Averaged Cross Section with an unprecedented few percent uncertainty. In both cases, the n_TOF results bore important Astrophysics implications.

The construction of the second experimental area at n_TOF, with the much larger neutron intensity, opened the way to measurements of neutron capture cross sections of isotopes with shorter half-life, and/or to samples of sub-mg masses. Three important s-process branching point isotopes with these characteristics were recently measured at n_TOF: ^{171}Tm, ^{147}Pm and the ^{204}Tl [9]. Table I lists the half-life of these isotopes, together with the mass of the samples used in the n_TOF measurements, and the measured or theoretically calculated Maxwellian Averaged Cross Section.

Table 1. Measurements on the capture cross section of branching point isotopes measured so far at n_TOF. Experimental results are reported with the respective error, while for calculations a range of predicted values is indicated.

Isotope	Half-life (years)	Sample mass (mg)	Experimental area	MACS at 30 keV (mb)
^{151}Sm	90.	200.	EAR1	3100±160 (Ref. [7])
^{63}Ni	100.	130.	EAR1	67±14 (Ref. [8])
^{204}Tl	3.8	11.	EAR1	97-328 (calculated)
^{171}Tm	1.9	3.6	EAR1 & EAR2	243-1332 (calculated)
^{147}Pm	2.6	0.3	EAR2	400-2000 (calculated) 709±100 (Ref. [9])

The ^{171}Tm branching point is independent of stellar temperature and can therefore provide a strong constraint on the s-process neutron density in low mass AGB stars, while ^{204}Tl, responsible for the production of ^{204,206}Pb (screened from the r-process by the stable ^{204}Hg and ^{203}Tl isotopes), is crucial for understanding the nucleosynthesis of heavy elements in AGB stars, as well as for providing hints on the stellar object that polluted the proto-solar nebula at the epoch of the solar system formation. Finally, the analysis of the branching point isotope ^{147}Pm in the Nd-Pm-Sm region is important for modeling the AGB star evolution and to put accurate constraints on the interplay between metallicity, initial stellar mass and mixing processes.

Measuring the neutron capture cross section of these three isotopes is a challenging task because of the high specific activity of these radionuclides, typically resulting in a poor signal-to-background ratio, and because of the difficulty in the procurement of a sufficient amount of material with adequate purity, as well as in sample preparation and characterization. Because of these problems, very scarce or no data have been collected so far on these isotopes, and at present one can only rely on highly uncertain theoretical calculations, as reported in Table I, or on a single activation measurement for ^{147}Pm [10]. The availability of intense neutron beams at n_TOF finally provided the unique opportunity to collect, for the first time, high quality data at a time-of-flight installation for these reactions. The radioisotopes were produced at the Institute Laue Langevin ILL (Grenoble, France) high flux reactor, by irradiating for several weeks under the thermal neutron beam the stable progenitors with thermal neutrons. The chemical separation and sample preparation was carried out at the Paul Scherrer Institute (PSI), Switzerland, as described in Ref. [11].

The analysis of these reactions is currently under way, but once completed it is expected to provide important experimental information on the capture cross section of these isotopes, in some cases for the first time ever, possibly leading to a better characterization of these branching points.

3.2 Studies of stellar neutron sources by neutron spectroscopy

One of the main issues in Nuclear Astrophysics regards the modelling of neutron generation in the stellar environment. Large efforts are currently being devoted in various

DOI: 10.1051/epjconf/201716501014

laboratories worldwide to determine the reaction rate of the two most important neutron sources in AGB stars, i.e. the $^{13}C(\alpha,n)^{16}O$ and $^{22}Ne(\alpha,n)^{25}Mg$ reaction. The accurate knowledge of the reaction rate for these sources is decisive for deriving information on neutron density, temperature, and pressure in the He-burning layers of AGB stars. At low projectile energy, direct measurements are difficult and yield, at present, highly uncertain results. An alternative approach to constrain the reaction rate at low energy is to measure indirect and transfer reactions. In particular, the study of neutron-induced reactions could provide useful information on the properties of the states above the neutron threshold of the compound nuclei formed in the reactions.

In this respect, neutron spectroscopy of ^{26}Mg states, of relevance for the $^{22}Ne(\alpha,n)^{25}Mg$ neutron source, was recently carried out using data from neutron capture and transmission measurements performed at n_TOF and GELINA, respectively [12]. This latter facility, located at the European Commission Joint Research Center in Geel, Belgium, is based on a linear electron accelerator and a mercury-cooled uranium target for neutron production [13]. The time properties of the electron beam and the small target size make GELINA particularly suited for high-resolution cross section measurements in the resonance region. A ^{6}Li-glass detector was used in the transmission measurement, performed in the experimental station placed at 50 m distance from the neutron-producing target.

High-resolution neutron capture yields for a high-purity, metallic ^{25}Mg sample measured in EAR1 at n_TOF, were combined with the transmission data collected at GELINA in a simultaneous R-matrix analysis performed on both data sets. This allowed to univocally assign the spin and parity of some important excited states of ^{26}Mg with natural parity just above the neutron separation energy, not easily accessible in the direct α+Ne reactions. Based on the n_TOF/GELINA data, a new, more reliable estimate of the $^{22}Ne(\alpha,n)$ cross section below 800 keV α-particle energy was obtained. The result, shown Figure 2, puts firmer constraints on the reaction rate of this important stellar neutron source. The use of neutron spectroscopy is now being considered at n_TOF for the primary neutron source as well, the $^{13}C(\alpha,n)^{16}O$ one. In this case, the Collaboration is planning a measurement with high resolution of the time-reversal reaction, the $^{16}O(n,\alpha)^{13}C$.

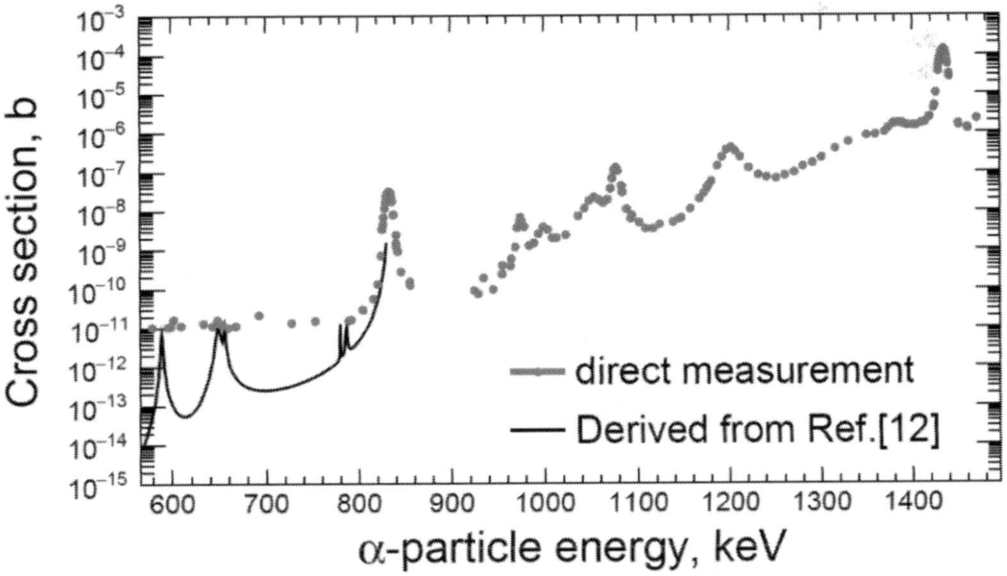

Fig. 2. Cross section of the ^{22}Ne(α,n)^{25}Mg reaction from a direct measurement (red symbols) and calculated on the basis of the neutron spectroscopy of Ref. [12].

3.3 Measurement of (n,cp) reactions in EAR2@n_TOF

The very high instantaneous neutron intensity in the second experimental area at n_TOF has also made feasible to measure (n, charged particle) reactions of low cross section and/or for isotopes of short-half life and/or with samples of extremely small mass. Two isotopes of high relevance in astrophysics reactions have been investigated so far at n_TOF: ^{7}Be and ^{26}Al.

Neutron-induced reactions on ^{7}Be ($t_{1/2}$=53.2 days) are of interest for Big Bang Nucleosynthesis, more specifically for the so-called "Cosmological Lithium problem" [2]. More details on the motivation and on the measurements can be find in [14-15]. Two reactions were measured on this isotope at n_TOF: the ^{7}Be(n,α)^{4}He and the ^{7}Be(n,p)^{7}Li reactions. For the first one, never measured before above the thermal neutron energy, two samples of ~3 μg total mass were used (corresponding to a total activity of 40 GBq). A sufficient amount of ^{7}Be was extracted from the SINQ spallation source of PSI, chemically separated and deposited on different backings using different techniques. More details on the sample preparation can be found in [16]. For the (n,p) reaction, characterized by a much higher cross section, it was sufficient to use a sample of mass as small as 90 ng. In this last case, however, a sample of high isotopic purity was produced by implantation of a ^{7}Be beam at ISOLDE. High performance experimental setups, based on solid state detectors [17-18], were specifically designed and built for the measurements. Combined with the characteristics of the EAR2 neutron beam, the world-class expertise in sample preparation and detector development has led the n_TOF Collaboration to collect high quality results on both reactions, which are expected to clarify once and for all the role of the neutron induced reactions on ^{7}Be in the Cosmological Lithium Problem.

Other successful measurements recently performed at n_TOF regard the ^{26}Al(n,p) and (n,α) reactions [19]. Precise satellite observations of this cosmic γ-ray emitter, whose relatively short lifetime (1 My) provides evidence of ongoing nucleosynthesis in our galaxy, offer the opportunity to improve calculations of the production and destruction rates of some key astrophysical reaction. However, the poor knowledge of the (n, cp) reactions on ^{26}Al, responsible for its destruction, are at present the major source of uncertainty in predicting the amount of ^{26}Al ejected into the interstellar medium by Wolf-Rayet stars in the stellar wind or following the supernova explosion. The challenge in this case is related to the availability of a sufficient amount of ^{26}Al, and to the need of identifying the emitted particles. At n_TOF, a sample of 10 μg mass, produced at Los Alamos National Laboratory (USA) [20] was irradiated under the high-flux neutron beam of the second experimental area, with the reaction product detected in a position-sensitive solid state telescope. While the analysis is still in progress, preliminary results indicate that the n_TOF measurement might provide accurate new data on the important n+^{26}Al reaction cross sections.

4 Conclusions and perspectives

The n_TOF Collaboration has been carrying out since several years an ambitious experimental program on Nuclear Astrophysics, with the aim of reducing the uncertainties on cross sections of various reactions involved in stellar and primordial nucleosynthesis. The unique features of the neutron beam in the two experimental areas have made possible to study neutron capture reactions on several unstable isotopes acting as branching points, to perform neutron spectroscopy relevant for stellar neutron sources and, more recently, to measure challenging (n, cp) reactions on short-lived radioisotopes or sub-mg samples. In all cases, a substantial reduction of the uncertainties on cross sections relevant for stellar and primordial nucleosynthesis has been achieved. New, challenging measurements are foreseen for the near future, whose results are expected to lead to significant advancement in the field of Nuclear Astrophysics.

We acknowledge the support by the Narodowe Centrum Nauki (NCN), under the Grant No. UMO-2012/04/M/ST2/00700.

References

1. F. Käppeler et al., Rev. Mod. Phys **83**, 157 (2011)
2. R.H. Cyburt et al., Phys. Rev. D **69**, 123519 (2004)
3. C. Rubbia et al, Tech Rep. CERN/LHAC/98-02 CERN (1998)
4. C. Guerrero et al., Eur. Phys J. A **49**, 27 (2013)
5. C. Weiss et al., Nucl. Instrum. Methods A **799**, 90 (2015)
6. F. Gunsing et al., Eur. Phys. J. Plus **131**, 371 (2016)
7. U. Abbondanno et al., Phys. Rev. Lett. **93**, 161103 (2004)
8. C. Lederer et al., Phys. Rev. Lett. **110**, 022501 (2013)
9. C. Guerrero et al., JPS Conf. Proc. **14**, 010903 (2016)
10. C. Arlandini, M. Heil, R. Reifarth, F. Käppeler, P.V. Sedyshev, Nucl. Phys. A **688**, 487c (2001)
11. S. Heinitz et al., Radiochemica Acta, DOI: 10.1515/ract-2016-2728
12. C. Massimi et al., Phys. Lett. B **768**, 1 (2017)
13. W. Mondelaers and P. Schillebeeckx, Notiziario Neutroni e Luce di Sincrotrone **11**, 19 (2006)
14. M. Barbagallo et al., Phys. Rev. Let. **117**, 152701 (2016);
15. M. Barbagallo et al., contribution to NPA VIII
16. E.A. Maugeri et al., J. Inst. **12**, P02016 (2017)
17. L. Cosentino et al., Nucl. Instrum. Methods A **830**, 197 (2016)
18. M. Barbagallo et al., submitted to Nucl. Instrum. Methods A
19. C. Lederer et al., CERN-INTC-20154-006/INTC-P-406
20. C. Ingelbrecht et al., Nucl. Instr. Meth. A **480**, 114 (2002)

EPJ Web of Conferences **165**, 01015 (2017) DOI: 10.1051/epjconf/201716501015
NPA8 2017

Cross section measurements in the $^{12}C+^{12}C$ system

S. Courtin[1,2,3], C.L. Jiang[4], G. Fruet[1,2], M. Heine[1,2], D.G. Jenkins[5], P. Adsley[6], L.G. Morris[5], P.H. Regan[7,8], M. Rudigier[7], D. Montanari[1,2,3], S. Della Negra[6], N. de Séréville[6], F. Haas[1,2], F. Hammache[6], O.S. Kirsebom[9], J. Lesrel[6], A. Meyer[6], D. Montanari[1,2,3], K. Auranen[4], M.L. Avila[4], A.D. Ayangeakaa[4], B.B. Back[4], S. Bottoni[4], M. Carpenter[4], C. Dickerson[4], B. DiGiovine[4], J.P. Greene[4], D.J. Henderson[4], C.R. Hoffman[4], R.V.F. Janssens[4], B.P. Kay[4], S.A. Kuvin[4], T. Lauritsen[4], R.C. Pardo[4], K.E. Rehm[4], D. Santiago-Gonzalez[4], J. Sethi[4], D. Seweryniak[4], R. Talwar[4], C. Ugalde[4], S. Zhu[4], C.M. Deibel[10], S.T. Marley[10], D. Bourgin[1,2], C. Stodel[11], A. Lefebvre-Schuhl[12], S. Almaraz-Calderon[13], X. Fang[14], X.D. Tang[15], M. Alcorta[16], B. Bucher[16], M. Albers[4], P. Bertone[4]

[1]IPHC, Université de Strasbourg, F-67037 Strasbourg, France
[2]CNRS, UMR7178, F-67037 Strasbourg, France
[3]USIAS, F-67083 Strasbourg, France
[4]Physics Division, Argonne National Laboratory, Argonne, IL 60439, USA
[5]Department of Physics, University of York, Heslington, York YO10 5DD, UK
[6]Institut de Physique Nucléaire d'Orsay, CNRS/IN2P3, Université Paris-Sud, Orsay, France
[7]Department of Physics, University of Surrey, GU2 7XH Guildford, United Kingdom
[8]AIR Division, National Physical Laboratory, TW11 0LW Teddington, United Kingdom
[9]Department of Physics and Astronomy, Aarhus University, DK-8000 Århus C, Denmark
[10]Department of Physics and Astronomy, Louisiana State University, Baton Rouge, USA
[11]Grand Accélérateur National d'Ions Lourds, CNRS/IN2P3, F-14076 Caen, France
[12]CSNSM IN2P3-CNRS and University of Paris Sud, F-91405 Orsay, France
[13]Department of Physics, Florida State University, Tallahassee, FL 32306, USA
[14]University of Notre Dame, Notre Dame, IN 46556, USA
[15]Institute of Modern Physics, Lanzhou, China
[16]TRIUMF, Vancouver, BC V6T 2A3, Canada
[16]Lawrence Livermore National laboratory, Livermore, CA 94551, USA

Abstract. The $^{12}C+^{12}C$ fusion reaction is one of the most important for nuclear astrophysics since it determines the carbon ignition in stellar environments. Two experiments which make use of the gamma-particle coincidence technique to measure the $^{12}C+^{12}C$ S-factors at deep sub barrier energies are discussed. Results are presented showing a decrease of the S-factor below $E_{c.m.}$ = 3 MeV.

1 Introduction

The $^{12}C+^{12}C$ fusion reaction is interesting from the point of view of nuclear structure since resonances have been observed in the cross section at the Coulomb barrier (CB) and below, possibly due to ^{12}C-^{12}C molecular configurations of ^{24}Mg. The persistence of resonances at the lowest energies is still an open question. It is also a very important reaction for nuclear astrophysics since carbon burning occurring essentially via the $^{12}C+^{12}C$ fusion in stellar environments, determines if a star will join to the heavy-ion burning branches following hydrogen and helium burning and if white dwarfs will evolve into type Ia supernovae. This reaction has thus been subject to numerous experimental studies from the CB to lower energy regions of astrophysical interest. The Gamow window for this reaction is centered at $E_0 = 2.42\times T_9^{2/3} \pm 0.75\times T_9^{5/6} = 1.5 \pm 0.3$ MeV at T = 5×10^8 K [1]. The present contribution discusses two experiments aiming at measuring the $^{12}C+^{12}C$ fusion cross-section using a direct method.

2 Experimental technique

Fusion of $^{12}C+^{12}C$ leads to the ^{24}Mg compound nucleus and the exit channels for this reaction are $^{12}C(^{12}C,\alpha)^{20}Ne$, $^{12}C(^{12}C,p)^{23}Na$ and $^{12}C(^{12}C,n)^{23}Mg$. Due to its negative Q-value (-2.62 MeV), the neutron evaporation exit channel contribution is negligible at energies relevant for astrophysics whereas the Q-values for the α and p channels are respectively 4.62 MeV, 2.24 MeV. Direct measurements of $^{12}C(^{12}C,^{24}Mg)$ fusion cross-sections where in the past essentially based on the measurement of charged particles (α and p) or gamma-rays of the evaporation residues ^{20}Ne and ^{23}Na. At the lowest energies, such a measurement is subject to very large background, essentially coming from proton and deuteron recoil particles and the d(^{12}C,p) reaction. This is due to the ubiquitous contamination of hydrogen and deuterium in the targets. The present article is based on the detection of charged particles and gamma-rays in coincidence, leading to a drastic reduction of the background [2]. In a first experimental campaign at the Argonne National Laboratory a ^{12}C beam of intensity 600 pnA was delivered by the ATLAS facility impinging on a highly enriched (99.9 %) ^{12}C target of thickness ~ 50 μg.cm^{-2}. Ten energy points have been measured ranging from $E_{c.m.}$ = 4.93 down to 2.68 MeV. Gamma transitions from the evaporation residues were measured using the Gammasphere 100 HPGe detectors array in coincidence with charged particles detected in 3 annular double-sided silicon strip detectors (DSSD) covering ~ 25% of 4π. Two surface barrier silicon monitor detectors where used to normalize the cross section results to Mott scattered ^{12}C nuclei at 45° in the forward direction. Characteristic gamma-rays from ^{20}Ne (1635 keV) and ^{23}Na (440 keV) were measured in the Gammasphere 100 Ge detectors. A further experimental campaign at the Andromede facility in Orsay has used the coincidence technique with the STELLA experimental station coupled to the FATIMA 30 $LaBr_3$ scintillator array [3] and a newly developed rotating target system. The experiment is described in another contribution to this conference [4].
The measured S-factors are presented in the next section.

3 Results and discussion

The cross sections measured during the Argonne campaign at $E_{c.m.}$ between 4.93 and 2.68 MeV converted in S-factors are presented in Fig. 1., together with the most recent results for a direct measurement in the system [5]. The region indicated in blue in this figure is the Gamow region for T = $5x10^{9}$ K and the yellow region represents the Gamow energy region for type IA supernova. At $E_{c.m.}$ = 2.68 MeV the S-factor corresponds to a cross section of 6 nb. On this figure, the presented extrapolations are based on standard potentials and use the sudden model or coupled-channel approach except the red one which takes into account the fusion hindrance phenomenon. This phenomenon has been observed in numerous fusion reactions of medium mass systems. It was first described by Jiang et al. [6]. At low energies, and for cross sections below ~ 0.1 mb, the measured cross sections were systematically smaller than predicted by standard coupled-channel calculations.

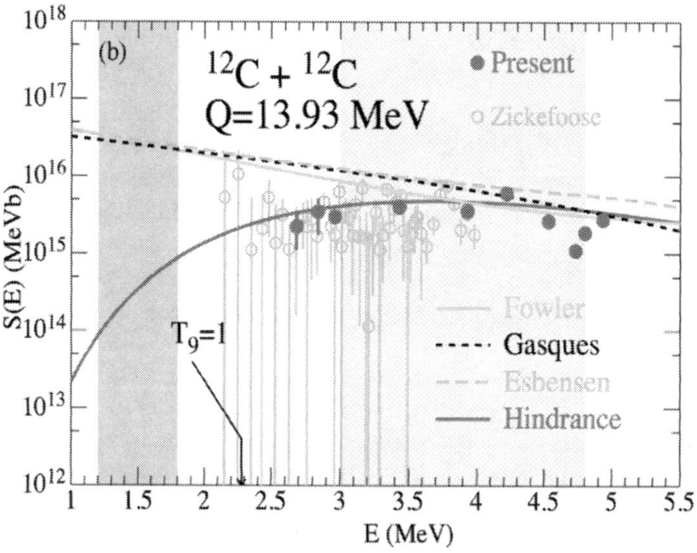

Fig. 1. (color online) S-factors measured in the Argonne $^{12}C+^{12}C$ fusion experiment (red points) as a function of the energy in the center of mass frame together with results from ref. [5]. The blue, black, red and green dashed lines correspond to extrapolations from Fowler [7], Gasques [8], Jiang [9] and Esbensen [10].

The data measured in this experiment are compatible at low energies with the extrapolation based on hindrance.

The coincidence technique is also used in the French-UK STELLA project whose commissioning has measured the $^{12}C+^{12}C$ fusion reaction around and below the Coulomb barrier. Fig. 2 shows preliminary particle spectra measured at E_{Lab} = 11 MeV. On the spectrum in coincidence with a characteristic gamma-ray of ^{23}Na, excellent identification of protons from the $^{12}C(^{12}C,p)^{23}Na$ reaction from p_0 to p_4 is achieved. The few counts remaining in the p_0 zone of the coincident spectrum are from random coincidences. From these spectra, an experimental value for the detection efficiency at E_γ = 440 keV is obtained by taking the ratio of integrated coincident and single p_1 peaks. This gives a result of 6%, which is in agreement with Geant4 simulation of the $LaBr_3$ setup developed within the STELLA project in the configuration which uses 28 $LaBr_3$ detectors [11].

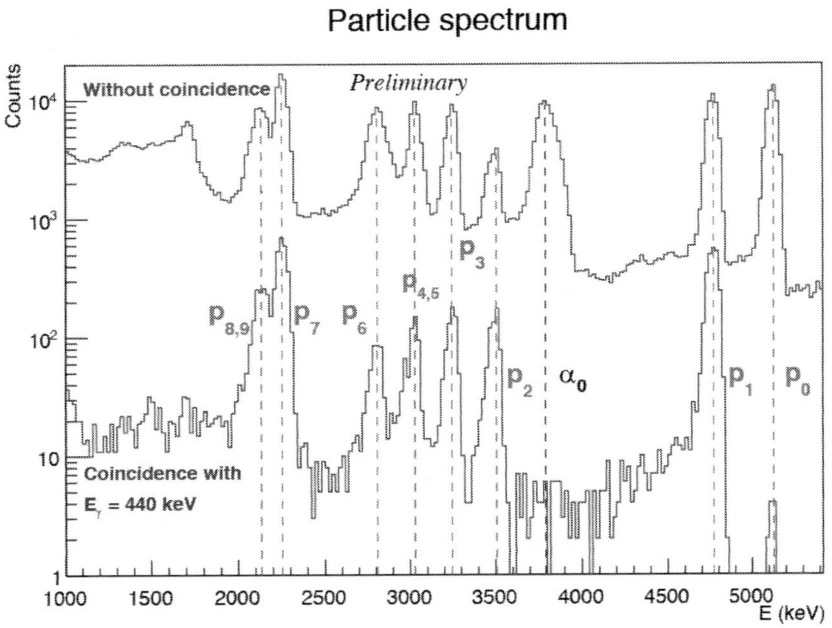

Fig. 2. (color online) Blue: $^{12}C+^{12}C$ fusion reaction particle spectrum measured in the STELLA project at E_{Lab} = 11 MeV. Purple: same spectrum in coincidence with the $5/2^+ \rightarrow 3/2^+$ transition of ^{23}Na, from $^{23}Na+p$ channel.

4 Conclusions

The gamma-particle coincidence method has been used to measure the direct $^{12}C(^{12}C,^{23}Mg)$ fusion reaction at sub barrier energies. Two different experimental campaigns are discussed, the first one took place at the Argonne National Laboratory using the Gammasphere array and the second one is taking data at Orsay using the STELLA+FATIMA setup. Results from the first campaign are presented. The $^{12}C+^{12}C$ S-factors have been measured down to $E_{c.m.}$ = 2.68 MeV with smaller error bars than the previous studies for this system. This measurement shows a decrease of the S-factors below $E_{c.m.}$= 3 MeV, after a broad maximum, which could indicate that fusion hindrance may play a role in this reaction and thus in the evolution of massive stars. The impact of fusion hindrance on stellar burning and astrophysics has been studied by Gasques *et al.* in [12]. The fusion hindrance phenomenon, as discussed in this article would indeed drastically reduce the corresponding reaction rates at temperatures below 3 to $10x10^8$K affecting the abundance of many species, in particular enhancing ^{26}Al and ^{60}Fe.

Acknowledgements

This work was supported by the US Department of Energy, Office of Nuclear Physics, under Contract No. DE-AC02- 06CH11357 and by USIAS, Strasbourg Fellowship of S. Courtin.

References

1. C. E. Rolfs, and W. S. Rodney, Cauldrons in the Cosmos, The University of Chicago Press, 1988.
2. C. L. Jiang *et al.*, NIM **A** 682 (2012) 12.
3. O.J. Roberts *et al.*, NIM **A** 748, 9195 (2014).
4. M. Heine *et al.*, contribution to this conference.
5. J. Zickefoose, PhD Thesis, U. Connecticut, (2011).
6. C.L. Jiang et al., Phys. Rev. Lett. 89, 052701 (2002).
7. W. Fowler, G. Caughlan and B. Zimmerman, Annu. Rev. Astrophys. 13, 69 (1975).
8. L.R. Gasques *et al.*, Phys.Rev. **C** 72, 025806(2005).
9. C.L. Jiang, K.E. Rehm, B.B. Back and R.V.F. Janssens, Phys. Rev. **C** 75, 015803 (2007).
10. H. Esbensen, X.D. Tang, and C.L. Jiang, Phys. Rev. **C** 84, 064613 (2011).
11. M. Heine *et al.*, J. Phys. Conf. Ser. 763, 012005 (2016).
12. L.R. Gasques *et al.*, Phys. Rev. **C** 76, 035802 (2007).

Constraining The Symmetry Energy (Far) Above Saturation Density Using Elliptic Flow

M.D. Cozma[1,⋆]

[1] *IFIN-HH, Reactorului 30, 077125 Măgurele-Bucharest, Romania*

Abstract. A QMD transport model that employs a modified momentum dependent interaction (MDI2) potential, supplemented by a phase-space coalescence model fitted to experimental multiplicities of free nucleons and light clusters, is used to study the density dependence of the symmetry energy above the saturation point by a comparison with experimental elliptic flow ratios measured by the FOPI-LAND and ASYEOS collaborations for ^{197}Au+^{197}Au collisions at 400 MeV/nucleon impact energy. Comparing theoretical predictions with experimental data for neutron-to-proton and neutron-to-charged particles elliptic flow ratios the following constraint is extracted for the slope L and curvature K_{sym} of symmetry energy at saturation: L=59±24(exp)±16(th)±10(sys) MeV and K_{sym}=0±370(exp)±220(th)±150(sys) MeV. Theoretical errors are the result of poorer known model ingredients. Systematical uncertainties are generated by the inability of the transport model to reproduce experimental light-cluster-to-proton multiplicity ratios. A more accurate value for L, free of systematical theoretical uncertainties, can be extracted from the neutron-to-proton elliptic flow ratio alone: L=63±18(exp)±14(th) MeV.

1 Introduction

Knowledge of the density dependence of the isospin dependent part of the equation of state (asy-EoS) of nuclear matter, commonly known as the symmetry energy (SE), is crucial for the understanding of the structure of rare isotopes, dynamics of heavy-ion collisions and properties of astrophysical objects such as neutron stars [1, 2]. Experimental studies of isospin diffusion, pygmy and giant dipole resonances, neutron skin thickness and other phenomena has made possible the extraction of constraints with satisfactory accuracy for the density dependence of the SE in the vicinity or below the saturation density (ρ_0) [3]. Similarly, recent advances in theoretical many-body simulations of nuclear matter have allowed increasingly more accurate predictions for the asy-EoS up to densities close to the saturation point [4]. The density dependence far above saturation, $\rho \geq 2\rho_0$, has remained up to now extremely uncertain. Its knowledge is mandatory for a proper understanding of properties of neutron stars such as radius and maximum allowed mass and may provide the key for a simple solution of the so called hyperon puzzle [5].

Intermediate energy heavy-ion collisions (HIC) provide the unique opportunity to study, in terrestrial laboratories, nuclear matter in the vicinity of twice saturation density [6]. To this end, elliptic

⋆e-mail: dan.cozma@theory.nipne.ro

flow ratios (EFR) of isospin partners have been shown to be sensitive to the density dependence of SE [7, 8]. Consequently, constraints for the asy-EoS stiffness have been extracted [7, 9–11] from a comparison of theoretical models with experimental data of the FOPI-LAND [12] and ASYEOS [9] collaborations. Due to the fact the neutron-to-proton (npEFR) and neutron-to-hydrogen (nhEFR) elliptic flow ratios probe on average different density regimes, 1.4-1.5ρ_0 and 1.0-1.1ρ_0 respectively [9], it is in principle possible to extract constraints for both the slope L and curvature K_{sym} of SE at saturation rather than an average value of the slope at an undetermined density.

The study of Ref. [13] had as main goal the extraction of constraints for the slope L and curvature K_{sym} from the FOPI-LAND npEFR and ASYEOS neutron-to-charged-particles elliptic flow ratio (nchEFR) experimental data. In this conference proceeding the main modifications to the transport model are briefly presented, supplemented by the main results of that study.

2 The model

Heavy-ion collision dynamics is simulated using an upgraded version [14, 15] of the Tübingen quantum molecular dynamics model (QMD) transport model [16, 17]. To allow for independent variations of the L and K_{sym} parameters, the potential component of the MDI Gogny-inspired version of the equation of state of nuclear matter [18] has been modified (and named MDI2) by the addition of a density dependent, momentum independent term proportional to a parameter denoted D

$$\begin{aligned} \frac{E}{N}(\rho,\beta) &= A_u(x,y)\frac{\rho(1-\beta^2)}{4\rho_0} + A_l(x,y)\frac{\rho(1+\beta^2)}{4\rho_0} + \frac{B}{\sigma+1}\frac{\rho^\sigma}{\rho_0^\sigma}(1-x\beta^2) \\ &+ \frac{D}{3}\frac{\rho^2}{\rho_0^2}(1-y\beta^2) + \frac{1}{\rho\rho_0}\sum_{\tau,\tau'} C_{\tau\tau'}\int\int d^3\vec{p}\,d^3\vec{p}\,'\frac{f_\tau(\vec{r},\vec{p})f_{\tau'}(\vec{r},\vec{p}\,')}{1+(\vec{p}-\vec{p}\,')^2/\Lambda^2}\,. \end{aligned} \tag{1}$$

In the above expression ρ, β and p denote the density, isospin asymmetry and momentum variables respectively. The parameters x and y have been introduced to allow independent adjustments of the stiffness parameters L and K_{sym} of symmetry energy. The free parameters of the equation of state are determined by using the following data as input: value of the optical potential at infinite momentum, effective nucleon mass in isospin symmetric matter, isovector effective mass difference, value of the saturation density of isospin symmetric matter, value of the binding energy at saturation, compressibility K_0 and skewness Q_0 parameters of isospin symmetric matter, value of the symmetry energy at a particular density $\tilde{\rho}$ and the slope L and curvature K_{sym} of symmetry energy at saturation. Additional details and values of the model parameters can be found in Ref. [13].

The final state spectra of HIC are determined by making use of a minimum spanning-tree coalescence algorithm. It is applied to identify all clusters with A≤15. A number of 23 additional clusters with A>15 that correspond to known stable or unstable isotopes of B, C, N and O are also identified. Clusters with lifetimes larger than 1 ms are considered as stable while the rest decay until clusters stable against strong interaction decays are reached (ex. ^{4}Li→p+^{3}He). For clusters with A≤15 that do not correspond to a known stable or unstable isotope, protons or neutrons are evaporated for proton and neutron rich clusters respectively until a known nucleus is reached (ex. 1p8n→6n+^{3}H). Nucleons belonging to clusters with other masses are discarded from final spectra.

The values of the r-space coalescence parameters are determined from a fit of cluster and free nucleon multiplicities to experimental data. For the case of Au+Au collisions at an impact energy of 400 MeV/nucleon the FOPI cluster multiplicity experimental data for central collisions ($b \leq 2.0$ fm) have been used [19], namely the multiplicities of p, n, ^{2}H, ^{3}H, ^{3}He, ^{4}He, Li, Be and C. The left panel of figure 1 presents the result of the fit for three values of the momentum space coalescence parameter δp. The experimental deuteron, Be, B and C multiplicities are closely reproduced, in opposition to

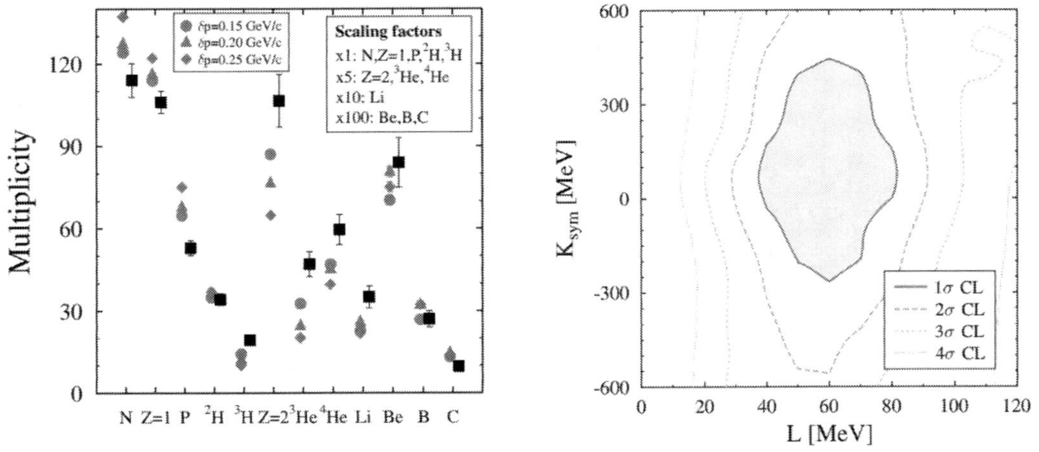

Figure 1. (Left panel) Multiplicities of free nucleons and clusters with $Z \leq 6$ for three values of the momentum space coalescence parameter compared to the experimental FOPI data [19] (full squares with error bars). The asy-EoS stiffness parameters have been set to L=60.5 MeV and K_{sym}=-81.0 MeV. The values of the r-space coalescence parameters for each case are listed in Ref. [13]. (Right panel) Constraint for the slope L and curvature K_{sym} extracted from FOPI-LAND npEFR and ASYEOS p_T dependent nchEFR. Curves for the 1, 2, 3 and 4 σ confidence levels are shown.

similar models reported in the literature [7]. The multiplicities of Li, He isotopes and triton are under-predicted, but the discrepancy with respect to experimental values is smaller as compared to results reported elsewhere. The reason behind this improvement has been identified to be the momentum dependence of the MDI2 interaction.

3 Selected results

By keeping the value of the SE at saturation and its slope L fixed the sensitivity of npEFR, nhEFR, nchEFR observables to the curvature K_{sym} parameters can be investigated. It was found [13] that npEFR and nhEFR (or alternatively nchEFR) show different sensitivity to variations of K_{sym}. This is a direct consequence of the fact that these observables probe, on average, different density regions [9]. The combination of observables FOPI-LAND npEFR and ASYEOS nchEFR presents the best opportunity for constraining both L and K_{sym} in view of the higher accuracy of the latter data set. The extracted values for the slope and curvature parameters read L=59±24 MeV and K_{sym}=88±372 MeV at 1 σ CL. The 1, 2, 3 and 4 σ CL contour curves are plotted in the right panel of figure 1. The minimum value for the quality-of-fit parameter is for the presented case χ^2_{min}/point=0.14.

The extracted constraints are systemically affected by the under-prediction, by the model, of experimental cluster-to-proton multiplicity ratios. Consequently, the value of charged-particles elliptic flow v_2^{ch} entering the expression of nchEFR needs to be corrected by a factor that has been estimated to lie in the interval $1.05 \leq f_{corr} \leq 1.15$. This correction leads to a softer value for K_{sym}. Additionally, the effect of poorer known model ingredients such as the values for the compressibility modulus of symmetric nuclear matter and isovector neutron-proton effective mass difference, scenarios used for in-medium nucleon-nucleon cross-sections or total energy conservation of the system and employed Pauli blocking algorithm on the extracted SE constraints has also been investigated. Their impact is summed up in quadrature to determine the theoretical uncertainty of the extracted constraint. The final result, obtained by using the more realistic value f_{corr}=1.10, is

$$\begin{aligned} L &= 59 \pm \ 24(\mathrm{exp}) \pm \ 16(\mathrm{th}) \pm \ 10(\mathrm{sys})\ \mathrm{MeV} \\ K_{sym} &= \ 0 \pm 370(\mathrm{exp}) \pm 220(\mathrm{th}) \pm 150(\mathrm{sys})\ \mathrm{MeV}\,. \end{aligned} \tag{2}$$

It is possible to extract a more precise constraint for the slope L, free of systematical uncertainties, from the FOPI-LAND npEFR data alone by taking into account the sensitivity of this observable to K_{sym}. By using a conservative range for the location of the real value of this parameter, $-300 \leq K_{sym} \leq 300$ MeV, the uncertainty it induces on the extracted value of L can be estimated. The extracted constraint for L reads

$$L = 63 \pm 18(\mathrm{exp}) \pm 14(\mathrm{th})\ \mathrm{MeV}\,. \tag{3}$$

Differences with respect to a previous study [10] were found to be equally due to a more accurate description of nucleon density profile of nuclei and usage of a coalescence algorithm rather than a density cut-off method to determine final spectra in the present study.

In conclusion, it has been shown that existing experimental data for neutron-to-proton and neutron-to-charged particles elliptic flow ratios allow the extraction of constraints for both the slope and curvature of the symmetry energy at saturation. For the former parameter the result is compatible with constraints extracted from nuclear structure and astrophysical experimental data and of similar accuracy. The accuracy of the latter parameter is rather poor but it is nevertheless compatible with a stiff symmetry energy at densities equal or larger than twice the saturation density hinting for a possible solution of the hyperon puzzle.

References

[1] B.A. Li, L.W. Chen, C.M. Ko, Phys. Rept. **464**, 113 (2008)
[2] J.M. Lattimer, M. Prakash, Phys. Rept. **442**, 109 (2007)
[3] M. Baldo, G.F. Burgio, Prog. Part. Nucl. Phys. **91**, 203 (2016)
[4] C. Drischler, A. Carbone, K. Hebeler, A. Schwenk, Phys. Rev. **C94**, 054307 (2016)
[5] I. Vidana, A. Polls, A. Ramos, L. Engvik, M. Hjorth-Jensen, Phys. Rev. **C62**, 035801 (2000)
[6] B.A. Li, Nucl. Phys. **A708**, 365 (2002)
[7] P. Russotto et al., Phys. Lett. **B697**, 471 (2011)
[8] M.D. Cozma, Phys. Lett. **B700**, 139 (2011)
[9] P. Russotto et al., Phys. Rev. **C94**, 034608 (2016)
[10] M.D. Cozma, Y. Leifels, W. Trautmann, Q. Li, P. Russotto, Phys. Rev. **C88**, 044912 (2013)
[11] Y. Wang, C. Guo et al., Phys. Rev. **C89**, 044603 (2014)
[12] Y. Leifels et al. (FOPI), Phys. Rev. Lett. **71**, 963 (1993)
[13] M.D. Cozma (2017), `arXiv:1706.01300`
[14] M.D. Cozma, Phys. Lett. **B753**, 166 (2016)
[15] M.D. Cozma, Phys. Rev. **C95**, 014601 (2017)
[16] D.T. Khoa, N. Ohtsuka, M.A. Matin, R.K. Puri, Nucl. Phys. **A548**, 102 (1992)
[17] V.S. Uma Maheswari, C. Fuchs, A. Faessler, L. Sehn, D.S. Kosov, Z. Wang, Nucl. Phys. **A628**, 669 (1998)
[18] C.B. Das, S.D. Gupta, C. Gale, B.A. Li, Phys. Rev. **C67**, 034611 (2003)
[19] W. Reisdorf et al. (FOPI), Nucl. Phys. **A848**, 366 (2010)

EPJ Web of Conferences **165**, 01017 (2017) DOI: 10.1051/epjconf/201716501017
NPA8 2017

Introduction of the new LUNA experimental setup for high precision measurement of the $^{13}C(\alpha,n)^{16}O$ reaction for astrophysical purposes

L. Csedreki[1,*], *G. F. Ciani*[1,3], *I. Kochanek*[1] *and A. Best*[2]

[1]INFN, Laboratori Nazionali del Gran Sasso, Via Giovanni Acitelli 22, 67100 Assergi, Italy
[2]Università di Napoli Federico II and INFN, Sezione di Napoli, Via Cintia, 80126 Napoli, Italy
[3]Gran Sasso Science Institute, Viale Francesco Crispi 7, 67100 L'Aquila, Italy

Abstract. The $^{13}C(\alpha,n)^{16}O$ reaction is the prevalent neutron source for the main *s*-process. The direct measurement of this reaction at stellar temperature (kT=8 keV) has so far not been possible due to the very low cross section at the corresponding energy. The extrapolation of the astrophysical *S*-factor of this reaction into the Gamow window ($E_{\alpha,c.m.}$=140-230 keV) is complicated by the large uncertainties of the low-energy experimental data and the existence of a state of ^{17}O near the α-threshold that can have a large effect on low energy cross section. The aim of this paper is to introduce the new LUNA experimental setup, dedicated to the investigation of $^{13}C(\alpha,n)^{16}O$ reaction below $E_{\alpha,lab}$=400 keV.

1 Introduction

The $^{13}C(\alpha,n)^{16}O$ reaction is very important in astrophysical context. This reaction is the dominant neutron source for the synthesis of the main *s*-process component of heavy elements in thermally pulsing, low-mass asymptotic giant branch (AGB) stars [1]. Several papers can be found in literature using direct [2-6] and indirect [7-10] methods to investigate the behaviour of the $^{13}C(\alpha,n)^{16}O$ reaction. However, the extrapolation of the astrophysical *S*-factor of this reaction into the Gamow window ($E_{\alpha,c.m.}$=140-230 keV) is challenging due to the contested effect of a resonance of ^{17}O located near the threshold and high uncertainties of the experimental data in the low-energy region.

The aim of the new project at the LUNA 400 kV accelerator is the determination of cross section of the $^{13}C(\alpha,n)^{16}O$ reaction approaching, in energy, the Gamow window with an accuracy of about 10% being performed in the Laboratori Nazionali del Gran Sasso (LNGS), Italy. The LNGS Underground Laboratory provides an ideal environment to establish the direct measurement of this reaction at low energy thanks to the reduction of neutron background with 3 orders of magnitude compared with other experiment performed on the Earth surface. This paper gives an overview of the design and of the features of the experimental setup. Moreover, the characterization of the environmental neutron and gamma-ray background and preliminary results of the efficiency calculation using Geant4 code [11] are presented.

*Corresponding author: laszlo.csedreki@lngs.infn.it

2 Experimental setup

The LUNA setup for the investigation of the $^{13}C(\alpha,n)^{16}O$ reaction is shown in figure 1. The setup includes 18 ^{3}He counters under a pressure of 10 atm (6 counters with 25 cm and 12 with 40 cm active length) with stainless steel body material. As ^{3}He has a very high cross section for capturing thermal neutrons through the reaction $^{3}He(n,p)^{3}H$ (σ=5330 barn, Q=764 keV), effective thermalisation of the emitted neutrons from the $^{13}C(\alpha,n)^{16}O$ reaction is required. Therefore, counters are embedded in a polyethylene moderator arranged in two cycles. Additionally the polyethylene block will be coated with borated-polyethylene for further reduction of count rates from the environmental neutron background.

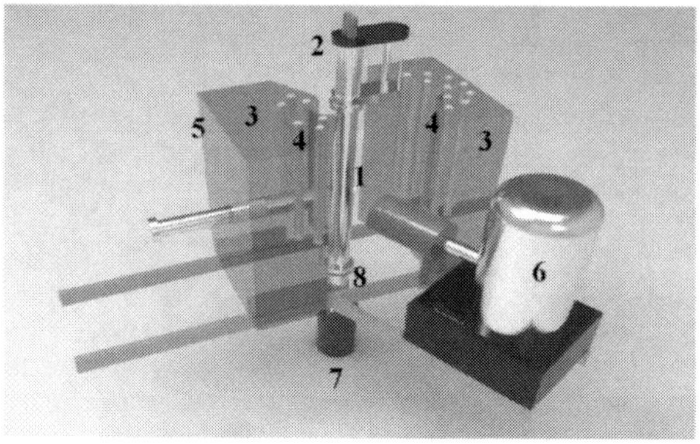

Fig. 1. Design of the experimental setup: 1-Vacuum chamber, 2-Linear motion feedthrough, 3-Polyethylene moderator composed of the two separable parts (left and right side with respect to the vacuum chamber), 4-^{3}He counters embedded in the moderator, 5-Borated-polyethilen shield, 6-HPGe detector, 7-Vacuum pump, 8-Electrical vacuum break.

In the energy range of the LUNA-400, the $^{13}C(p,\gamma)^{14}N$ reaction is suitable for the monitoring of the target degradation before and after the alpha irradiation. Therefore, the setup is designed to allow the use of a High-Purity Germanium (HPGe) detector in close geometry during the experiment. In regular intervals the $^{13}C(\alpha,n)^{16}O$ measurement can be interrupted, the moderator opened and the target degradation can be checked observing the $^{13}C(p,\gamma)^{14}N$ reaction with the HPGe detector.

Multi-stage target holder is applied to minimize the possible contamination and simplify the target changing procedure during the experiment. Moreover, cooling loop is integrated in the target holder for effective removing the dissipated beam power.

2.1. Characterization of environmental background

Left and right panels of figure 2 present the comparison of measured gamma-ray and neutron environmental background on surface and in LNGS Underground Laboratory. The violet bars include the region of the thermal neutron peak. Signals beyond this region are due to alpha intrinsic background. Underground neutron measurement using the counters agrees with the most recent literature value [12]. The intrinsic background of ^{3}He counter is in the range of 10^{-8} cm^{-2} s^{-1}, which is essential for the measurement of the $^{13}C(\alpha,n)^{16}O$ reaction at low bombarding energy range.

2.2. Simulation of neutron detection efficiency

The geometry of the experimental setup is optimized, using Geant4 simulation, to reach the highest neutron detection efficiency in the energy region of our interest ($E_{n,lab}$~2.1-2.5 MeV). The estimated neutron detection efficiency of the setup is **ε_{tot-n}~40%.** However for

DOI: 10.1051/epjconf/201716501017

the validation of the Geant4 simulation and the determination of absolute neutron detection efficiency, the measurement using calibrated neutron source and/or particle induced neutron threshold nuclear reaction are required.

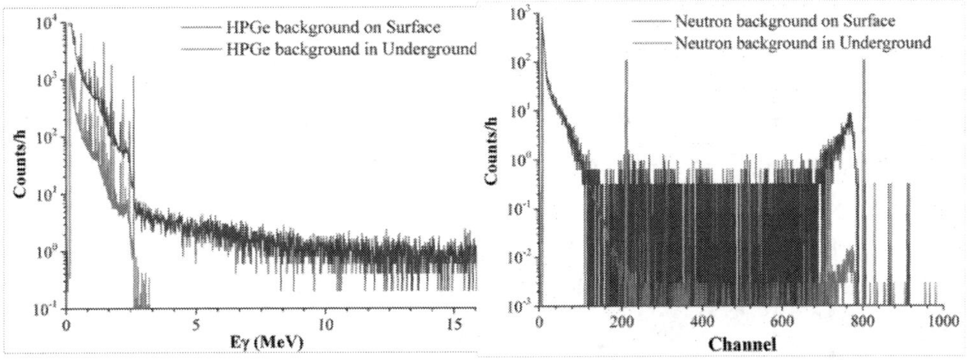

Fig. 2. Left panel: Gamma-ray environmental background measured on surface (blue line) and in the LNGS Underground Laboratory (red line) with HPGe detector – Right panel: Environmental neutron background measured on surface (blue line) and in the LNGS Underground Laboratory (red line) with a ^{3}He counter. The violet bars include the region of the thermal neutron peak. Signals beyond this region are due to alpha intrinsic background.

3 Summary and outlook

In the LNGS Underground Laboratory, a new high efficiency setup is designed and developed to investigate the low energy yield of the ^{13}C(α,n)^{16}O reaction close to its Gamow window. Neutron and gamma-ray environmental background are characterized and the comissioning phase of the experimental arrangement is in progress.

The authors are gratefull to D. Orlandi and M. Paris for technical assistance on chamber designing.

References

1. R. Gallino, C. Arlandini, M. Busso, M. Lugaro, C. Travaglio, O. Straniero, A. Chieffi, M. Limongi, Astrophys. J. **497** (1998)
2. S. Kellogg, R. Vogelaar, and R. Kavanagh, Bull. Am. Phys. Soc. **34** (1989)
3. C. R. Brune, I. Licot, and R. W. Kavanagh, Phys. Rev. C **48** (1993)
4. H. Drotleff, A. Denker, H. Knee, M. Soiné, G. Wolf, J. W. Hammer, Astrophys. J. **414** (1993)
5. M. Heil, R. Detwiler, R. E. Azuma, A. Couture, J. Daly, J. Görres, F. Kappeler, R. Reifarth, P. Tischhauser, C. Ugalde, and M. Wiescher, Phys Rev. C **78** (2008)
6. Y. Xu, K. Takahashi, S. Goriely, M. Arnould, M. Ohta, H. Utsunomiya, Nucl. Phys. A **918** (2013)
7. B. Guo et al., Astrophys. J. **756** (2012)
8. La Cognata et al. Phys. Rev. Lett. **109** (2012)
9. M.L. Avila, G.V. Rogachev, E. Koshchiy, L.T. Baby, J. Belarge, K.W. Kemper, A.N. Kuchera, D. Santiago-Gonzalez, Phys. Rev. C **91** (2015)
10. A. M. Mukhamedzhanov, Shubhchintak and C. A. Bertulani, Phys. Rev. C **96** (2017)
11. J. Allison et al., NIMA **835** (2016)
12. A. Best, J. Görres, M. Junker, K.-L. Kratz, M. Laubenstein, A.Long, S. Nisi, K. Smith, M. Wiescher, NIMA **812** (2012)

The Trojan Horse Method application on the ^{10}B(p,α_0)^{7}Be reaction cross section measurements

A. Cvetinović[1], *C.* Spitaleri[1,2], *R.* Spartá[1], *G. G.* Rapisarda[1], *S. M.* Puglia[1], *M.* La Cognata[1], *S.* Cherubini[1,2], *G. L.* Guardo[1], *M.* Gulino[1,3], *L.* Lamia[1], *R. G.* Pizzone[1], *S.* Romano[1,2], and *M. L.* Sergi[1]

[1] *INFN-Laboratori Nazionali del Sud, Catania, Italy*
[2] *Dipartimento di Fisica e Astronomia, University of Catania, Catania, Italy*
[3] *Università degli Studi di Enna "Kore", Enna, Italy*

Abstract. The ^{10}B(p,α_0)^{7}Be reaction cross section has been measured in an wide energy range from 2.2 MeV down to 3 keV in a single experiment applying THM. Optimized experimental set-up ensured good energy resolution leading to a good separation of α_0 and α_1 contributions to the cross section coming from the ^{7}Be ground and first excited state, respectively.

1 Introduction

Since excitation function of the ^{10}B(p,α)^{7}Be reaction is dominated by strong s-wave resonance at 10 keV (J^π=3$^+$, 8.699 MeV ^{11}C level) lying exactly at the energy corresponding to the Gamow peak (EG) [2], the cross section function extrapolation (that is usually carried out from direct experimental data measured at higher energies) should be avoided due to significant uncertainties that can be introduced in this way. Therefore the precise study of the ^{10}B(p,α)^{7}Be reaction cross section at astrophysical energies applying indirect approach such as Trojan Horse Method (THM) ([1] and references therein) is well suited.

The ^{10}B(p,α)^{7}Be reaction plays an important role in prediction of boron abundance in stars because the ^{10}B burning process mostly proceeds via the (p,α) reaction. Since depletion of light elements such as Li, Be and B occurs at different depths, this reaction can be used as probe for internal stellar structure, so, analysis of the resulting atmospheric abundances of these elements can help in understanding and confining the mixing processes [3–5].

Up to now, several experiments were performed applying direct [6–14] and indirect [15, 16] approaches in order to investigate the ^{10}B(p,α)^{7}Be reaction. The biggest problem in the precise cross section analysis comes from the significant differences between the existing experimental data sets [6], probably caused by unknown systematic uncertainties.

2 Applied technique

In order to extract a two-body ^{10}B(p,α_0)^{7}Be reaction cross section free of suppression due to penetration thought the Coulomb barrier and electron screening effects, the THM was applied by selecting the Quasi-Free (QF) contribution to the ^{2}H(^{10}B,α ^{7}Be)n reaction cross section at energies above the Coulomb barrier [17]. Deuteron has been selected as a Trojan Horse (TH) nucleus due to its $p \oplus n$ cluster configuration. After ^{2}H breakup in the collision with the projectile, ^{10}B interacts only with the transferred particle p, while n does not participate in the process and acts as a spectator.

In the Plane Wave Impulse Approximation (PWIA), the three-body reaction cross section can be calculated [18–20] as:

$$\frac{d^3\sigma}{dE_\alpha d\Omega_\alpha d\Omega_{^7\mathrm{Be}}} \propto KF \cdot \phi^2(p_n) \cdot \left(\frac{d\sigma(E_{cm})}{d\Omega}\right)^{HOES} . \quad (1)$$

Here KF represents the kinematical factor containing the final state phase-space factor and it is dependent on the masses, momenta, and angles of the outgoing particles. $\phi^2(p_n)$ represents the squared Fourier transform of the radial wave function describing the p-n inter-cluster motions, usually given for deuterium by the Hulthén function. The factor $(d\sigma(E_{cm})/d\Omega)^{HOES}$ stands for the Half-Off-Energy-Shell (HOES) differential cross section of two-body reaction at the center-of-mass energy E_{cm}=$E_{^2\mathrm{H}-p}$=$E_{^7\mathrm{Be}-\alpha}$-Q_{2b} [21].

3 Experiment

The experimental study of the ^{2}H(^{10}B,α ^{7}Be)n reaction was performed by bombarding a 56 μg/cm^2 self-supported deuterated polyethylene (CD_2) target by the 28 MeV ^{10}B ion beam produced with the SMP Tandem Van de Graaff accelerator at the *Laboratori Nazionali del Sud* in Catania, Italy. The beam intensity was around 1 enA. The detection system consisted of four 500 μm thick Position Sensitive Silicon Detectors (PSD) of which two were used to measure the residual energy of the detected particles (E stage) as a part of two ΔE-E telescope systems. As for ΔE stage, two Ionisation Chambers were used with 1.5 μm thick Mylar entrance and exit windows, filled with butane gas (C_4H_{10}) at a pressure of 110 mbar. Two PSDs used as a part of telescope systems were placed at a distance d=400 mm from the target and centred at opposite sides of the beam axis at the laboratory angle of 14.5°. Another two position sensitive detectors were placed at a distance d=200 mm from the target and centred at opposite sides of the beam axis at the laboratory angle of 28°. The trigger for the event acquisition was given by the coincidence of signals of two PSDs.

4 Results

Analysing the ΔE-E plot, it was possible to identify and select only events corresponding to the three-body ^{2}H(^{10}B,α ^{7}Be)n reaction. In following, the Q-value spectrum for selected ^{7}Be nuclei and α-particles detected in coincidence was reconstructed by assuming that mass is equal to 1 for the undetected third particle (neutron). The reconstructed Q-value spectrum is reported in the panel (a) of Fig.1. For the experimentally deduced Q-values corresponding to the α_0 (Q=-1.079 MeV) and the α_1 (Q=-1.54 MeV) channels have been found to be in good agreement with the theoretical values (marked by arrows in this figure). Two peaks are very well separated thus allowing accurate selection of events corresponding only to the ^{2}H(^{10}B,α_0 ^{7}Be)n reaction.

The ^{2}H(^{10}B,α_0 ^{7}Be)n excitation function has been extracted in an energy range from 3 keV to 2.2 MeV. For the experimental energy resolution, evaluated from FWHMs of the ^{11}C levels at energies of 8.654 MeV (E_{cm}=-35 keV) and 8.699 MeV (E_{cm}=10 keV) [22], was found to be $\delta\epsilon$=17±1 keV. This is notable improvement compared to the previous THM experiments on the same reaction, where the experimental resolution was found to be 31±3 keV in [15] and 87±5 keV in [16]. Since the behaviour of the cross section at low energies is strongly dominated by the ^{11}C resonant state at the energy

DOI: 10.1051/epjconf/201716501018

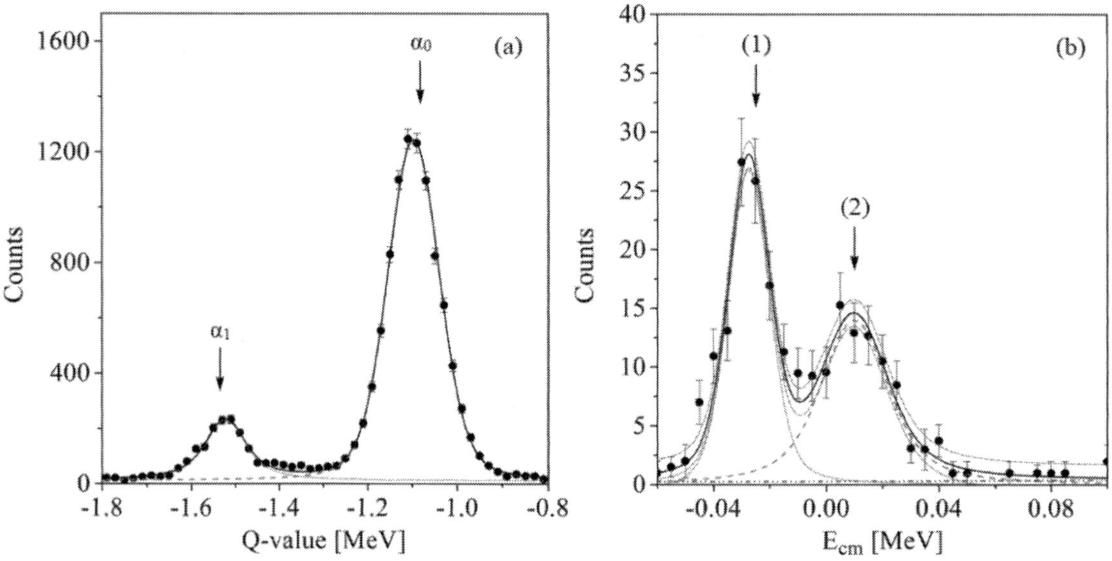

Figure 1. (a) Reconstructed three-body Q-value spectra. The theoretical values $Q^{th}_{\alpha_0}$=-1.079 MeV and $Q^{th}_{\alpha_0}$=-1.54 MeV are marked by arrows. (b) Contributions of the ^{11}C resonances at energies of 8.654 MeV (1) (dotted red line), 8.699 MeV (2) (dashed green line) and 9.200 MeV (dash-dotted line). Cumulative fit is displayed as a full black line, while blue dash-dotted lines give the upper and lower 95% confidence limits to the fit.

of 8.699 MeV, falling just in the Gamow peak energy region, the good separation of this resonance from sub-threshold contribution is crucial for accurate determination of the astrophysical S-factor and so, electron screening potential U_e. The subtraction of the 8.654 MeV level was done as described in [15, 16]. Namely, these two resonances were fitted with sum of three smeared incoherent Breit-Wigner functions, describing the contributions of resonances at energies of 8.654 MeV and 8.699 MeV and the contribution of the tail of the resonance at the energy of 9.200 MeV (see panel (b) of Fig.1).

In order to obtain the bare-nucleus THM S-factor in absolute units, the THM data has been normalised to the direct data from [14] in an energy range from 0.2 MeV up to 1.2 MeV. This procedure yielded an error affecting normalization constant of 2.8%. As for values of the astrophysical S-factor at center-of-mass energy of 10 keV and the electron screening potential (S_b(10 keV)=2950±206 MeV b and U_e=265±46 eV, respectively) was found to be in a good agreement with ones observed in [16] (S_b(10 keV)=2942±395 MeV b and U_e=240±50 eV).

References

[1] C Spitaleri et al., Eur. Phys. J. A **52**, 77 (2016)
[2] C. Rolfs, *Cauldrons in the Cosmos* (The University of Chicago 1988)
[3] A. M. Boesgaard et al., Astr. Phys Journ. **621**, 991 (2005)
[4] L. Lamia et al., The Astrophys. Jour. **768**, 65 (2013)
[5] L. Lamia et al., Astrophys. Jour. **811**, 99 (2015)
[6] M. Wiescher et al., Phys. Rev. C **95**, 044617 (2017)
[7] J. Szabo et al., Nucl. Phys. A **195**, 527 (1972)
[8] N. A. Roughton et al., Nucl. Phys. **23**, 177 (1979)

[9] M. Wiescher et al., Phys. Rev. C **28**, 1431 (1983)
[10] M. Youn et al., Nucl. Phys. A **533**,321 (1991)
[11] C. Angulo et al., Z. Phys. A **345**, 231 (1993)
[12] T. Rauscher et al., Phys. Rev. C **53**, 2496 (1996)
[13] I. Lombardo et al., J. Phys. G: Nucl. Part.Phys. **43**, 045109 (2016)
[14] A. Caciolli et al., The Europ. Phys. Jour. A **52**, 136 (2016)
[15] C. Spitaleri et al., Phys. Rev. C **90**, 035801 (2014)
[16] C. Spitaleri et al., Phys. Rev. C **95**, 035801 (2017)
[17] G. Baur, Phys. Lett **178**, 135 (1986)
[18] V. G. Neudatchin et al., At. Energy Rev. **3**, 157 (1965)
[19] G. Jacob and Th. A. J. Maris, Rev. Mod. Phys. **38**, 121 (1966)
[20] JI. Slaus et al., Phys. Lett. **67**, 43 (1977)
[21] H. R. Weller et al., Phys. Rev. Lett **32**, 177 (1974)
[22] J. H. Kelley et al., Nucl. Phys. A **880**, 88 (2012)

Nuclear reactions in AGB nucleosynthesis: the ^{19}F(α, p)^{22}Ne at energies of astrophysical relevance

G. D'Agata[1,2], *R. G.* Pizzone[1], *M.* La Cognata[1], *I.* Indelicato[1], *C.* Spitaleri[1,2], *S.* Blagus[3], *S.* Cherubini[1,2], *P.* Figuera[1], *L.* Grassi[3], *G. L.* Guardo[1], *M.* Gulino[4], *S.* Hayakawa[1,5], *R.* Kshetri[1,6], *L.* Lamia[1], *M.* Lattuada[1,2], *T.* Mijatovič[3], *M.* Milin[7], *Đ.* Miljanič[3,†], *L.* Prepolec[3], *G. G.* Rapisarda[1], *S.* Romano[1,2], *M. L.* Sergi[1], *N.* Soić[3], *V.* Tokić[3], *A.* Tumino[1,4], and *M.* Uroić[3]

[1] *INFN - Laboratori Nazionali del Sud, Via Santa Sofia 62, 95123, Catania, Italy*
[2] *Dipartimento di Fisica e Astronomia, Università degli Studi di Catania, Via S. Sofia, 64, 95123, Catania, Italy*
[3] *Ruđer Bošković Institute, Bijenicka cesta, 54, 10000, Zagreb, HR*
[4] *Facoltà di Ingegneria ed Architettura, Kore University, Viale delle Olimpiadi, 1, I-94100 Enna, Italy*
[5] *RIKEN, CNS, 2-1 Hirosawa, Wako, Saitama 351-0198, Japan*
[6] *Department of Physics, Sidho-Kanho-Birsha University, Purulia - 723104, WB, India*
[7] *Department of Physics, University of Zagreb, Bijenicka 32, Zagreb, Croatia*

Abstract. The abundance of ^{19}F in the universe is strictly related to standard and extra-mixing processes taking place inside AGB-stars, that are considered to be the most important sites for its production. Nevertheless the way in which it is destroyed is far from being well understood. For this reason we studied the ^{19}F(α,p)^{22}Ne reaction, that is supposed to be the main destruction channel in the Helium-rich part of the star. In this experiment, the reaction has been studied in the energy range of relevance for astrophysics (0÷1 MeV) via the Trojan Horse Method (THM), using the three-body reaction ^{6}Li(^{19}F,p^{22}Ne)d.

1 Introduction

The only Fluorine stable isotope - ^{19}F - has been clearly observed in AGB stars [1], where it can be produced in the He-intershell region through the chain of reactions ^{18}O(p,α)^{15}Na(α,γ)^{19}F. Due to the high abundance of alpha particles in the He-intershell, the ^{19}F(α,p)^{22}Ne reaction is expected to dominate over the competitive reaction channels in this region. Fluorine can in fact be destroyed by the ^{19}F(n,γ)^{20}F, reaction triggered by the neutrons produced in ^{13}C(α,n)^{16}O and ^{22}Ne(α,n)^{25}Mg reactions, and via the ^{19}F(p,α)^{16}O reaction (already studied using THM by [2] and [3]). It is now clear how Fluorine abundance is really sensitive to the physical condition of the stars, and can be used as a probe to clarify if stellar interior nucleosynthesis is well understood or not [4, 5]: in this case Fluorine abundance can not be reproduced by the up-to-date models. A possible reason of this fact is the large uncertaities at Helium burning temperatures ($0.2 \leq T_9 \leq 0.8$), due to the lack of experimental data about the cross-section in the energy region of astrophysical interest. In particular, before this measurement (results partially published in [6, 7]), there were no experimental data below

†Deceased

0.660 MeV in the center-of-mass reference frame, being the Gamow window between 0.200 and 1.200 MeV. This is caused by the presence of the Coulomb barrier. In cases like this one, the use indirect methods, such as the Trojan Horse Method (THM) [8] can be rather useful to overcome the technical difficulties related to the height of the Coulomb barrier.

2 The Experiment

To study the $^{19}F(\alpha,p)^{22}Ne$ reaction, the THM was applied by using a ^{6}Li beam (6 MeV energy, 5 enA intensity) and that can be considered as a cluster $\alpha \oplus d$, impinging on a ^{7}LiF target (150 μm thick), with the aim to induce the $^{6}Li(^{19}F,p^{22}Ne)d$ three-body reaction to study the $^{19}F(\alpha,p)^{22}Ne$ two-body one: in this reaction, the α particle is considered as the participant to the two-body reaction, while the deuteron continues its course undisturbed, and is therefore considered the spectator for the process of interest. Following the THM prescriptions, the beam energy was chosen to measure the $^{19}F(\alpha,p)^{22}Ne$ cross-section in the energy region of interest for astrophysics.

The experiment was performed at Ruđer Bošković Institute (Zagreb, Croatia) and the set-up was

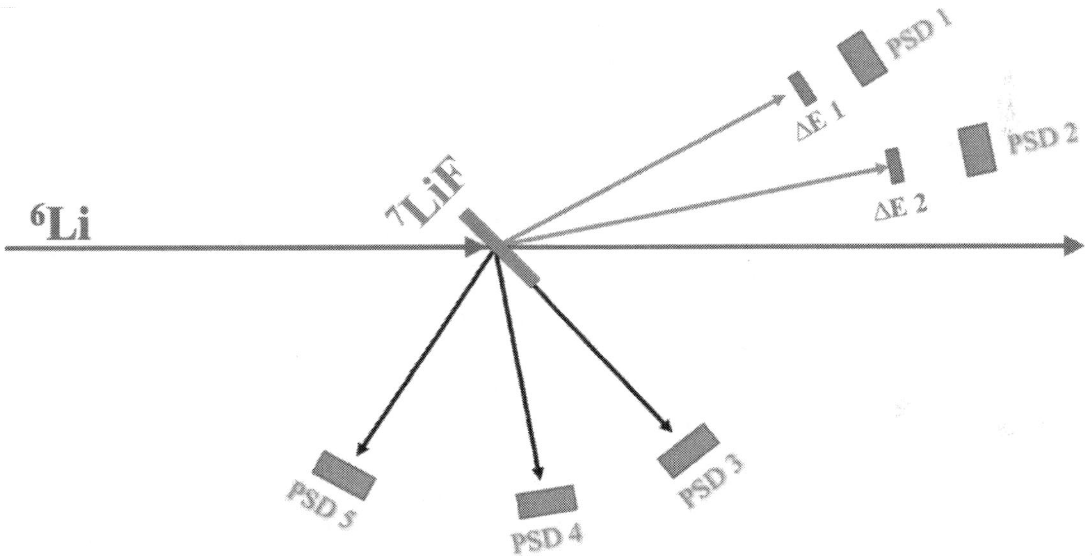

Figure 1. Sketch of the experimental set-up

composed by two ΔE-E telescopes (placed at 12.3°±7° and 32.3°±7°) made up using thin silicon detectors (82 mm^2 active surface, 15μm and 9μm thickness respectively) as ΔE stage and thick (500μm each) Position Sensitive Detectors (PSDs) as E stage, both meant for deuteron particles detection. On the opposite side of the beam, another 3 PSDs with the same specifics are placed at 37.7°±12°, 81°±9° and 119.9°±11° and are devoted to proton detection (Fig. 1). The necessity to have detectors sensible to the position is given by the fact that one of the three particles in the exit channel of the $^{6}Li(^{19}F,p^{22}Ne)d$, in this case ^{22}Ne, is not experimentally detected, but its characteristics are reconstructed form the energies and positions of the two measured ones.

About the THM application in this case, in this case the Trojan Horse (TH) nucleus is in the beam and we chose to reveal the spectator (deuterons) along with the light outgoin particle (protons). This last fact is due to the characteristics of the ^{22}Ne particles coming from the $^{19}F(\alpha,p)^{22}Ne$ reaction, that

are emitted at very forward angles ($\vartheta_{Ne} \leq 6°$) - where the elastic scattering ^{6}Li + ^{19}F (circa 4kHz) would eventually break down the detectors - and at energies (3.5 MeV) that are not high enough to emerge from the target.
In order to maximize the statistics on PSD1 and PSD2, it was also decided to tilt the target at 45° with respect to the beam direction in order to maximize deuteron detection. This is crucial because of the low reaction rate for the reaction and the low energy ($E_d \leq$ 5MeV, determined by means of a proper Monte Carlo simulation) of the deuteron particles.

3 Results and Conclusions

Once deuterons are identified by means of the ΔE-E technique, the Q-value for the three-body process can be isolated and the quasi-free contribution can be separated from the sequential decay (procedure explained in [7]). The half-of-energy-shell binary cross-section of interest can be expressed, in quasi-free conditions, as a function of the measured three-body differential cross-section (in the simplest approach) as follows:

$$\left(\frac{d\sigma}{d\Omega}\right)^{HOES} \propto \left(KF\,|\Phi(p_s)|^2\right)^{-1} \cdot \frac{d^3\sigma}{dE_{CM}d\Omega_d\Omega_p} \tag{1}$$

where KF is a kinematic factor and $|\Phi(p_s)|^2$ is tied to the relative motion of the TH nucleus, with p_s momentum of the spectator particle (deuteron) inside the TH nucleus (^{6}Li). Dividing the measured three-body triple differential cross-section for KF and $|\Phi(p_s)|^2$ it is therefore possible to obtain the two-body one of interest in arbitrary units. Those data were then fitted by means of the *Modified R-Matrix* procedure [2] and a first evaluation of the widths of the involved resonances has been performed [6, 7]. The measured cross-section was then normalized to direct data [9] in the overlap region (0.6÷0.9 MeV in the center-of-mass reference frame), obtaining in the end the absolute units cross-section for the ^{19}F(α,p)^{22}Ne reaction. A comparison between the existing reaction rate [9] and the new one was also attempted, and it resulted in an enhancement - in the temperature window of astrophysical interest - up to a factor of four for the destruction rate of ^{19}F via the (α,p) reaction (Tab. 1).

Temperature [10^9K]	Rate $\left[\frac{cm^3}{mol \times sec}\right]$	$\frac{R_{THM}}{R}$
0.10	2.59×10^{-22}	1.08
0.20	6.00×10^{-14}	1.81
0.30	4.67×10^{-10}	3.71
0.40	1.11×10^{-7}	3.29
0.50	5.09×10^{-6}	1.66
0.60	1.14×10^{-4}	1.12

Table 1. Reaction rate and ratio between experimental THM results (R_{THM}) and the one parametrized from the results of [9] (R), $\frac{R_{THM}}{R}$, for several temperatures in units of 10^9 K

In conclusion, with this experiment it was possible, for the first time, to measure the ^{19}F(α,p)^{22}Ne cross-section inside the energy range of astrophysical relevance, and its impact on the reaction rate has been evaluated. An investigation on the effect of this new rate on fluorine nucleosynthesis is still in progress [7].

References

[1] Lucatello, S., Masseron, T., Johnson, J. et al. 2011, ApJ, **729**, 40
[2] La Cognata M., Mukhamezhanov A., Spitaleri C., et al., 2011, ApJ L, **739**, L 54
[3] Indelicato, I., La Cognata, M., Spitaleri, C., et al., 2017, ApJ, **845**, 1
[4] Lugaro, M., Ugalde, C., Karakas, A. I., et al. 2004, ApJ, **615**, 934
[5] Cristallo, S., Di Leva, A., Imbriani, G. et al. 2014, A&A, **570**, 46
[6] Pizzone, R. G., D'Agata, G., La Cognata M., Indelicato I., et al. 2017, ApJ, **836**, 57
[7] D'Agata, G., Pizzone, R. G., La Cognata M., Indelicato I., et al., ApJ, **submitted**
[8] Spitaleri, C., La Cognata, M., Lamia, L., et al. 2016, Eur. Phys. J. A, **52**, 77
[9] Ugalde, C., Azuma, R. E., Coutre, A., et al. 2008, Phys. Rev. C, **77**, 035801

A new high-precision upper limit of direct α-decays from the Hoyle state in ^{12}C

D. Dell'Aquila[1,2,3,★], *I.* Lombardo[4], *G.* Verde[3,4], *M.* Vigilante[1,2], *L.* Acosta[5], *C.* Agodi[6], *F.* Cappuzzello[7,6], *D.* Carbone[6], *M.* Cavallaro[6], *S.* Cherubini[6,7], *A.* Cvetinovic[6], *G.* D'Agata[7,6], *L.* Francalanza[2], *G.L.* Guardo[6], *M.* Gulino[8,6], *I.* Indelicato[6], *M.* La Cognata[6], *L.* Lamia[7], *A.* Ordine[2], *R.G.* Pizzone[6], *S.M.R.* Puglia[6], *G.G.* Rapisarda[6], *S.* Romano[6], *G.* Santagati[6], *R.* Spartà[6], *G.* Spadaccini[1,2], *C.* Spitaleri[7,6], and *A.* Tumino[8,6]

[1] *Dip. di Fisica "E. Pancini", Università di Napoli Federico II, I-80126 Napoli, Italy*
[2] *INFN-Sezione di Napoli, I-80126 Napoli, Italy*
[3] *Institut de Physique Nucléaire, CNRS-IN2P3, Univ. Paris-Sud, Université Paris-Saclay, 91406 Orsay Cedex, France*
[4] *INFN - Sezione di Catania, Via S. Sofia, I-95125 Catania, Italy*
[5] *Universidad Nacional Autónoma de México, A.P. 20-364, Cd.Mx, D.F. 01000 México*
[6] *INFN - Laboratori Nazionali del Sud, Via S. Sofia, I-95125 Catania, Italy*
[7] *Dip. di Fisica e Astronomia, Università di Catania, Via S. Sofia, I-95125 Catania, Italy*
[8] *Facoltà di Ingegneria ed Architettura, Università Kore, I-94100 Enna, Italy*

Abstract. The Hoyle state in ^{12}C ($E_x = 7.654$MeV) is characterized by a pronounced 3α cluster configuration. It is involved in the so-called 3α process in stars, that is responsible of ^{12}C nucleosynthesis. We studied the decay path of the Hoyle state by using the ^{14}N(d, α_2)^{12}C(7.654) reaction at 10.5MeV incident energy. We found, with a precision higher of a factor 5 than any other previous experiment, an almost total absence of direct decays by-passing the ground state of ^{8}Be. A new upper limit of such a decay width is placed at 0.043% (95% C.L.). Astrophysical 3α process reaction rate calculations have to be consequently revised.

1 Introduction

The origin of elements in our universe, as well as the evolution of stars, reflect the properties of nuclei which are involved in the reactions taking place in stellar environments [1]. Studying nuclear reactions at low energy is therefore important to understand how and to what extent elements are produced in stars [2–5]. A particularly remarkable case is the 3α process, which occurs during the helium burning stage of stellar evolution. This process is responsible of the creation of one of the major constituent of the universe and human being: the carbon. The 3α process proceed with the initial fusion of two α particles into the ^{8}Be, followed by the radiative capture of a third α, α+^{8}Be→^{12}C+γ [6, 7]. Anyway, the lifetime of the intermediate ^{8}Be, formed in the initial fusion, is extremely short ($\tau \approx 10^{-16}s$) and the corresponding reaction rate is therefore limited by the almost instantaneous decay of the unbound

★e-mail: dellaquila@na.infn.it

^{8}Be. To explain the ratio of carbon and oxygen abundances observed in our universe, which amounts at about 0.6 [8]), Fred Hoyle formulated in 1953 an interesting hypothesis: the radiative capture of the third α takes place via a resonant process through a $J^\pi = 0^+$ state of ^{12}C [9, 10]. This state should be close to the α-threshold of ^{12}C in a way to introduce a strong enhancement in the reaction cross section. The existence of this state was then confirmed, a few years later, by Cook and collaborators [11], lying at an excitation energy of 7.654MeV, and it is now named as the *Hoyle state* of ^{12}C [12]. The amount of carbon produced in such a process so depend on the competition between the Hoyle state radiative decays and the α-ones. This can be calculated, for an environment of temperature higher than 10^8 K, as follows [12]:

$$< \sigma v > \propto \frac{\Gamma_\alpha \Gamma_{rad}}{\Gamma} e^{-\frac{E_R}{k_B T}} \quad (1)$$

where E_R is the resonance energy of the Hoyle state, k_B the Boltzmann constant and Γ the total level width. The latter is fully dominated by the α decay width, i.e. $\Gamma = \Gamma_\alpha + \Gamma_{rad} \approx \Gamma_\alpha$, and so it results in a dynamical equilibrium ^{4}He + ^{8}Be $\leftrightarrows$ ^{12}C* only broken by the small leakage to the ^{12}C ground state given by the radiative decays.

The reaction rate of the 3α process is therefore strongly affected by the decay properties of the Hoyle state [13]. However, while for star temperatures greater than 10^8 K it is dominated by the *sequential* (resonant) process through the ground state of ^{8}Be, for lower temperatures another mechanism plays a role in the reaction rate determination: the so-called *direct* (non-resonant) process, where the formation of the Hoyle state happens by-passing the ^{8}Be ground state level. While the first process is affected by the *sequential decay* (SD) width of the Hoyle state, the second one is determined by the *direct decay* (DD) one. As an example, for a temperature of $T = 0.02$GK the 3α reaction rate associated to the direct process is found to be 7-20 orders of magnitude greater than the one calculated via a resonant process [14, 15].

Recently, a quite large number of experiments have been performed in order to measure the DD width of the Hoyle state [16–21]. Anyway, the picture resulting from these experiments is quite complicated, see Ref. [21], and no definitive conclusions can be drawn regarding the possible existence of a non-vanishing DD width.

For these reasons, we performed a new high-precision experiment with the aim of solving the present ambiguities in the experimental knowledge of the DD width of the Hoyle state in ^{12}C.

2 Experimental results

The experiment was performed at the INFN-Laboratori Nazionali del Sud (LNS) of Catania, Italy. A 10.5 MeV deuteron beam was delivered to the scattering chamber on a $C_3N_6H_6$ (40 μg/cm^2) composite target, deposed on a C backing ($\approx$ 10 μg/cm^2). To produce ^{12}C*(Hoyle) we used the ^{14}N(d,α) reaction. A ΔE-E silicon telescope is placed at an angle $\theta_{lab} = 125°$ to detect the ejectiles from the reaction of interest. A strong reduction of the background due to contaminant reactions is achieved by using the anti-coincidence telescope technique, as described in Ref. [22], where a detailed sketch of the present experimental apparatus is presented. By measuring energy and directions of the α ejectiles we are able to detect the recoiling direction of the emitted ^{12}C residuals and their excitation energy. A ^{12}C excitation energy spectrum obtained from the anti-coincidence spectrum measured in the backward telescope is shown in figure 1 (black spectrum). States populated in ^{12}C are indicated by labels. A peak corresponding to the energy of the Hoyle state is clearly observed. It can be used to tag the recoiling ^{12}C*(Hoyle). In kinematical coincidence with these recoils we placed a hodoscope of 64 silicon independent cells [23], in order to detect the three-body disintegration of the produced and

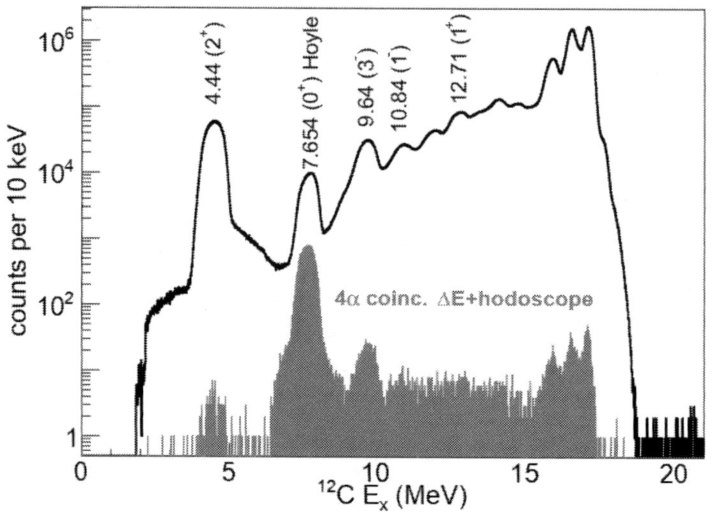

Figure 1. ^{12}C excitation energy spectrum obtained from the single-particle spectrum measured by the ΔE-E anti-telescope (black spectrum). It collapses into the filled one by including only 4-particles fully reconstructed events

tagged $^{12}C^{\star}$(Hoyle). Such an apparatus is extremely sensitive to the disintegration of the Hoyle state. Indeed, if one requires that 3 particle are detected in the hodoscope in coincidence with the backward detected ejectile, the spectrum collapses into the filled one of figure 1. Here contributions due to ^{12}C states different from the Hoyle state are almost totally suppressed, while the peak associated to the Hoyle state is prominent.

Invariant mass reconstruction (see for example [24, 25]) of the 3α particles detected by the hodoscope allow to study the decay pattern of the Hoyle state. The total energy of the Hoyle state decay is normally shared among the three α-particles. If one calls with E_1, E_2 amd E_3 these values, so the normalized energy of the decay are defined as $\varepsilon_{1,2,3} = E_{1,2,3}/(E_1 + E_2 + E_3)$. It is obvious that $\varepsilon_1 + \varepsilon_2 + \varepsilon_3 = 1$. In the case of a sequential decay, the first α particle is emitted with the maximum energy in the center of mass. In other words, as discussed more in detail for example in [19, 21], in order for the total momentum to be conserved, the first α-particle should bring an amount of kinetic energy almost equal to the half of the total energy of the $^{12}C^{\star} \rightarrow 3\alpha$ decay. Indicating with i the first emitted α-particle, one should observe $\epsilon_i \approx 0.5$ for a SD, while any value from 0.33 to 0.67 is possible, according to the available phase space, in the presence of a DD.

Figure 2 shows the experimental distribution of ε_i, i.e. the largest normalized energy in the decay, obtained in the present experiment. A peak is present at around 0.5 which represents events associated to a SD. A dashed line is the result of a detailed simulation of the experiment which assumed a full SD path of the Hoyle state. Only a few events fall outside of the SD region, and they are not reproduced by our simulation. By using the Feldman and Cousin's method for evaluating small signal in the presence of small background, we are able to indicate that such amount of counts is statistically non-significant. The corresponding upper limit of DD is placed at 0.043% with a confidence limit (C.L.) of 95%.

DOI: 10.1051/epjconf/201716501020

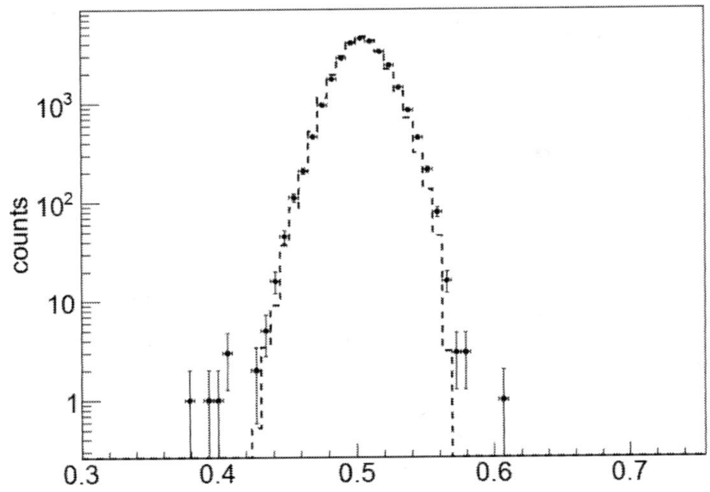

Figure 2. Distribution of the largest normalized energy ε_i in the decays of the Hoyle state. They are obtained by means of invariant mass technique on experimental Hoyle state decay data measured in the hodoscope.

3 Conclusions

We have measured the DD of the Hoyle state in ^{12}C by means of a high-precision experiment. We used the ^{14}N(d, α_2)^{12}C(7.654) reaction at 10.5MeV incident energy to tag ^{12}C nuclei, whose decay is studied via invariant mass techniques of the 3α decay. To measure the three particles of the decay we used a hodoscope of 64 silicon independent cells, thus avoiding background typical of the use of a single strip detectors. We do not evidence any DD contribution, placing a upper limit for direct decay branches of the Hoyle state of 0.043% (95% C.L.). This result, recently published in [26], has a precision of about a factor 5 better than any other previous investigation. An analogous result has been also obtained by a very recent experiment performed by the group of Birmingham [27].

In the next future, an experiment will be performed by the CHIMERA group of LNS to increase the precision in the knowledge of the γ decay width of the Hoyle state. The experiment will benefit of the coupling of a large acceptance detector [28] and a high-granularity array [29]. A more accurate knowledge of the radiative decay width of the Hoyle state, together with the present improved limit of the DD width will help in better understanding the structure of the Hoyle state and the rate of the 3α process in low temperature stars.

References

[1] C.E. Rolfs, W.S. Rodney, *Cauldrons in the Cosmos: Nuclear Astrophysics*, 5th edn. (The University of Chicago Press, Chicago and London, 1997)
[2] C. Spitaleri et al., Phys. Rev. **C95**, 035801 (2017)
[3] I. Lombardo et al., J. Phys. G. **43**, 45109 (2016)
[4] I. Lombardo et al., Phys. Lett. B **748**, 178 (2015)
[5] I. Lombardo et al., J. Phys. G **40**, 125102 (2013)

[6] E. Opik, Proc. R. Irish Acad. A **54**, 49 (1951)
[7] E. Salpeter, Phys. Rev. **88**, 547 (1952)
[8] M. Freer, Rep. Prog. Phys. **70**, 2149 (2007)
[9] F. Hoyle et al., Phys. Rev. **92**, 1095c (1953)
[10] F. Hoyle, Astrophys. J. Suppl. Ser. **1**, 121 (1954)
[11] C. Cook, W. Fowler, C. Lauritsen, T. Lauritzen, Phys. Rev. **107**, 508 (1957)
[12] M. Freer, H. Fynbo, Prog. Part. Nuc. Phys. **78**, 1 (2014)
[13] K. Ogata, M. Kan, M. Kamimura, Prog. Theor. Phys. **122**, 1055 (2009), 0905.0007
[14] N.B. Nguyen, F.M. Nunes, I.J. Thompson, E.F. Brown, Phys. Rev. Lett. **109**, 141101 (2012), 1112.2136
[15] N.B. Nguyen, F.M. Nunes, I.J. Thompson, Phys. Rev. **C87**, 054615 (2013), 1209.4999
[16] M. Freer et al., Phys. Rev. C **49**, R1751 (1994)
[17] A. Raduta et al., Phys. Lett. B **705**, 65 (2011)
[18] J. Manfredi, R.J. Charity, K. Mercurio, R. Shane, L.G. Sobotka, A.H. Wuosmaa, A. Banu, L. Trache, R.E. Tribble, Phys. Rev. C **85**, 037603 (2012)
[19] O. Kirsebom et al., Phys. Rev. Lett. **108**, 202501 (2012)
[20] T. Rana et al., Phys. Rev. C **88**, 021601(R) (2013)
[21] M. Itoh et al., Phys. Rev. Lett. **113**, 102501 (2014)
[22] D. Dell'Aquila et al., J. Phys. Conf. Ser. **876**, 012006 (2017)
[23] D. Dell'Aquila et al., Nucl. Instrum. Meth. Phys. Res. A **877**, 227 (2018)
[24] D. Dell'Aquila et al., Phys. Rev. C **93**, 024611 (2016)
[25] D. Dell'Aquila, Il Nuov. Cim. C **39**, 272 (2016)
[26] D. Dell'Aquila et al., Phys. Rev. Lett. **119**, 132501 (2017)
[27] R. Smith, T. Kokalova, C. Wheldon, J.E. Bishop, M. Freer, N. Curtis, D.J. Parker, Phys. Rev. Lett. **119**, 132502 (2017)
[28] G. Cardella et al., Nucl. Instr. and Meth. A **799**, 64 (2015)
[29] E.V. Pagano et al., EPJ Web of Conf. **117**, 10008 (2016)

EPJ Web of Conferences **165**, 01021 (2017)
NPA8 2017
DOI: 10.1051/epjconf/201716501021

New direct measurement of the ^{10}B(p,α)^{7}Be reaction with the activation technique

Rosanna Depalo[1,2,⋆], *Antonio* Caciolli[1,2], *Carlo* Broggini[2], *Marco* La Cognata[3], *Livio* Lamia[4], *Roberto* Menegazzo[2], *Liliana* Mou[5], *Sebastiana Maria Regina* Puglia[3], *Valentino* Rigato[5], *Stefano* Romano[3,4], *Carlos* Rossi Alvarez[5], *Maria Letizia* Sergi[3], *Claudio* Spitaleri[3,4], and *Aurora* Tumino[3,6]

[1] *Dipartimento di Fisica e Astronomia, Universitá degli Studi di Padova, Padova, Italy*
[2] *INFN - Sezione di Padova, Padova, Italy*
[3] *INFN, Laboratori Nazionali del Sud, Catania, Italy*
[4] *Dipartimento di Fisica e Astronomia, Universitá degli Studi di Catania, Catania, Italy*
[5] *INFN, Laboratori Nazionali di Legnaro, Legnaro, Italy*
[6] *Facoltá di Ingegneria e Architettura, Universitá degli Studi di Enna "Kore", Enna, Italy*

Abstract. Boron plays an important role in astrophysics and, together with lithium and beryllium, is a probe of stellar structure during the pre-main sequence and main-sequence phases. In this context, the ^{10}B(p,α)^{7}Be reaction is of particular interest.
The literature data show discrepancies in the energy range between 100 keV and 2 MeV. This also poses a normalization problem for indirect data obtained with the Trojan Horse Method.
A new measurement of the ^{10}B(p,α)^{7}Be reaction cross section was performed at Legnaro National Laboratories (LNL). At LNL, the cross section was determined with the activation technique by measuring the activated samples at a low-background counting facility. The analysis of that experiment is now complete and the results are here presented.

1 Introduction

Boron-burning nuclear reactions are of great interest in many different fields ranging from nuclear physics and astrophysics to plasma physics.
The ^{11}B(p,α)^{8}Be $\longrightarrow$ $\alpha\alpha$ reaction is one of the main candidates for aneutronic energy production in plasma-induced fusion reactors [1]. Natural boron consists of ^{11}B (80.1%) and ^{10}B (19.9%) and the (p,α) reaction on ^{10}B represents a contaminant for aneutronic fusion, since it produces the radioactive nucleus ^{7}Be. In order to estimate the impact of ^{10}B contaminations in future fusion reactors, the cross section of the ^{10}B(p,α)^{7}Be reaction should be accurately known.
In astrophysics, boron is used together with lithium and beryllium as a tracer of stellar structure and evolution during the pre-main sequence and main sequence phases. In stars, lithium, beryllium and boron are burned at different temperatures (T > 2.5 MK for Li, T > 3.5 MK for Be and T > 5 MK for B). As a consequence, when material is mixed from the surface down to the inner layers of a star,

⋆e-mail: rdepalo@pd.infn.it

each element survives to different depths [2]. Moreover, the relative abundances of the stable boron isotopes ^{10}B and ^{11}B in pre-MS stars provides additional information on early stellar evolution: the N(^{11}B)/N(^{10}B) ratio for different stellar masses and metallicities shows an increase with time for stellar masses in the range 0.1 - 0.3 solar masses [3]. Therefore, by measuring the amount of each isotope on the surface of a star it is possible to shed light on stellar mixing in the early phases of stellar evolution, provided that the cross sections of all the reactions which may affect the elemental abundances are known.

2 State of the art

The ^{10}B(p,α)^{7}Be reaction has been widely studied in the past, with experimental data available in the energy range from 5 to 6000 keV [4] - [13]. At energies below 100 keV, the direct data in ref. [10] show an enhancement of the astrophysical S-factor S(E). This enhancement is produced by the combination of a resonance at $E_{CM} \sim$ 10keV and by the electron screening effect.
In the energy range between 100 and 400 keV, experimental data show great discrepancies, as the data of ref. [9] had to be multiplied by a factor of 1.83 to match the results of ref. [10].
To overcome the difficulties related to the electron screening effect and the suppression of the cross section at ultra-low energies, recent measurements of ^{10}B(p,α_0)^{7}Be with the Trojan Horse Method (THM) [12] have provided the bare-nucleus S(E) down to 5 keV without the needs of extrapolation procedures, providing the only experimental dataset at energies below the 10 keV resonance. However, the mismatch between direct data above 100 keV, where electron screening is negligible, makes it difficult to choose a reliable normalization for the Trojan Horse data.
Together with the present experiment, new results from different groups have been published recently in the 0.630–1.028 MeV energy range [14, 16]. Here we report on the results of a measurement of the ^{10}B(p,α)^{7}Be cross section performed at Legnaro National Laboratories [17].

3 Experimental setup

The experiment was performed at the AN2000 accelerator of Legnaro National Laboratories (LNL).
A beam energy calibration was performed exploiting the ^{27}Al(p,γ)^{28}Si narrow resonances at 632 keV and 992 keV [18] and through Rutherford Backscattering Spectroscopy on a SiO_2 sample.
The ^{10}B(p,α_0)^{7}Be cross section has been measured through activation technique. A sketch of the setup used for the target irradiation is shown in fig. 1 and an extensive description can be found in [17]. The beam was collimated by slits positioned at a distance of 1 meter from the target, then entered the target chamber through a collimator of 9 mm diameter.
Targets were made by evaporating ^{10}B powder (isotopic enrichment up to 93%) on thin carbon layers (~ 20 μg/cm^2). Three different target thicknesses were chosen: 25, 60, and 100 μg/cm^2. Targets with the same thickness were evaporated at the same time in order to ensure their uniformity. Moreover, two samples of each group were not irradiated and used for off-line target analysis.
The reaction kinematics is such that ^{7}Be ions are emitted at all angles and their kinetic energy is high enough to escape the target. Therefore, in order to collect all ^{7}Be nuclei produced by the reaction, the target was surrounded with a 0.2 mm thick aluminum catcher. The catcher was designed to cover 99.8% of the total solid angle.
Each target was irradiated at a different beam energy and the duration of the irradiation was decided according to the beam energy and expected cross section. Once the irradiation was completed, the ^{7}Be activity in the targets and in the catchers was measured at the low-background γ counting

DOI: 10.1051/epjconf/201716501021

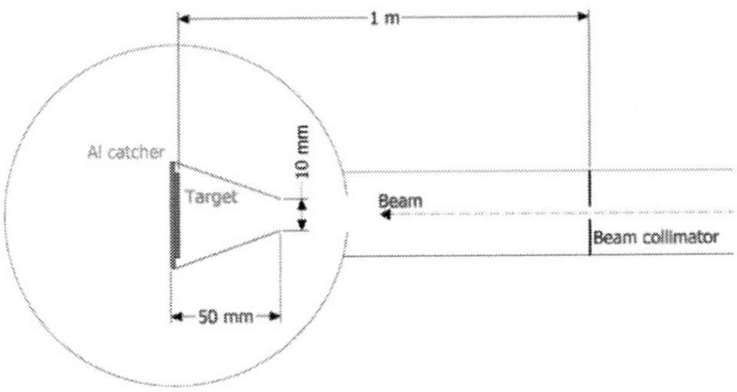

Figure 1. Experimental setup used for the irradiation of ^{10}B samples [17].

facility of LNL [19]. The counting facility consists of two HPGe detectors of 80% relative efficiency facing each other. Both detectors are enclosed in a copper and lead lining serving as a shield against environmental background from naturally occurring radioactive isotopes. Samples were positioned between the two detectors, at a distance of 2.5 cm from each endcap.
^{7}Be decays with (10.44 ± 0.04)% probability to the first excited state of ^{7}Li. The de-excitation of ^{7}Li to the ground state produces a 478 keV γ ray which was used to count the number of ^{7}Be decays per unit time.

4 Results and conclusions

The ^{10}B(p,α_0)^{7}Be cross section has been measured in the energy range from 249 up to 1182 keV in the center-of-mass system, an energy range poorly explored in previous works.
The number of ^{7}Be nuclei produced in each irradiation was deduced from the analysis of the activation spectra. Targets and catchers samples were analysed separately and the results were then summed to obtain the total amount of ^{7}Be produced at each irradiation energy.
Targets thickness and elemental composition (including oxygen and nitrogen contaminants) were measured performing Rutherford backscattering spectrometry and nuclear reaction analysis at the AN2000 accelerator.
The astrophysical S-factor from the present experiment is reported in fig. 2, compared to some of the literature S-factors. At the lowest energies investigated, the present data show a discrepancy of about a factor of two with respect to Youn et al. [9], matching the results from Angulo et al. [10]. At energies above 500 keV, our results are in fair agreement with literature data from ref. [4] and [7], as well as with the more recent results from Lombardo et al. [14], while some disagreement exists with the recently published results from Wiescher et al. [16].
A simplified R-matrix analysis of the present-work data was performed (see fig. 2 and ref. [17] for more details). The fit successfully reproduces our experimental data, with the exception of one point at E = 647 keV, which significantly deviates from the calculation and might point to the existence of a new resonance. Further investigations with a smaller energy step may help clarify this point.

The results from the present work have been used in combination with [7] and [14] for the normalization of both low energy [12] and high energy [15] Trojan Horse results, which provide consistent data in the energy range between 5 keV and 1.5 MeV [15].
Some disagreement still exists between the results from different direct experiments and therefore

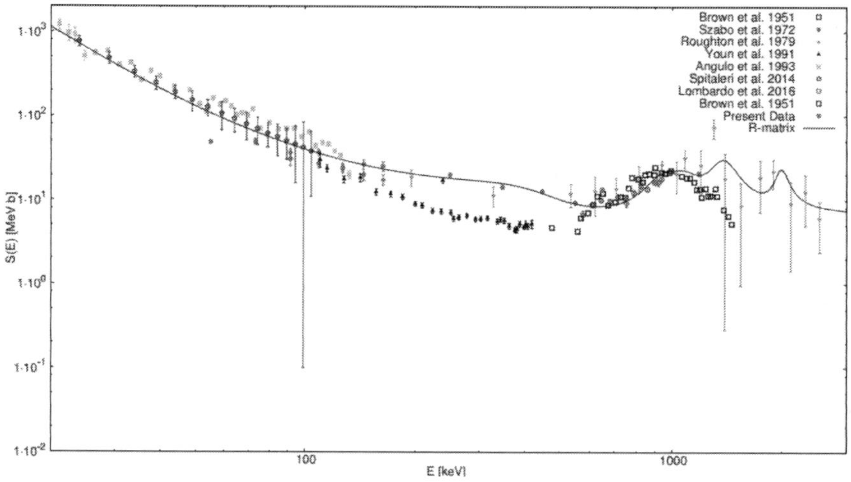

Figure 2. Astrophysical S-factor of the $^{10}B(p,\alpha_0)^7Be$ reaction. Present results are compared with literature data.

new direct investigations covering a broader energy range are still needed.

Following the measurement of the $^{10}B(p,\alpha_0)^7Be$ cross section, a new experiment was performed at LNL to measure the cross section of the (p,α_1) channel in the energy range between 300 and 2000 keV. The data analysis is still ongoing and results will be presented in a forthcoming publication.

References

[1] M. C. Spraker et al., Journal of Fusion Energy **31**, 357 (2011)
[2] A. M. Boesgaard et al., The Astrophysical Journal **621**, 991 (2005)
[3] L. Lamia et al., The Astrophysical Journal **811**, 99 (2015)
[4] A. B. Brown et al., Physical Review **82**, 159 (1951)
[5] J. W. Cronin et al., Physical Review **101**, 298 (1956)
[6] J. Szabó et al., Nuclear Physics A **195**, 527 (1972)
[7] N. Roughton et al., Atomic Data and Nuclear Data Tables **23**, 177 (1979)
[8] M. Wiescher et al., Physical Review C **28**, 1431 (1983)
[9] M. Youn et al., Nuclear Physics A **533**, 321 (1991)
[10] C. Angulo et al., Zeitschrift für Physik A **345**, 231 (1993)
[11] A. Kafkarkou et al., Nuclear Instruments in Physics Research B **316**, 48 (2013)
[12] C. Spitaleri et al., Physical Review C **90**, 035801 (2014)
[13] A. Lagoyannis et al., Nuclear Instruments in Physics Research B **342**, 271 (2015)
[14] I. Lombardo et al., Journal of Physics G: Nuclear and Particle Physics **43**, 045109 (2016)
[15] C. Spitaleri et al., Physical Review C **95**, 035801 (2017)
[16] M. Wiescher et al., Physical Review C **95**, 044617 (2017)
[17] A. Caciolli et al., European Physical Journal A **52**, 136 (2016)
[18] S. Harissopulos et al., European Physical Journal A **9**, 479 (2000)
[19] G. Xhixha et al, Journal of Radioanalytical and Nuclear Chemistry **295**, 445 (2013)

DOI: 10.1051/epjconf/201716501022

Determination of alpha spectroscopic factors for unbound ^{17}O states

Nicolas de Séréville[1,*], *Anne* Meyer[1], *Faïrouz* Hammache[1], *Alison M.* Laird[2], and *Marco* Pignatari[3]

[1] *Institut de Physique Nucléaire d'Orsay, UMR8608, IN2P3-CNRS, Université Paris sud 11, 91406 Orsay, France*
[2] *Department of Physics, University of York, York YO10 5DD, United Kingdom*
[3] *E. A. Milne Centre for Astrophysics, Deparment of Physics and Mathematics, University of Hull, HU6 7RX, United Kingdom*

Abstract. It has been recently suggested that hydrogen ingestion into the helium shell of massive stars could lead to high ^{13}C and ^{15}N excesses when the blast of a core collapse supernova (ccSN) passes through its helium shell. This prediction questions the origin of extremely high ^{13}C and ^{15}N abundances observed in rare presolar SiC grains which is usually attributed to classical novae. In this context the ^{13}N(α,p)^{16}O reaction plays an important role since it is in competition with ^{13}N β^+-decay to ^{13}C. As a first step to the determination of the ^{13}N(α,p)^{16}O reaction rate, we present a study aiming at the determination of alpha spectroscopic factors of ^{17}O states which are the analog ones to those in ^{17}F, the compound nucleus of the ^{13}N(α,p)^{16}O reaction.

1 Introduction

Primitive meteorites hold several types of dust grains that condensed in stellar winds or ejecta of stellar explosions. These grains carry isotopic anomalies which are used as a signature of the stellar environment in which they formed. As such, extreme excesses of ^{13}C and ^{15}N in rare presolar SiC grains have been considered as a diagnostic of an origin in classical novae [1], however an origin in ccSNe has also been recently proposed [2]. In the context of ccSNe, explosive He shell burning can reproduce the high ^{13}C and ^{15}N abundances if H was ingested into the He shell and not fully destroyed before the explosion [3]. The supernova shock will then produce an isotopic pattern similar to the hot-CNO cycle signature obtained in classical novae. It has been shown that a variation of a factor of five for the ^{13}N(α,p)^{16}O reaction rate induces several orders of magnitude uncertainty in the production of ^{13}N which β^+-decays to ^{13}C.

Currently the ^{13}N(α,p)^{16}O reaction rate is calculated using a statistical model or the time reverse reaction and these determinations have large uncertainties. The goal of this work is to put the ^{13}N(α,p)^{16}O reaction rate on a firmer basis. Given that the alpha emission threshold (S_α = 5.819 MeV) is much higher than the proton emission threshold (S_p = 0.600 MeV) in the compound nucleus ^{17}F, the resonance strength of individual resonances is directly proportional to their alpha

*e-mail: deserevi@ipno.in2p3.fr

widths. We report on the analysis of $^{13}C(^{7}Li,t)^{17}O$ data populating the analog states of the states of interest in ^{17}F. After a DWBA analysis of the measured differential cross sections, alpha spectroscopic factors are extracted and alpha widths are deduced.

2 Data reduction

An analysis of existing data from the alpha-transfer $^{13}C(^{7}Li,t)^{17}O$ experiment performed at the Tandem-ALTO facility in Orsay, France, was undertaken. All the experimental details can be found in Ref. [4] which focused on the study of the sub-threshold state at 6.356 MeV in ^{17}O relevant for the $^{13}C(\alpha,n)^{16}O$ reaction and its role in the main *s*-process. Tritons from the $^{13}C(^{7}Li,t)^{17}O$ reaction were momentum analyzed by an Enge Split-Pole magnetic spectrometer, and detected and identified at the focal plane. Figure 1 shows the triton focal-plane position spectrum obtained in the case of a spectrometer angle of 7° covering ^{17}O excitation energies between 6.2 and 7.4 MeV. Calibration of the focal-plane position detector at the same magnetic field using the natural carbon target and known ^{16}O levels populated through the $^{12}C(^{7}Li,t)^{16}O$ reaction was used to identify the ^{17}O states. The best least-square fit of the triton spectrum for states above the α+^{13}C threshold (S_α = 6.359 MeV) is also represented in Fig. 1 together with the individual contribution of each ^{17}O state. While narrow states were described by a skewed Gaussian function needed to account for the low energy tail of each triton peak, a Voigt function was used for the broad ^{17}O state at 7.202 MeV (Γ = 280 (30) keV [5]). Since most of the levels have a natural width much smaller than the experimental width, a common width was taken for each Gaussian function describing ^{17}O states in the fitting procedure. A width of ≈ 60 keV (FWHM) is obtained which reflects the experimental resolution. The previous procedure has been repeated for each of the eleven spectrometer angles where the measurement was performed.

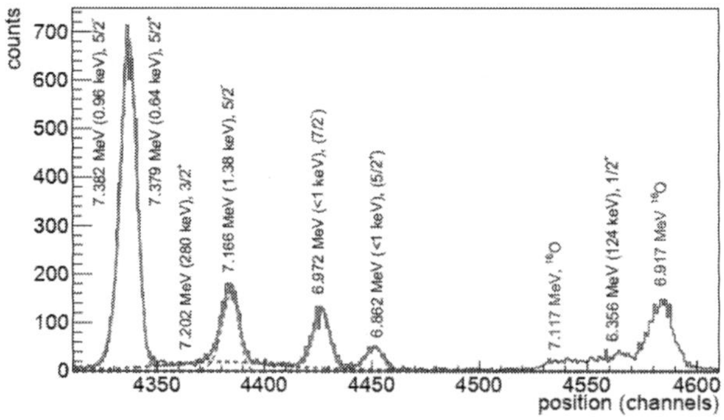

Figure 1. (Color online) Tritons magnetic rigidity spectrum at spectrometer angle of 7°. Excitation energies in ^{17}O between 6.2 and 7.4 MeV are covered and the best fit of the spectrum is represented for states above the α+^{13}C threshold. Note the presence of an experimentally unresolved doublet including ^{17}O states at 7.379 and 7.382 MeV. Note also the presence of two ^{16}O contamination states away from the region of interest.

3 Angular distributions and DWBA analysis

The differential cross sections corresponding to ^{17}O states were obtained by normalizing the number of tritons determined at each detection angle to the target thickness, solid angle and accumulated charge. As an example, the differential cross section of the 7.382 MeV ($5/2^-$) state is displayed in Fig. 2. Finite-range DWBA calculations were performed with the FRESCO code [6]. Several combinations of entrance and exit optical potentials have been tried as inputs of the DWBA calculations together with two values of the number of radial nodes N of the α-wave function in ^{17}O (see Fig. 2). The most sensitive parameter is the number of radial nodes and N = 2 was selected for the 7.382 MeV state. For this state the different combinations of optical potential give similar angular distribution shapes even if the spread on the normalization factor accounts for ±20%. The best compromise for describing all differential cross sections at the same time is with potential III from Schumacher et al. [7] for the entrance channel and with potential I f7/2 from Garrett et al. [8] for the exit channel, in line with the study of the sub-threshold 6.356 MeV state [4].

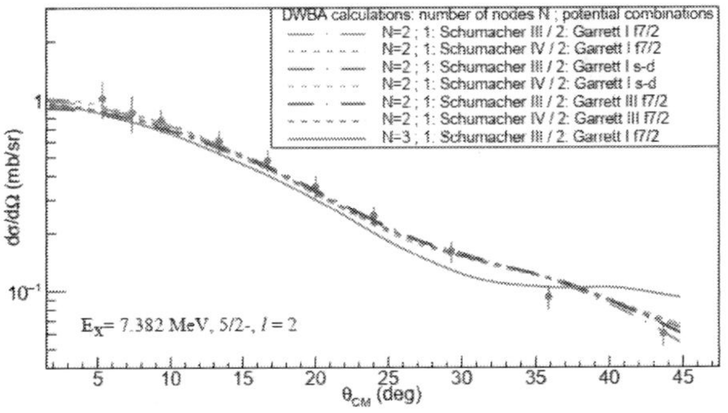

Figure 2. (Color online) Experimental cross section of the ^{17}O state at 7.382 MeV populated with the ^{13}C(^{7}Li,t)^{17}O reaction. Finite-range DWBA calculations are represented for different sets of entrance and exit optical potentials. The effect of the number of radial nodes of the α-wave function in ^{17}O is also explored.

Assuming that the overlap between the ^{7}Li and α+t systems is equal to one (see discussion in Ref. [4]), the normalization factor between the experimental and DWBA differential cross sections is the ^{17}O alpha spectroscopic factor (C^2S_α). The procedure to extract C^2S_α for unbound ^{17}O states follows the prescription from Ref. [9]. For low transferred angular momentum ($\ell < 2$) such as for the ^{17}O state at 7.202 MeV, the alpha spectroscopic factor is determined at several binding energies and extrapolated at the actual α-separation energy (BE = -844 keV) as shown in Fig. 3. For higher transferred angular momentum, C^2S_α is determined for an hypothetical ^{17}O state bound by 100 keV since the α-cluster is quasi-bound due to the large centrifugal barrier. Alpha spectroscopic factor of 0.40 is obtained in the case of the 7.202 MeV broad state. Concerning the experimentally unresolved doublet including ^{17}O states at 7.379 and 7.382 MeV, alpha spectroscopic factors are 0.28 and 0.42, respectively, assuming all the strength is on one or the other state.

The alpha spectroscopic factor can be used to derive the corresponding partial width in case of unbound states using the following formula [10]: $\Gamma_\alpha = 2P_l(r,E)\frac{\hbar^2 r}{2\mu}C^2S_\alpha\,|\phi(r)|^2$, where $P_l(r,E)$ is the penetrability for transferred angular momentum ℓ, and $|\phi(r)|$ is the radial part of the α+^{13}C wave function. This formula should be evaluated at the interaction radius r where the α+^{13}C wave function reaches an asymptotic behavior. This radius was determined by comparing the alpha reduced width γ_α^2 with the corresponding Whittaker function (see Fig. 4) and a value of r = 6.5 fm is obtained. This procedure was done for all states and a similar radius was obtained. Alpha widths obtained for the 7.202 MeV and 7.379 MeV states are 6.5×10^{-2} and 7.8×10^{-3} keV, respectively. This is in very good agreement (within a factor of two) with alpha widths of 0.07 and 0.01 keV, respectively, reported in the literature [5].

4 Conclusions

Data from the ^{13}C(^{7}Li,t)^{17}O reaction were analysed with the goal to determine the alpha spectroscopic factors of ^{17}O states relevant for the study of the ^{13}N(α,p)^{16}O reaction. Details of the analysis in the case of the ^{17}O states at 7.202 MeV and 7.382 MeV are given. Results for other ^{17}O states will be presented in a forthcoming paper together with a new ^{13}N(α,p)^{16}O reaction rate and its astrophysical implication.

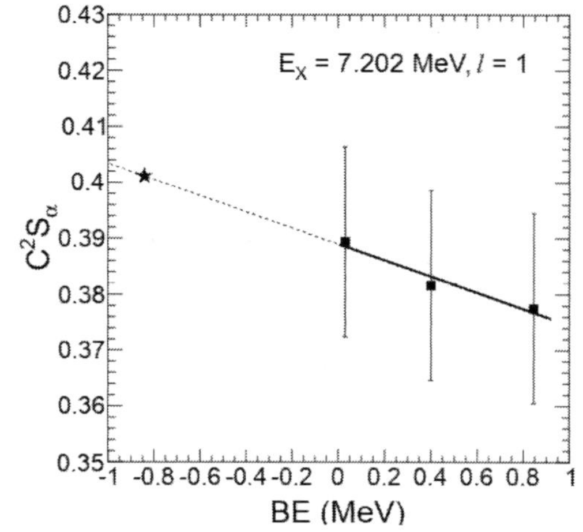

Figure 3. Determination of the α-spectroscopic factor for low transferred angular momentum (here the ^{17}O state at 7.202 MeV). DWBA calculations at several binding energies are performed, and the spectroscopic factors are extrapolated to the actual α-separation energy (filled star).

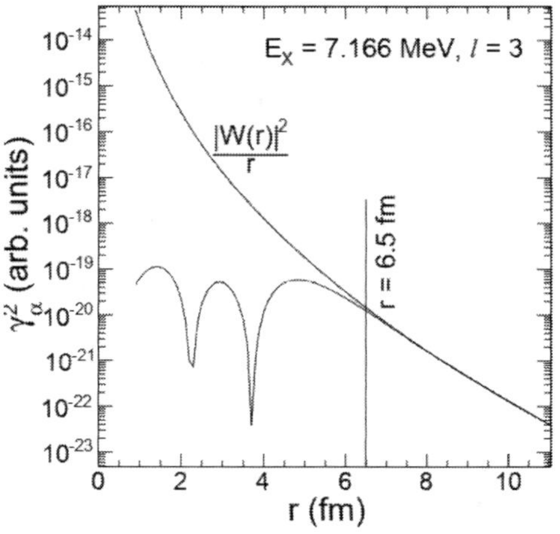

Figure 4. Alpha reduced width (γ_α^2) as a function of the interaction radius for the 7.166 MeV state in ^{17}O. The asymptotic behavior of γ_α^2 is given by ($|W(r)|^2/r$) where $W(r)$ represents the Whittaker function. The interaction radius is defined when both functions have the same radius dependance.

Acknowledgments

A. M. L. acknowledges the support of the Science and Technology Facilities Council.

References

[1] S. Amari, X. Gao, L.R. Nittler, E. Zinner, J. José, M. Hernanz, R.S. Lewis, Astrophys. J. **551**, 1065 (2001)

[2] N. Liu, L.R. Nittler, C.M. O'D. Alexander, J. Wang, M. Pignatari, J. José, A. Nguyen, Astrophys. J. **820**, 140 (2016)

[3] M. Pignatari, E. Zinner, P. Hoppe, C.J. Jordan, B.K. Gibson, R. Trappitsch, F. Herwig, C. Fryer, R. Hirschi, F.X. Timmes, Astrophys. J. Lett. **808**, L43 (2015)

[4] M.G. Pellegriti, F. Hammache, P. Roussel, L. Audouin, D. Beaumel, P. Descouvemont, S. Fortier, L. Gaudefroy, J. Kiener, A. Lefebvre-Schuhl et al., Phys. Rev. **C77**, 042801 (2008)

[5] D.R. Tilley, H.R. Weller, C.M. Cheves, Nuclear Physics A **564**, 1 (1993)

[6] I.J. Thompson, Computer Physics Report **7**, 167 (1988)

[7] P. Schumacher, N. Ueta, H.H. Duhm, K.I. Kubo, W.J. Klages, Nuclear Physics A **212**, 573 (1973)

[8] J.D. Garrett, O. Hansen, Nuclear Physics A **212**, 600 (1973)

[9] F.D. Becchetti, E.R. Flynn, D.L. Hanson, J.W. Sunier, Nuclear Physics A **305**, 293 (1978)

[10] C. Iliadis, *Nuclear Physics of Stars* (Wiley-VCH, 2008), ISBN 9783527406029

EPJ Web of Conferences **165**, 01023 (2017) DOI: 10.1051/epjconf/201716501023
NPA8 2017

Neutron capture cross sections of Kr

Stefan Fiebiger[1,★], *Bayarbadrakh* Baramsai[2], *Aaron* Couture[2], *Milan* Krtička[3], *Shea* Mosby[2], *René* Reifarth[1], *John* O'Donnell[2], *Gencho* Rusev[2], *John* Ullmann[2], *Mario* Weigand[1], and *Clemens* Wolf[1]

[1] *Goethe University Frankfurt, Germany*
[2] *Los Alamos National Laboratory, USA*
[3] *Charles University, Czech Republic*

Abstract. Neutron capture and β^--decay are competing branches of the s-process nucleosynthesis path at ^{85}Kr [1], which makes it an important branching point. The knowledge of its neutron capture cross section is therefore essential to constrain stellar models of nucleosynthesis. Despite its importance for different fields, no direct measurement of the cross section of ^{85}Kr in the keV-regime has been performed. The currently reported uncertainties are still in the order of 50% [2, 3]. Neutron capture cross section measurements on a 4% enriched ^{85}Kr gas enclosed in a stainless steel cylinder were performed at Los Alamos National Laboratory (LANL) using the Detector for Advanced Neutron Capture Experiments (DANCE). ^{85}Kr is radioactive isotope with a half life of 10.8 years. As this was a low-enrichment sample, the main contaminants, the stable krypton isotopes ^{83}Kr and ^{86}Kr, were also investigated. The material was highly enriched and contained in pressurized stainless steel spheres.

1 Introduction

Most of the elements with proton numbers higher than iron are produced through neutron-induced processes (neutron captures). There are two major processes, the rapid neutron-capture process (r-process) and the slow neutron capture process (s-process) [4]. Only a few isotopes on the proton-rich side of the valley of stability get significant contributions from other processes.

If the conditions in the star make the rates for neutron capture comparable to the rate of β-decay for a particular isotope, the s-process path branches at that isotope: a fraction of the isotope transforms via neutron capture, while the other fraction β-decays. The branching ratio, or relative likelihood, for the different reactions depends on the physical conditions in the interior of the star – temperature, neutron density and electron density. At higher neutron densities with all other conditions equal, more nuclei of a given isotope capture a neutron before having the chance to β-decay. Thus, the branching ratios deduced from the isotopic ratios observed in stellar material provide the tools to effectively constrain modern models of the stars where the nucleosynthesis occurs. Therefore the fundamental rates, hence cross sections, for neutron capture and β-decay are essential.

A particularly interesting branching of the s-process occurs in the region around ^{85}Kr (Figure 1) [5]. The main path is indicated by green arrows while the red arrows show the path if ^{85}Kr captures a

★e-mail: fiebiger@iap.uni-frankfurt.de

 DOI: 10.1051/epjconf/201716501023

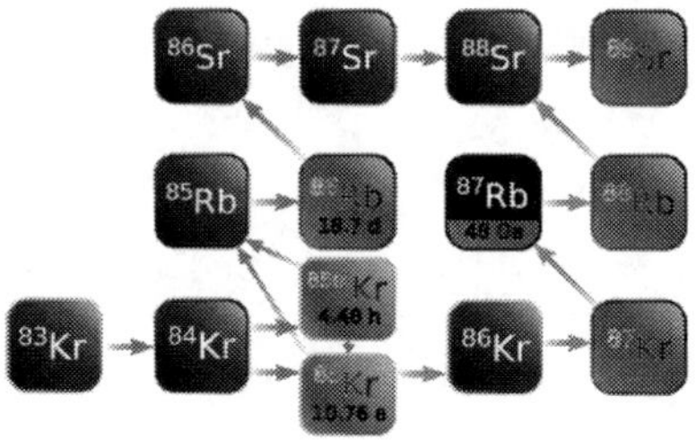

Figure 1. Reaction network during s-process nucleosynthesis around the important branch point ^{85}Kr.

neutron. There is isomeric state with half life of 4.48 h which always decays before neutron capture – either through β-decay or through internal conversion to the ground state.

2 Samples

As Kr is gaseous and in the case of ^{85}Kr radioactive, preparing suitable samples for the DANCE setup proved to be a major challenge. Two types of samples were designed for ^{85}Kr and ^{83,86}Kr respectively (Figure 2). Both had to fit in the existing beam pipe while simultaneously minimizing exposure of the sample holders. The first sample was a reactor produced ^{85}Kr sample containing fractions of ^{83,84,85,86}Kr. Except for ^{83}Kr, which has a higher Q-value (10.52 MeV) than ^{85}Kr (9.86 MeV), all other Q-values are smaller than for ^{85}Kr which is important to distinguish between the isotopes. The gas is enclosed in a stainless steel cylinder with a height of 70 mm and a diameter of 3.7 mm. The pressure inside the sample is about 0.05 bar (Figure 2, left).

Figure 2. Left: Target holder for ^{85}Kr. Right: Target holder for ^{83}Kr/^{86}Kr.

Both the ^{83}Kr and ^{86}Kr samples were pressurized gas spheres with a diameter of 10.6 mm that were prepared with a special gas filling system at Goethe University Frankfurt. Gas pressure was 11 bar for the ^{83}Kr sample and 200 bar for the ^{86}Kr sample. This corresponds to a total mass of 0.02 g / 0.45 g respectively (Figure 2, right).

3 Experiment Analysis

The neutron capture cross sections are investigated using the time of flight technique. The analysis consists of several steps. As the gaseous samples were inside their respective containers, separate target holder measurements were also performed to characterize their contributions to the background. Furthermore corrections include pile-up and neutron scattering. A gold sample was used to determine the neutron flux.

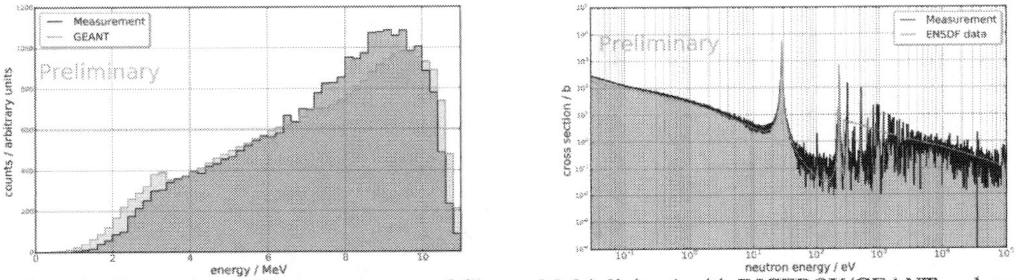

Figure 3. Left: Comparison of Esum spectrum of Cluster Multiplicity 4 with DICEBOX/GEANT and measurement. Right: Preliminary ^{83}Kr cross section.

Extensive DICEBOX/GEANT [6, 7] simulations were performed to reliably estimate the detector efficiency. The DICEBOX code was used to generate γ-ray cascades within the statistical model using known decay pattern below 3.55 MeV and models of γ-strength functions and nuclear level density at higher excitation energies. The resulting γ-cascades are than utilized in GEANT to simulation the response of the complete DANCE setup. A comparison of simulated and measured spectra is shown in Figure 3 (left). The difference between both spectra might be due to the fact that the utilized DICEBOX model does not fully describe the target nucleus.

As the purest sample, ^{83}Kr exhibits very little background from other isotopes. The preliminary cross section, shown in Figure 3 (right), is in very good agreement with recommended ENDF/B-VII.1 data [8]. Determination of the cross sections of the other isotopes is currently in progress.

Acknowledgments

This project was supported by the European Research Council under the European Union's Seventh Framework Programme (FP/2007-2013)/ERC Grant Agreement n. 615126. Milan Krtička acknowledges the support by Grant No. 13-07117S of the Czech Science Foundation.

References

[1] C. Abia et al., *Astrophysical Journal*, 559:1117 (2001)
[2] R. Raut et al. Cross-Section Measurements of the 86Kr(g,n) Reaction to Probe the s-Process Branching at 85Kr (2013)
[3] Z. Y. Bao et al. *Atomic Data Nucl. Data Tables*, 76:70 (2000)
[4] R. Reifarth et al. *Journal of Physics G Nuclear Physics*, 41(5):053101 (2014)
[5] C. Abia et al. *Astrophysical Journal*, 559:1117 (2001)
[6] F. Bečvář, *Nucl. Instr. Meth. A*, 417:434 (1998)
[7] J. Apostolakis, CERN program library long writeup, W5013 (1993)
[8] M. B. Chadwick et al., *Nuclear Data Sheets*, 112:2887 (2011)

A new analysis technique to measure fusion excitation functions with large beam energy dispersions

P.Figuera[1], A.Di Pietro[1], M.Fisichella[1], M.Lattuada[1,2], A.C. Shotter [3], C. Ruiz[4] and M.Zadro[5]

[1]INFN Laboratori Nazionali del Sud, Catania, Italy
[2]Dipartimento di Fisica ed Astronomia Universita' di Catania, Catania, Italy
[3]University of Edinburgh, Edinburgh, UK
[4]TRIUMF, Vancouver, Canada
[5]Ruder Boskovic Institute, Zagreb, Croatia

Abstract. Peculiar nuclear structures of two colliding nuclei such has clustering, neutron halo/skin or very low breakup thresholds can affect the reaction dynamics below the Coulomb barrier and this may also have astrophysical consequences. In order to have a better understanding of this topic, in the last decade, several experiments were performed. A typical experimental challenge of such studies is the need to measure excitation functions below the Coulomb barrier, having a strong energy dependence, with rather large beam energy dispersions inside the target. This may easily lead to ambiguities in associating the measured cross section with a proper beam energy. In this paper a discussion on this topic is reported and a new technique to deal with the above problem will be proposed.

1 Introduction

In the last decade many efforts have been devoted to a better understanding of structure effects on reaction dynamics below the Coulomb barrier (see e.g. [1] and references therein). Besides being a challenging topic in itself, both from an experimental and theoretical point of view, a better understanding of structure effects on the dynamics can also improve the quality of calculations in certain astrophysical scenarios. As an example, one of the uncertainties when attempting to calculate fusion of extremely neutron rich ions in X-ray superbursts is the poorly known effect of the neutron skin on the dynamics e.g. [2]. It has also been recently suggested that clustering affects the dynamics below the barrier influencing the extraction of the electron screening potential [3] and a better understanding of such effects can shed some light on the so called 'electron screening puzzle'.

A typical experimental challenge of such studies, often performed with low intensity radioactive beams (RIBs), is the need to measure excitation functions below the barrier, having a strong energy dependence, with rather large beam energy dispersions inside the target due to a combination of target thickness/uniformity and incoming beam energy spread. This problem usually becomes very important when measuring cross sections with activation techniques irradiating stacks of several targets, which is a common procedure when using low intensity RIBs (see e.g. [4] and references therein). The beam energy spread inside the target may easily lead to ambiguities in associating the measured cross section to a proper beam energy. In this paper a discussion on this topic is reported and a new technique to deal with the above problem will be discussed.

2 Unfolding procedures

Many fusion reactions induced by low intensity RIBs around and below the barrier have been studied by irradiating stacks of several thick targets and measuring on-line or off-line the radiation emitted in the decay of the evaporation residues. As discussed in [4], in some of these experiments the authors had very large beam energy distributions inside the targets in the range 2-6 MeV FWHM. In these activation experiments, the fusion excitation function has been determined by associating the cross-section σ_{mean}, measured in each target, either with the average energy in the considered target $\overline{E}$ (see e.g. [5,6]) or with an effective energy E_{eff}, (see e.g. [7,8]) defined as follows:

$$E_{eff} = \frac{\int_0^\infty E\sigma(E)D(E,t_0)dE}{\int_0^\infty \sigma(E)D(E,t_0)dE} \qquad (1)$$

where $D(E, t_0)$ represents the probability that a beam particle will have energy E inside the considered target of thickness t_0. This has been calculated with different degrees of approximation by various authors. However, it can be shown already analytically [4] that the above methods of relating σ_{mean} to $\overline{E}$ or to E_{eff} can result in misinterpretation of the real excitation function. As an example, in figure 1, the result of a numerical simulation concerning the study of the $^{9}Li+^{120}Sn$ fusion reaction is reported [4]. It has been supposed that a 28 MeV ^{9}Li beam impinged on a stack of five targets each composed of one ^{120}Sn foil with an average thickness of 5 mg/cm^2, followed by a ^{93}Nb catcher/degrader foil 1.5 mg/cm^2 thick. The simulations were performed for the case of uniform and non uniform targets and degrader foils. The $^{9}Li+^{120}Sn$ fusion excitation function has been supposed to be known and to have a Wong-like [10] behavior with realistic parameters, resulting in the black line of figure 1. The beam energy distributions inside the targets $D(E,t_0)$ have been calculated exactly with the help of the SRIM code [9] as discussed in [4]. In this way, for each target, the mean cross-section that would be measured in it has been calculated by using the following relation:

$$\sigma_{mean} = \frac{\int_0^\infty \sigma(E)D(E,t_0)dE}{\int_0^\infty D(E,t_0)dE} \qquad (2)$$

DOI: 10.1051/epjconf/201716501024

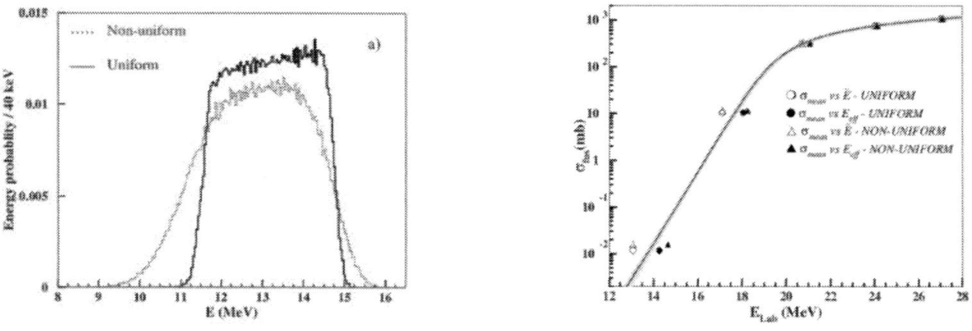

Fig. 1 . Results of the simulations for the study of the $^{9}Li+^{120}Sn$ fusion reaction. (a) Beam energy probability distribution $D(E,t_0)$ inside the last target of the stack assuming uniform and non uniform foils. (b) The continuous black line shows the real fusion excitation function assumed for the collision. The open and closed symbols are the experimental results that would be obtained plotting the measured σ_{mean} versus $\overline{E}$ and E_{eff} (circles uniform foils triangles non uniform foils). The blue shaded area represents the result of our deconvolution procedure with its associated uncertainty. See text for details.

In Fig. 1 the σ_{mean} obtained by applying Eq. (2) is plotted versus the mean energy $\overline{E}$ (open symbols), and versus the effective energy E_{eff} (closed symbols). As one can see, at energies below the Coulomb barrier, the real excitation function (black line) is not reproduced by any of the two representations. Plotting the measured cross-sections versus $\overline{E}$ or E_{eff} generates differences up to a factor of 5 with respect to the real fusion excitation function.

As a possible alternative solution, we propose the following unfolding procedure. The task is to deduce the continuous excitation function $\sigma(E)$ from a finite number of cross sections $\sigma_{mean}i$ measured in the considered experiment. This may be done by choosing a suitable function $\hat{g}(E,\mu)$ (with μ parameters to be determined) which shows the same expected energy behavior as $\sigma(E)$, and then minimizing the following expression S with respect to μ:

$$S = \sum_i \left(\frac{\sigma_{mean,i} - \hat{g}_{mean,i}}{\beta_i} \right)^2 \qquad \text{where} \qquad \hat{g}_{mean,i} = \frac{\int_0^\infty \hat{g}(E,\mu) D(E,t_0) dE}{\int_0^\infty D(E,t_0) dE} \qquad (3)$$

Here β_i is the experimental error associated with σ_{mean}, i. Thus the goal of the proposed deconvolution procedure is the deduction of the best curve $\hat{g}(E,\mu)$ which, when convoluted with the $D(E,t_0)$ functions, for each target, reproduces the value of the experimental mean cross sections. The blue shaded area in figure 1 represents the result of the above unfolding procedure applied using a 3 parameter Wong-like function [10] as $\hat{g}(E,\mu)$ and its error.

A problem in the suggested approach is that, given a functional form, it will always be possible (from a mathematical point of view) to apply the proposed procedure and obtain a deconvoluted excitation function, which may differ from the one obtained plotting the measured cross sections as a function of the mean energy. It is therefore important to check if the above-mentioned differences are simply due to the fact that the chosen functional form is not able to reproduce the real excitation function, or the differences are, instead, to be attributed to the real effect of the energy spread of the beam. A self consistency check that can be done, to verify if the chosen $\hat{g}(E,\mu)$ is a good representation of the fusion excitation function, is the following. One can calculate the ratio between the $\sigma_{calc\text{-}mean,i}$, obtained from Eq. (2) averaging the best functional form $\hat{g}(E,\mu)$ (deduced by the deconvolution procedure) over the beam energy distribution in the target $D_i(E,t_0)$, and the measured cross section $\sigma_{mean,i}$. If this ratio is equal to 1 within the error in the whole energy range, then the procedure is self consistent. As an example, we recently applied this approach in [11].

In our study of the above discussed problems, we also verified that another way to extract a correct excitation function when dealing with large beam energy dispersion is the recursive procedure suggested in [12]. This is true, however, only if the correct $D(E,t_0)$ is used in the procedure, whereas in [12] the authors use an approximation, taking only into account the effects of the energy loss.

3 Summary and conclusions

In the present study we have shown that the traditional way to represent an experimental fusion cross section, as a function of the average energy inside the target or as a function of an effective energy based on a weighted average, may lead to a wrong determination of the excitation function in the case of large beam energy dispersions and an alternative unfolding procedure was proposed and discussed. We believe that, in future publications where such kind of data are presented, a complete discussion on the beam energy spread and its effects on the deduced excitation functions should be presented.

References

1. L.F. Canto et al., Phys Rep. **596**, 1, (2015) and references therein
2. C.J. Horowitz et al., Phys. Rev. C **77**, 045807, (2008)
3. C. Spitaleri et al., Phys. Lett. B **755**, 275, (2016)
4. M. Fisichella et al., Phys. Rev. C **92**, 064611, (2015)
5. Y.E. Penionzhkevich et al., Phys. Rev. Lett. **96**, 162701, (2006)
6. Y.E. Penionzhkevich et al., Eur. Phys. Jour. A **31**, 185, (2007)
7. R. Wolksi et al., Eur. Phys. Jour. A **47**, 111, (2011)
8. D. Shapira et al. Nucl. Instr. and Meth. A **551**, 330, (2005)
9. J.F. Ziegler, 'SRIM code' www.srim.org
10. C. Y. Wong, Phys. Rev. Lett. **31**, 766, (1973)
11. M. Fisichella et al., Phys Rev. C **95**, 034617, (2017)
12. J.F. Liang et al. Phys. Rev. C **75**, 054607, (2007)

EPJ Web of Conferences **165**, 01025 (2017) DOI: 10.1051/epjconf/201716501025
NPA8 2017

The *r*-process nucleosynthesis and related challenges

Stephane Goriely[1,⋆], *Andreas* Bauswein[2], *Hans-Thomas* Janka[3], *Oliver* Just[4], and *Else* Pllumbi[3]

[1] *Institut d'Astronomie et d'Astrophysique, Université Libre de Bruxelles, CP 226, 1050 Brussels, Belgium*
[2] *Heidelberger Institut fr Theoretische Studien, Schloss-Wolfsbrunnenweg 35, 69118 Heidelberg, Germany*
[3] *Max-Planck-Institut für Astrophysik, Postfach 1317, 85741 Garching, Germany*
[4] *Astrophysical Big Bang Laboratory, RIKEN, 2-1 Hirosawa, Wako, Saitama, 351-0198, Japan*

Abstract. The rapid neutron-capture process, or *r*-process, is known to be of fundamental importance for explaining the origin of approximately half of the $A > 60$ stable nuclei observed in nature. Recently, special attention has been paid to neutron star (NS) mergers following the confirmation by hydrodynamic simulations that a non-negligible amount of matter can be ejected and by nucleosynthesis calculations combined with the predicted astrophysical event rate that such a site can account for the majority of *r*-material in our Galaxy. We show here that the combined contribution of both the dynamical (prompt) ejecta expelled during binary NS or NS-black hole (BH) mergers and the neutrino and viscously driven outflows generated during the post-merger remnant evolution of relic BH-torus systems can lead to the production of *r*-process elements from mass number $A \gtrsim 90$ up to actinides. The corresponding abundance distribution is found to reproduce the solar distribution extremely well. It can also account for the elemental distributions observed in low-metallicity stars. However, major uncertainties still affect our understanding of the composition of the ejected matter. These concern (i) the β-interactions of electron (anti)neutrinos with free neutrons and protons, as well as their inverse reactions, which may affect the neutron-richness of the matter at the early phase of the ejection, and (ii) the nuclear physics of exotic neutron-rich nuclei, including nuclear structure as well as nuclear interaction properties, which impact the calculated abundance distribution. Both aspects are discussed in the light of recent hydrodynamical simulations of NS mergers and microscopic calculations of nuclear decay and reaction probabilities.

1 Introduction

Among the various fields in nuclear astrophysics, nucleosynthesis is clearly the one the most closely related to nuclear physics, the nuclear physics imprint being found in the origin of almost all nuclides produced in the Universe [1]. Impressive progress has been made for the last decades in the various fields related to nucleosynthesis, especially in experimental and theoretical nuclear physics, as well as in ground-based or space astronomical observations and astrophysical modellings. In spite of these achievements, major problems and puzzles remain. Among the various nuclear astrophysics problems, one specific nucleosynthesis process remains extremely difficult to solve. It concerns the rapid neutron-capture process, or *r*-process, invoked to explain the production of the stable (and some

⋆e-mail: sgoriely@astro.ulb.ac.be

long-lived radioactive) neutron-rich nuclides heavier than iron that are observed in stars of various metallicities, as well as in the solar system (for a review, see Ref. [2]). In recent years, nuclear astrophysicists have developed more and more sophisticated *r*-process models, trying to explain the solar system composition in a satisfactory way by adding new astrophysical or nuclear physics ingredients. The *r*-process remains the most complex nucleosynthetic process to model from the astrophysics as well as nuclear-physics points of view. The site(s) of the *r*-process is (are) not identified yet, all the proposed scenarios facing serious problems. Complex—and often exotic—sites have been considered in the hope of identifying astrophysical conditions in which the production of neutrons is large enough to lead to a successful *r*-process.

Progress in the modelling of type-II supernovae and γ-ray bursts has raised a lot of excitement about the so-called neutrino-driven wind environment [2–4]. However, until now a successful *r*-process has not been obtained *ab initio* without tuning the relevant parameters (neutron excess, entropy, expansion timescale) in a way that is not supported by the most sophisticated existing models [2, 3]. Early in the development of the theory of nucleosynthesis, an alternative to the *r*-process in high-temperature supernova environments was proposed. It concerns the decompression of cold neutron star (NS) matter that was found to be favorable for strong *r*-processing, as discussed below.

2 NS mergers as a potential *r*-process site

Recently, special attention has been paid to NS mergers following the confirmation by hydrodynamic simulations that a non-negligible amount of matter, typically about 10^{-3} to $10^{-2} M_{\odot}$, can be ejected. In contrast to the supernova site, investigations with growing sophistication have confirmed NS merger ejecta as viable sites for strong *r*-processing [5, and references therein]. In particular, recent nucleosynthesis calculations [5] show that the combined contribution of both the dynamical (prompt) ejecta expelled during the binary NS-NS or NS-black hole (BH) merger and the neutrino and viscously driven outflows generated during the post-merger remnant evolution of the relic BH-torus systems lead to the production of *r*-process elements from $A \gtrsim 90$ up to thorium and uranium with an abundance distribution that reproduces extremely well the solar distribution (Fig. 1), as well as the elemental distribution observed in low-metallicity stars [6]. In this specific *r*-process scenario, during the dynamical phase, the number of free neutrons per seed nucleus can reach a few hundreds. With such a neutron richness, heavy fissioning nuclei can be produced. As a consequence, the final composition of the ejecta is rather insensitive to details of the initial abundances and the astrophysical conditions, namely the mass ratio of the two NSs, the quantity of matter ejected, and the equation of state [7]. This robustness, which is compatible with the uniform solar-like abundance pattern of the rare-earth elements observed in metal-poor stars [6] supports the possible creation of these elements by fission recycling in NS merger ejecta.

The ejected mass of *r*-process material, combined with the predicted astrophysical event rate (around $10\,\mathrm{Myr}^{-1}$ in the Milky Way [8]) can account for the majority of *r*-material in our Galaxy [9]. Recent studies [10, and references therein] have also reconsidered the galactic or cosmic chemical evolution of *r*-process elements in different evolutionary contexts, and although they do not converge towards one unique quantitative picture, most of them got to the conclusion that double compact star mergers may be the major production sites of *r*-process elements. However, the recent observation of the kilonova GW170817 after the present conference presents the first clear evidence regarding the significant contribution of binary NS mergers in the *r*-process enrichment of the Galaxy [11].

Despite the success shown by the NS merger models, one major question still concerns the impact neutrino reactions can have on the predictions. In particular, relativistic NS-NS merger simulations [12, 13] found that neutrino reaction can significantly affect the electron fraction in the dynamical ejecta of systems with delayed collapse of the merger remnant. In this case, in addition to heavy

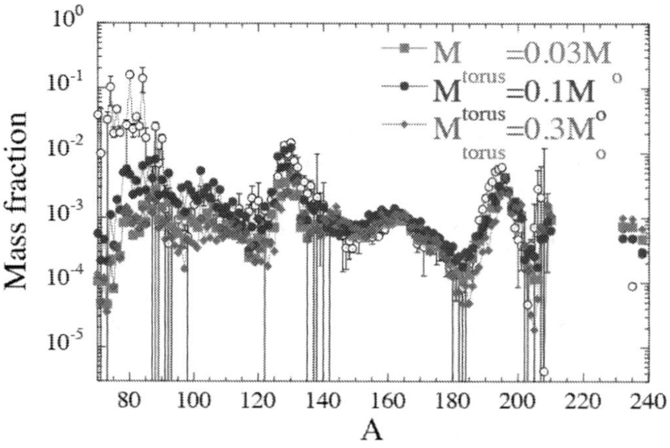

Figure 1. Abundance distribution as functions of the atomic mass A for three combined systems (merger model plus remnant model) corresponding to models with torus masses $M_{\rm torus} = 0.03, 0.1$ and $0.3M_\odot$. All distributions are normalized to the $A = 196$ r-abundance in the Solar System (dotted circles). See Ref. [5] for more details.

r-process elements ($A > 140$), nuclei with lower mass numbers may also be created in the dynamical ejecta. It is still unclear whether the observed effects apply similarly strongly to all high-density equation of states and all binary systems that lead to a delayed collapse of the merger remnant.

The accurate inclusion of neutrino interactions in hydrodynamical simulations remains a highly complex task. For this reason, a simple, parametric study was conducted by Goriely et al. [14] in order to quantify the potential impact of weak interactions on the electron-fraction evolution in merger ejecta and thus to explore the consequences of charged-current neutrino-nucleon reactions for the nucleosynthesis and possible r-processing in these ejecta. More specifically, Goriely et al. [14] investigated the influence of β-interactions of electron neutrinos (ν_e) and electron antineutrinos ($\bar{\nu}_e$) with free n and p and of their inverse reactions,

$$\begin{aligned} &\nu_e + n \rightleftharpoons p + e^- \\ &\bar{\nu}_e + p \rightleftharpoons n + e^+ \quad , \end{aligned} \tag{1}$$

on the Y_e distribution and r-process nucleosynthesis at conditions representative of the dynamical ejecta expelled by hydrodynamical forces during NS-NS mergers. Only recently these reactions have been included in r-process nucleosynthesis studies [12, 13]; however, their quantitative effects depend on the details of the merger dynamics, the strength of the shock heating during the merging, and therefore on the adopted equation of state. The role of weak interactions for the electron fraction and the corresponding implications for r-process nucleosynthesis, however, demand further exploration in more detail, in particular also by basic, parametric modeling, because a multitude of uncertainties will prevent rigorous, self-consistent solutions of the full problem in the near future. Such uncertainties are associated with, for example, the extreme complexities of 3D energy-dependent neutrino transport in relativistic environments, with the neutrino opacities of dense, potentially highly magnetized matter, and with neutrino-flavor oscillations at rapidly time-variable, largely aspherical conditions of neutrino emission.

The major effect of the neutrino interactions on free n and p (Eq. 1) is to increase the electron fraction Y_e that consequently may affect the efficiency of the r-process nucleosynthesis. The corresponding asymptotic value of Y_e can be approximated by [14]

$$Y_e^{\infty} \simeq \frac{L_{\nu_e}\langle E_{\nu_e}\rangle f_{\nu_e}^{\rm mr}}{L_{\nu_e}\langle E_{\nu_e}\rangle f_{\nu_e}^{\rm mr} + L_{\bar{\nu}_e}\langle E_{\bar{\nu}_e}\rangle f_{\bar{\nu}_e}^{\rm mr}}, \tag{2}$$

where $L_{\nu_e/\bar{\nu}_e}$ are the (anti)neutrino luminosities, $\langle E_{\nu_e/\bar{\nu}_e}\rangle$ the mean energies of the radiated neutrinos, and $f_{\nu_e/\bar{\nu}_e}^{\rm mr}$ corresponds to the weak magnetism and recoil corrections [15]. For the present study, we consider a given representative neutrino luminosity and angle-averaged mean energies [12, 16], namely $L_{\nu_e} = 0.6\ 10^{53}$ erg/s, $\langle E_{\nu_e}\rangle = 12$ MeV and $\langle E_{\bar{\nu}_e}\rangle = 16$ MeV, but various values of the antineutrino luminosity to study their impact on nucleosynthesis. All of them are assumed to remain constant in time. Electron and positron captures on free nucleons are also included in these cases. For these mean energies, $f_{\bar{\nu}_e}^{\rm mr}/f_{\nu_e}^{\rm mr} \simeq 0.69$. The impact of the neutrino interaction is essentially to increase Y_e, as illustrated in Fig. 2, where the Y_e distribution at the density $\rho_{\rm net}$ (at which the temperature drops below 10 GK; see [14] for more details) is illustrated for different values of the antineutrino luminosities and compared to the case where all weak interactions on nucleons have been neglected. The larger the antineutrino luminosity, the more efficient the *r*-process nucleosynthesis, as shown in Fig. 3. For $L_{\bar{\nu}_e} \gtrsim 3\ 10^{53}$ erg/s (*i.e.* about 5 times the neutrino luminosity), Y_e^{∞} drops below 0.20 and the ejected *r*-abundance distribution is seen to match relatively well the solar one for all nuclei with $A \gtrsim 90$. In contrast, for decreasing antineutrino luminosities, a weaker *r*-process is obtained with a second (first) *r*-process peak produced for $L_{\bar{\nu}_e}$ at least 3 (2) times higher than the neutrino luminosity[1].

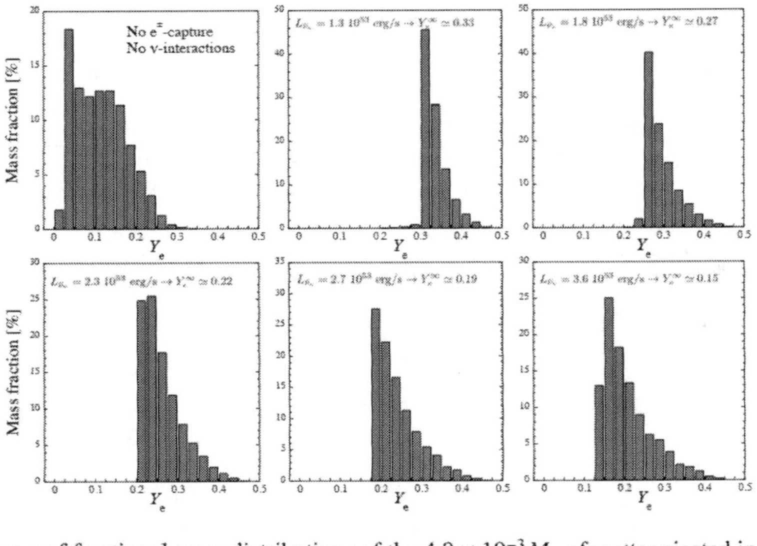

Figure 2. Histograms of fractional mass distributions of the $4.9 \times 10^{-3} M_{\odot}$ of matter ejected in the 1.35–1.35$M_{\odot}$ NS-NS merger model during the dynamical phase as a function of Y_e at density $\rho_{\rm net}$ without any weak interaction of free nucleons and for five different values of the antineutrino luminosity. For these 5 cases, $L_{\nu_e} = 0.6\ 10^{53}$ erg/s, $\langle E_{\nu_e}\rangle = 12$ MeV and $\langle E_{\bar{\nu}_e}\rangle = 16$ MeV. Also given is each panel is the approximated value of Y_e^{∞} as given by Eq. 2. See Ref. [14] for more details.

It should be mentioned that our assumption of time-independent (anti)neutrino luminosities can be questioned, since the neutrino-ejecta interaction is a highly time-dependent problem, where the

[1]Note that the $L_{\bar{\nu}_e} = 1.3\ 10^{53}$ erg/s case corresponds to Case 1 of Ref. [14], but differ in the nucleosynthesis predictions with the published version. This is due to a bug found in the nucleosynthesis code that has now been corrected and only affects Fig. 7 of Ref. [14].

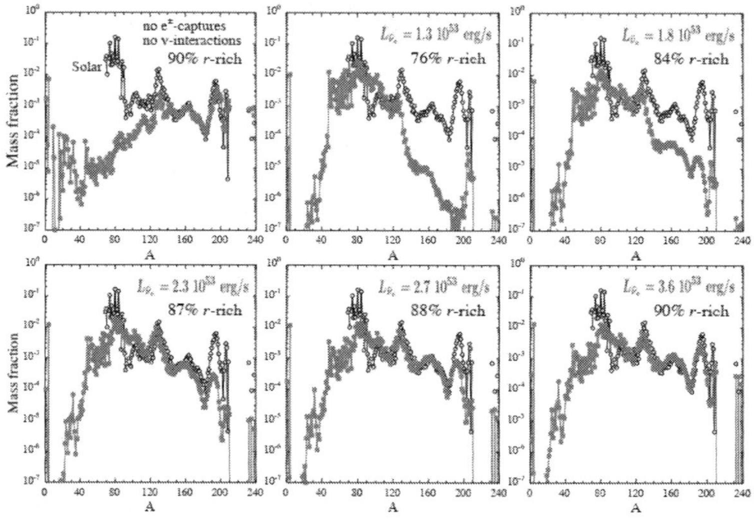

Figure 3. Impact of the neutrino interactions on the abundance distribution after ejection for the same 1.35-1.35$M_\odot$ NS merger model as in Fig. 2. The upper left panel is obtained without any weak interaction of free nucleons, the others for five different values of the antineutrino luminosity $L_{\bar{\nu}_e}$, together with L_{ν_e} = 0.6 10^{53} erg/s, $\langle E_{\nu_e}\rangle$ = 12 MeV and $\langle E_{\bar{\nu}_e}\rangle$ = 16 MeV. Each panel is labelled by the mass fraction of $A > 69$ r-nuclei in the ejected material. See Ref. [14] for more details.

relative time between the growth of the neutrino emission and the mass ejection matters. However, this simple approximation is sufficiently good to demonstrate the impact of neutrino processes on the time evolution and mass distribution of the electron fraction and the corresponding consequences for the r-process. A predictive assessment of neutrino effects on the nucleosynthesis in merger ejecta would also have to take into account variations of the neutrino emission with different directions. Matter expelled towards the polar directions is exposed to different neutrino conditions than matter that leaves the system along equatorial trajectories.

3 Nuclear Physics

R-process nucleosynthesis calculations require a reaction network including about 5000 species from protons up to Z $\gtrsim$ 110 lying between the valley of β-stability and the neutron drip line. All charged-particle fusion reactions on light and medium-mass elements that play a role when the nuclear statistical equilibrium freezes out need to be included in addition to radiative neutron captures and photodisintegrations. On top of these reactions, β-decays as well as β-delayed neutron emission probabilities and α-decay rates need to be taken into account, but also fission processes, including neutron-induced, spontaneous, β-delayed and photofission, together with the corresponding fission fragment distribution for all fissioning nuclei. All rates are based on experimental information whenever available, but since only a extremely small amount of data are known experimentally, theoretical models are fundamental in providing the various predictions.

For such applications, the necessary ingredients (properties of cold and hot nuclei, nuclear level densities, optical potentials, γ-ray strength functions, fission properties, β-strength functions) should ideally be derived from *global*, *universal* and *microscopic* models. The large number of nuclides involved in the modelling of some nucleosynthesis mechanisms demands the use of global models.

On the other hand, a universal description of all nuclear properties within a unique framework for all nuclei involved ensures the essential coherence of the predictions of all unknown data. Finally, a microscopic description provided by a physically sound theory based on first principles ensures extrapolations away from experimentally known energy or mass regions that are likely to be more reliable than predictions derived from more or less parametrized approaches of various types and levels of sophistication. Nowadays, microscopic models can be tuned to the same level of accuracy as the phenomenological models, and therefore could replace them in practical applications.

Today, due to our ignorance on the exact conditions in which the r-process takes place, it remains difficult to estimate the precision with which the various relevant rates need to be estimated. In particular, it strongly depends if an (n,γ)-(γ,n) equilibrium would be reached during the neutron irradiation or if, instead, a competition between neutron captures and β-decays would be responsible for the nuclear flow and final shaping of the r-abundance distribution. Much more work on the astrophysical modelling [2] is needed before providing such constraints that could shed light on the precision required from nuclear physics. In the meantime, a first educated guess would require the reaction rates to be estimated within a factor of 2 and β-decay rates within 50% for all nuclei that may be direct progenitors of r-nuclei, *i.e.* before the final β-decay cascade at the neutron freeze-out. Concerning the more exotic nuclei up to the neutron drip-line, depending if fission efficiently recycles material, *i.e.* depending on the number of neutrons per seed available, less stringent constraints could be envisioned. It also remains of first importance to estimate the statistical as well as systematic uncertainties affecting the predictions far away from the experimentally known region. Such a difficult task has been started regarding mass predictions [17], but remains to be performed for the reaction as well as β-decay rates. Our capacity to predict the fundamental nuclear ingredients for reaction models, namely nuclear masses, optical potentials, γ-ray strength functions, nuclear level densities, fission barriers, as well as the predictive power of the reaction and β-decay models are discussed in Ref. [17].

References

[1] M. Arnould, K. Takahashi, Rep. Prog. Phys. **62**, 395 (1999)
[2] M. Arnould, S. Goriely, K. Takahashi, Phys. Repts. **450**, 97 (2007)
[3] H.T. Janka, Ann. Rev. of Nucl. Part. Sci. **62**, 407 (2012)
[4] S. Wanajo, H.T. Janka, B. Müller, Astrophys. J. **726**, L15 (2011)
[5] O. Just, A. Bauswein, R. Ardevol Pulpillo, S. Goriely, H.T. Janka, MNRAS **448**, 541 (2015)
[6] I. Roederer, Astrophys. J. **732**, L17 (2011)
[7] A. Bauswein, S. Goriely, H.T. Janka, Astrophysical J. **773**, 78 (2013)
[8] M. Dominik, K. Belczynski, C. Fryer, D. Holz, E. Berti, T. Bulik, I. Mandel, R. O'Shaughnessy, Astrophys. J. **759**, 52 (2012)
[9] A. Bauswein, R. Ardevol Pulpillo, H.T. Janka, S. Goriely, Astrophysical J. Lett **773**, L9 (2014)
[10] E. Vangioni, S. Goriely, F. Daigne, P. Francois, K. Belczynski, MNRAS **455**, 17 (2016)
[11] B. Abbott, R. Abbott, T. Abbott et al., Phys. Rev. Lett. **119**, 161101 (2017)
[12] S. Wanajo, Y. Sekiguchi, N. Nishimura, K. Kiuchi, K. Kyutoku, M. Shibata, Astrophys. J. Lett **789**, L39 (2014)
[13] Y. Sekiguchi, K. Kiuchi, K. Kyutoku, M. Shibata, Phys. Rev. D **91**, 064059 (2015)
[14] S. Goriely, A. Bauswein, O. Just, E. Pllumbi, H.T. Janka, MNRAS **452**, 3894 (2015)
[15] E. Pllumbi, I. Tamborra, S. Wanajo, H.T. Janka, L. Hüdepohl, Astrophys. J. **808**, 19 (2015)
[16] M. Ruffert, H.T. Janka, Astron. Astrophys **380**, 544 (2001)
[17] S. Goriely, Eur. Phys. J. A **51**, 172 (2015)

Nuclear Astrophysics at ELI-NP: the ELISSA prototype tested at Laboratori Nazionali del Sud

Giovanni Luca Guardo[1,2,⋆], *Antonello* Anzalone[1], *Dimiter* Balabanski[2], *Svetlana* Chesnevskaya[2], *Walter* Crucillá[1], *Dan* Filipescu[2], *Marisa* Gulino[1,3], *Marco* La Cognata[1], *Dario* Lattuada[2], *Catalin* Matei[2], *Rosario Gianluca* Pizzone[1], *Giuseppe* Rapisarda[1], *Stefano* Romano[1,4], *Claudio* Spitaleri[1,4], *Alessandra* Taffara[1], *Aurora* Tumino[1,3], and *Yi* Xu[2]

[1] *INFN, Laboratori Nazionali del Sud, Catania, Italy*
[2] *ELI-NP/ IFIN-HH, Magurele, Romania*
[3] *Facoltá di Ingegneria e Architettura, Universitá Kore, Enna, Italy*
[4] *Dipartimento di Fisica e Astronomia, Universitá degli Studi, Catania, Italy*

Abstract. The Extreme Light Infrastructure-Nuclear Physics (ELI-NP) facility, under construction in Magurele near Bucharest in Romania, will provide high-intensity and high-resolution gamma ray beams that can be used to address hotly debated problems in nuclear astrophysics, such as the accurate measurements of the cross sections of the ^{24}Mg(γ,α)^{20}Ne reaction, that is fundamental to determine the effective rate of ^{28}Si destruction right before the core collapse and the subsequent supernova explosion.
For this purpose, a silicon strip detector array (named ELISSA, acronym for Extreme Light Infrastructure Silicon Strip Array) will be realized in a common effort by ELI-NP and Laboratori Nazionali del Sud (INFN-LNS), in order to measure excitation functions and angular distributions over a wide energy and angular range.
A prototype of ELISSA was built and tested at INFN-LNS in Catania (Italy) with the support of ELI-NP. In this occasion, we have carried out experiments with alpha sources and with a 11 MeV ^{7}Li beam. Thanks to our approach, the first results of those tests show up a very good energy resolution (better than 1%) and very good position resolution, of the order of 1 mm. Moreover, a threshold of 150 keV can be easily achieved with no cooling.

1 Introduction

1.1 The ELI-NP facility

The ELI-NP project aims at establishing a European Research Infrastructure Centre for ultra-high intensity lasers, laser-matter interactions, nuclear and material science, using laser-driven radiation sources [1]. This multidisciplinary facility will provide new opportunities to study fundamental processes that occur in ultra-intense laser fields during laser-matter interactions.
ELI-NP will benefit also from a very bright gamma beam system (GBS). The system was conceived as a dedicated device based on the inverse Compton backscattering of high-intensity laser pulses off

⋆e-mail: luca.guardo@eli-np.ro

high brightness relativistic electron bunches and able to provide gamma beams with unprecedented features: continuous tuneable gamma-ray energy over a broad range (from 200 keV to 19.5MeV), high spectral density (about 10^4 photons/s/eV), small relative energy bandwidth (lower than 0.5%), high degree of linear polarization (higher than 95%), and high peak brilliance [2].
Such a facility will open new experimental perspectives for studies in the field of photonuclear physics. Moreover, thanks to these excellent features, the ELI-NP facility will provides unique opportunities for nuclear astrophysics research.

1.2 Nuclear Astrophysics at ELI-NP

Using the gamma radiation beam with unique characteristics at ELI-NP, we will be able to respond to the need of Nuclear Astrophysics to perform accurate measurements of small cross sections (order of μb or even less) of nuclear reactions of the hydrogen and helium burning processes and hence of the astrophysical S-factor that are essential for stellar evolution modeling. Measuring capture reactions by means of the inverse photodisintegration reaction, besides being inherently low-background measurements, have the advantage of having a different systematic uncertainty than those of characteristic charged particle induced reactions measured at low energies of astrophysical interest. Such systematic issues, for example, involving the target and its deterioration, (effective) beam energy definition, etc., lead to different systematic errors and thus may allow us to resolve conflicting data.

1.3 A physical case: the ^{24}Mg(γ,α)^{20}Ne reaction

Silicon burning sets the chemical composition of the star right before the core collapse and the subsequent supernova explosion, thus constituting a key process for the understanding of core-collapse supernovae [3]. When ^{16}O is depleted at the conclusion of core oxygen burning, the most abundant nuclei are ^{28}Si and ^{32}S. The stellar core contracts and the temperature increases, reaching values as large as T= 2.8-4.1 GK, depending on the stellar mass. Fusion reactions such as ^{28}Si+^{28}Si or ^{28}Si+^{32}S are unlikely to occur owing to the Coulomb barrier between interacting nuclei, even at such high temperatures [4]. Instead, nucleosynthesis takes place through photodisintegration of less bound nuclei and radiative captures of the dissociated light particles (protons, neutrons, and α-particles) to create gradually heavier and more tightly bound nuclei [5]. Therefore, many photodissociation cross sections have to be known with adequate accuracy to model silicon burning. However, sensitivity studies show that the ^{24}Mg(γ,α)^{20}Ne reaction governs the downward flow from ^{24}Mg to ^{4}He, thus determining the effective rate of ^{28}Si destruction, making its reaction rate critically important to stellar models [3].
At present, the ^{24}Mg(γ,α)^{20}Ne reaction rate has been calculated from the ^{20}Ne(α,γ)^{24}Mg rate. In the temperature range of interest, around $3.9{\cdot}10^9$ K, the ^{20}Ne(α,γ)^{24}Mg reaction rate may be subject to systematic errors of the order of a factor of 2, as can be seen from the different results reported in [6]. A direct ^{24}Mg photodissociation measurement using gamma beams of energies 10-12 MeV will allow us to determine a much more accurate cross section to be used in nuclear reaction network calculations to improve the knowledge of the pre-supernova chemical composition.

2 The ELISSA detector

Generally speaking, photodissociation reactions are not so hard to study as the phase space factor enhances the (γ,p) and (γ,α) cross sections with respect to the inverse process, provided that a high quality gamma beam is available, such as the one delivered by the ELI-NP facility. In addiction, it is worth noting that the emitted fragments have low energies, ranging from few hundreds keV to few

 DOI: 10.1051/epjconf/201716501026

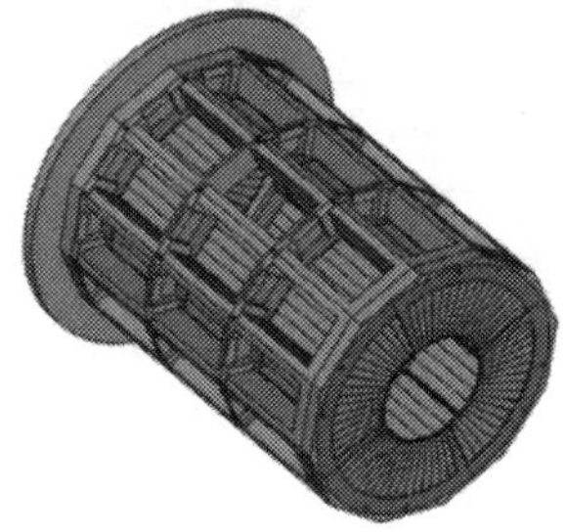

Figure 1. Final design of the ELISSA detector. It will be made of n. 3 rows of PSD detectors to form a barrel plus n. 8 end cap detectors to extend the angular coverage.

MeV, making it necessary to use a low-threshold detector. Moreover, excitation functions and angular distributions should be measured over a wide range allowing for a better understating of reaction mechanism. Indeed, taking into account the presence of beam induced background at forward angles, an angular coverage as large as possible should be covered in order to ensure a range large enough to perform an effective nuclear spectroscopy. Finally, silicon strip detectors would couple enhanced angular and energy resolution with high detection efficiency. High granularity is not a crucial aspect as a low counting rate is expected, making the probability of multiple hit very small. This type of array has already been successfully applied for nuclear astrophysics studies, e g. ORRUBA (Oak Ridge Rutgers University Barrel Array) or ANASEN (Array for Nuclear Astrophysics Studies with Exotic Nuclei) [7, 8].
Thus, the Extreme Light Infrastructure Silicon Strip Array (ELISSA) is under construction [9]. The performed Monte Carlo simulation using a code based on GEANT4 tracking libraries and the n-body event generator of ROOT, as described in [10], proved that the barrel configuration is particularly suited as it guarantees a very good resolution and granularity, ensuring also a compact detection system (useful as it would allow a simple integration with ancillary detectors, such as neutron arrays) and a limited number of electronics channel [11]. So that, the final design of ELISSA will consist of 3 rings of 12 X3 position-sensitive detectors produced by Micron Semiconductor Ltd. [12], in a barrel like configuration, ensuring a total angular coverage in the laboratory system of about 100. The angular coverage is extended down to about 20 (160 at backward angles) by using end cap detectors such as the assembly of four QQQ3 segmented detectors by Micron Semiconductor Ltd [12]. A sketch of the final expected setup is shown in fig. 1.

3 Tests on ELISSA prototype

To test the performance of the X3 detectors and make sure that their energy and angular resolutions are suitable for the realization of the ELISSA detector, we have carried out experiments with alpha sources and with a 11 MeV ^{7}Li beam at the INFN-LNS tandem in Catania (Italy), in collaboration with ELI-NP (Romania). In this occasion, a prototype of the ELISSA detector realized at INFN-LNS was installed at the "Camera 2000" scattering chamber. The tests were performed on February 3-4, 2016. The prototype is made up of a single X3 detector and a single QQQ3 detector, so it replicates in a smaller scale the whole detector, and their readout is performed by means of standard analog electronics. While QQQ3 are DSSSD, so their position resolution is fixed by the strip size and energy resolution is constant to a good approximation, X3 detectors showed more peculiar features. So that,

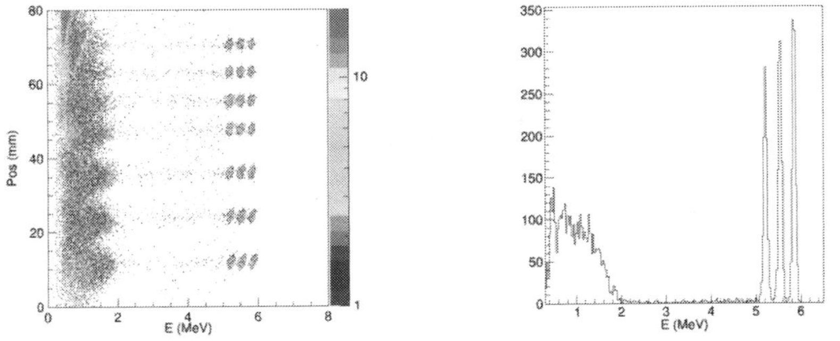

Figure 2. Left panel: Position vs. Energy plot of the X3 detector. See the text for details. Right panel: Energy spectrum for a fixed position.

in the following we focus on the X3 detector. To measure the detection threshold, we have performed a run using a standard 3-peak alpha source and a ^{241}Am source shielded by a 17 μm thick Al foil. This degrader shifted the energy peak to 1 MeV and, due to energy straggling, the energy range reached energy as low as zero. The result of this run is reported in the left panel of fig. 2 showing the position Vs. energy plot obtained using a mask (asymmetric respect the center of the detector and with 1mm and equally spaced slits in each side) in front of the detector. In the right panel the energy spectrum is shown for a fixed angle, corresponding to the slit at about 57mm. From the plot it is clear that a threshold of 150 keV can be easily achieved with no special expedient.
Energy and position resolutions were measured detecting scattering and reactions induced by the ^{7}Li beam on Au, C and CD_2 targets. In this way, we determined an energy resolution of $\leq$1% at high energy (scattering off gold) and $\leq$8% in the case of ^{7}Li scattered off ^{12}C (energy resolution measured at about 1 MeV), essentially due to straggling in the target and target inhomogeneity. At energies of about 11 MeV, the position resolution is better than 1 mm while at lower energies, below about 1 MeV, a position resolution of about 6 mm was found [9]. Notwithstanding such position resolution deterioration, the detector performances are still good enough for the measurement of angular distribution and the kinematical identification of the reactions induced on the target by gamma beams, since it corresponds to an angular resolution of about 1.5 degrees.
To sum up, these tests allows us to say that the X3 detectors, as well the standard QQQ3 detectors, are perfectly suited for nuclear astrophysics studies with ELISSA.

Acknowledgments

This work is supported in part by the Extreme Light Infrastructure Nuclear Physics (ELI-NP) Phase II, a project cofinanced by the Romanian Government and the European Union through the European Regional Development Fund - the Competitiveness Operational Programme (1/07.07.2016, COP, ID 1334).

References

[1] D. Filipescu et al., Eur. Phys. J. **A 51**, 185 (2015)

DOI: 10.1051/epjconf/201716501026

[2] D. L. Balabanski et al., EPL **117**, 28001 (2017)
[3] C. Iliadis, *Nuclear Physics of Stars*, (Weinheim, Wiley-VCH Verlag, 2007)
[4] A.J. Howard et al., Astrophys. J. **188**, 131 (1974)
[5] W.R. Hix et al., Astrophys. J. **460**, 869 (1996)
[6] C. Angulo et al., Nucl. Phys. **A 656**, 3 (1999)
[7] S. D. Pain et al., Nucl. Instr. Meth. Phys. Res. B 261, 1122 (2007)
[8] M. Matos et al., PoS NIC XI, 226 (2010)
[9] M. La Cognata et al., JINST 12 C03079, (2017)
[10] D. Lattuada et al. this volume
[11] O. Tesileanu et al., Rom. Rep. in Phys. **68**, S699 (2016)
[12] Micron Semiconductor, General cata logue, http://www.micronsemiconductor.co.uk/pdf/cat.pdf

EPJ Web of Conferences **165**, 01027 (2017) DOI: 10.1051/epjconf/201716501027
NPA8 2017

Target characterizations for a $^{14}N(p,\gamma)^{15}O$ cross section measurement

Gy. Gyürky[1,⋆], *A.* Csik[1], *Zs.* Mátyus[1,2], *Zs.* Fülöp[1], *Z.* Halász[1], *G.G.* Kiss[1], *T.* Szücs[1], and *L.* Wagner[3,4]

[1] *Institute for Nuclear Research (Atomki), Debrecen, Hungary*
[2] *University of Debrecen, Hungary*
[3] *Helmholtz-Zentrum Dresden-Rossendorf, Germany*
[4] *Technische Universität Dresden, Germany*

Abstract. The $^{14}N(p,\gamma)^{15}O$ reaction controls the rate of CNO cycle hydrogen burning in various astrophysical sites and it is therefore one of the most important reactions in nuclear astrophysics. An experimental program is in progress to measure the $^{14}N(p,\gamma)^{15}O$ cross section in a wide energy range using a novel approach. A crucial quantity for the cross section determination is the number of N atoms in the target. In this paper the results of different experiments used for N target characterization are presented.

1 Introduction

Hydrogen burning is inevitably the most important energy source of the universe as this burning makes most of the visible stars shine. The pp-chains and the CNO cycles are the two main processes of hydrogen burning. The first CNO cycle is dominant in main sequence stars more massive than about 1.3 solar masses. The rate of the cycle determines therefore the energy generation and evolution of massive stars. But the small contribution of the CNO cycle to the hydrogen burning of our Sun makes this cycle important also for the better understanding of e.g. the solar composition.

The slowest reaction of the CNO cycle is $^{14}N(p,\gamma)^{15}O$ which determines therefore the rate of the whole cycle. The knowledge of the $^{14}N(p,\gamma)^{15}O$ reaction rate is thus necessary for the stellar models. The rate is obtained from the reaction cross section which must be known at low, stellar energies. Typically these energies are not accessible experimentally due to the extremely low cross sections. Nuclear theory guided extrapolations are therefore needed which are based on and constrained by higher energy experimental cross sections.

The cross section of $^{14}N(p,\gamma)^{15}O$ was measured many times in the past focusing mostly to energies as low as possible (see the review [1] for the list of experiments and ref. [2] for a recent study). Nevertheless, the precision of the $^{14}N(p,\gamma)^{15}O$ reaction rate based on the experimental cross sections does not reach the level required for the stellar models. Further experimental study of this reaction is therefore necessary.

The aim of the present work is to measure the $^{14}N(p,\gamma)^{15}O$ cross section with a technique never applied for this reaction. The cross section is to be determined with the activation method, i.e. by the measurement of the decay of ^{15}O having roughly 2 minutes half life. An important parameter

⋆e-mail: gyurky@atomki.mta.hu

for the cross section calculations is the number of N target atoms. In this paper some details of the measurements carried out for the determination of this quantity as well as their results are presented.

2 Target characterizations

Since nitrogen in elemental form is gaseous, for the cross section measurement either a gas target must be used, or a suitable nitrogen compound must be chosen for a solid target experiment. Several previous experiments proved that titanium nitride (TiN) is a well suited solid nitrogen compound for the purposes of cross section measurement. In the present work TiN target were therefore used.

The targets were produced by reactive sputtering of TiN onto thick Ta backings at the Helmholtz-Zentrum Dresden-Rossendorf, Germany using a setup similar to the one described in ref. [3]. In the first phase of the experimental campaign six targets were used with three different nominal thicknesses of 100, 200 and 300 nm. In the next subsections the measurements carried out for the determination of the Ti:N atomic ratio and the actual thicknesses are described.

2.1 Ti:N ratio measurement by X-ray diffraction

Although TiN is known to have well defined Ti:N ratio of 1:1, technological errors during the sputtering process cannot be excluded, so the experimental proof of the ratio is necessary for a reliable determination of the number of N target atoms using the methods shown below. For the Ti:N ratio measurement the X-ray diffraction (XRD) method was used. As the target on a Ta backing is not suitable for a precise XRD analysis, TiN layers onto Si single crystals were also deposited along with the real targets. Owing to the same geometry, the thicknesses of these layers are equal to the layer thicknesses on the Ta backings.

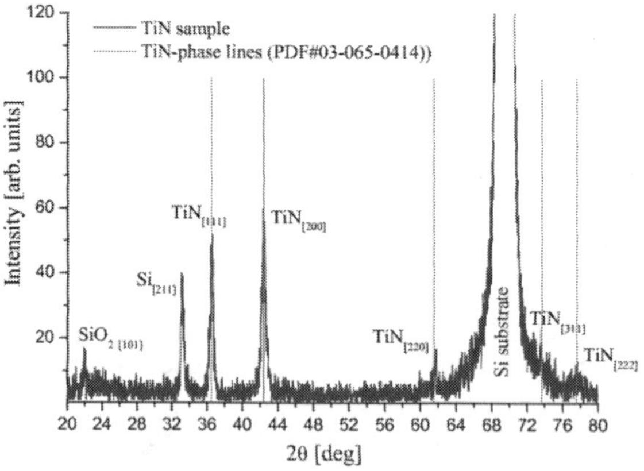

Figure 1. The high angle XRD pattern of the TiN sample. Vertical lines correspond to the TiN phase taken from Powder Diffraction File (PDF) database.

The XRD measurement was carried out by using a diffractometer of Atomki equipped with a Siemens type Cu-anode X-ray tube (operated at 40 kV/30 mA, λ=1.54056 Å) and a horizontal goniometer with graphite monochromator. A high angle symmetric scan was measured in Bragg-Brentano θ-2θ geometry and the standard Powder Diffraction File (PDF) database has been used for identification of diffraction lines. Scan was collected from 20° to 80° 2θ-angle, with a step width of $2\theta = 0.02°$ and a sampling time of 14 sec/step. A piece of the clean Si single crystal was additionally measured for identification of the substrate lines in diffraction pattern. As it can be seen in Fig. 1 the lines from the standard card of Powder Diffraction File (#03-065-0414) fit well the measured data, which means that the TiN layer has Ti:N ratio of 1:1. All other visible diffraction peaks were identified as Si or SiO lines from substrate.

2.2 Target thickness measurement by SNMS

The mass spectrometry method also was used to determine the composition of the TiN layers. As in the case of the XRD, the layers deposited onto the Si wafers were used for these measurements. Quantitative depth profile analyses of the samples were carried out by Secondary Neutral Mass Spectrometer (SNMS) INA-X type (SPECS GmbH, Berlin) of Atomki [4, 5]. The surface bombardment and post ionization were performed by low pressure Electron Cyclotron Wave Resonance (ECWR) argon plasma. The SNMS was operated in high frequency mode which insured effective charge compensation against surface charge accumulation and thus results in a homogeneous ion bombardment. 350 V sputtering potential at 100 kHz frequency with 80 % duty cycle was applied on a sample surface. The investigated area on the sample surface was confined within a circle of 2 mm diameter by a Ta mask. The lateral homogeneity of ion bombardment was checked by measuring the shape of the sputtered crater with an AMBIOS XP-1 type profilometer. The profilometer was also used to determine the sputtering rate by measuring the crater depth as a function of sputtering time and at the same time to determine the absolute thickness of the samples. Figure 2 shows a typical depth profile of the measured samples. As it can be seen, the top part of the sample contains a low amount of oxygen, nevertheless predominant part of the TiN layer has Ti:N ratio of 1:1.

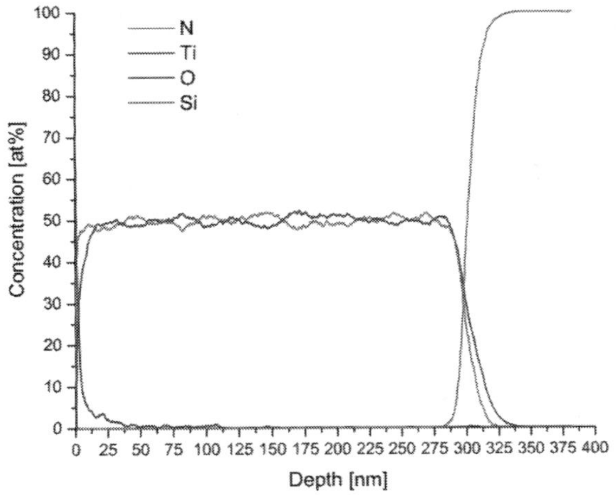

Figure 2. The typical depth profile of the TiN samples measured by Secondary Neutral Mass Spectrometry (SNMS) method.

2.3 Target thickness measurement by nuclear resonance analysis

With a suitable nuclear resonance with small natural width the target thickness can also be measured by Nuclear Resonance Analysis (NRA). As the TiN targets were produced using natural isotopic composition material, the 0.36 % abundance of ^{15}N allows the measurement of the strong resonance in the $^{15}N(p,\alpha\gamma)^{12}C$ reaction at $E_p = 897$ keV proton energy [6]. The NRA measurements were carried out at the new Tandetron accelerator of Atomki using a setup described in ref. [7]. The resonance profile of a 300 nm nominal thickness target is shown in Fig. 3. Knowing the Ti:N ratio and the stopping power of protons in Ti and N, the width of the resonance profile provides the target thickness. Taking into account the typically 4 % uncertainty of the stopping power and the 3 % statistical uncertainty of the fit, the target thicknesses could be determined with a precision of 5 % using the NRA method.

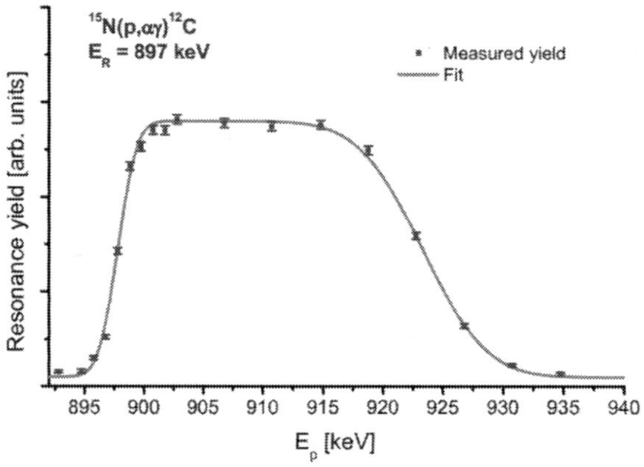

Figure 3. Profile of the E_p = 897 keV resonance in $^{15}N(p,\alpha\gamma)^{12}C$ reaction measured on a 300 nm thick TiN target

3 Conclusions

The results of the TiN target thickness measurement carried out with the two methods described above are in perfect agreement and reproduce well the nominal thicknesses from the sputtering process. Just to give an example: the thickness of a nominal 300 nm target was found to be 302 ± 11 nm with the SNMS technique and 295 ± 15 nm with NRA. Based on the XRD and SNMS methods the Ti:N ratio of 1:1 was confirmed with 1.2 % accuracy. The number of N target atoms in the targets used for the $^{14}N(p,\gamma)^{15}O$ cross section measurements is known with an uncertainty of typically 4 %. The cross section measurements are in progress.

Acknowledgments

This work was supported by OTKA K108459 and K120666. XRD and SNMS measurements were carried out in frame of the GINOP-2.3.2-15-2016-00041 project, which is co-financed by the European Union and the European Regional Development Fund. L.W. acknowledges support for the TU Dresden Graduate Academy. The authors thank Mario Steinert (HZDR Ion Beam Center) for the target production.

References

[1] E.G. Adelberger *et al.*, Rev. Mod. Phys. **83** 195 (2011)
[2] Q. Li *et al.*, Phys. Rev. C **93** 055806 (2016)
[3] V. Rigato *et al.*, Surface and Coating Technology **142-144** 943 (2001)
[4] H. Oechsner, Applied Surface Science **70-71** 250 (1993)
[5] K. Vad, A. Csik, G.A. Langer, Spectroscopy Europe **21** 13 (2009)
[6] M. Marta *et al.*, Phys. Rev. C **81** 055807 (2010)
[7] Gy. Gyürky *et al.*, Phys. Rev. C **95** 035805 (2017)

EPJ Web of Conferences **165**, 01028 (2017) DOI: 10.1051/epjconf/201716501028
NPA8 2017

First results of total and partial cross-section measurements of the ^{107}Ag(p,γ)^{108}Cd reaction

Felix Heim[1,⋆], *Jan* Mayer[1], *Philipp* Scholz[1], *Mark* Spieker[1], and *Andreas* Zilges[1]

[1] *Institute for Nuclear Physics, University of Cologne*

Abstract. The γ process is assumed to play an important role in the nucleosynthesis of the majority of the p nuclei. Since the network of the γ process includes so many different reactions and - mainly unstable - nuclei, cross-section values are predominantly calculated in the scope of the Hauser-Feshbach statistical model. The values heavily depend on the nuclear-physics input parameters. The results of total and partial cross-section measurements are used to improve the accuracy of the theoretical calculations. In order to extend the experimental database, the ^{107}Ag(p,γ)^{108}Cd reaction was studied via the in-beam method at the high-efficiency HPGe γ-ray spectrometer HORUS at the University of Cologne. Proton beams with energies between 3.5 MeV and 5.0 MeV were provided by the 10 MV FN-Tandem accelerator leading to the determination of four new total cross-section values. After slight adjustments of the nuclear level density and γ-ray strength function an excellent agreement between theoretical calculations and experimentally deduced values for both total and partial cross sections has been obtained.

1 Introduction

About 30 to 35 neutron-deficient nuclei are bypassed by neutron-capture processes and the question about their detailed origin is still open. There are several different processes in various astrophysical scenarios proposed which might contribute to the synthesis of the so-called p nuclei [1, 2]. However, the main contribution is assumed to stem from the γ-process, a huge network of photodisintegration reactions on approx. 2,000 mainly unstable nuclei. In order to predict reaction rates or cross-section values, one often has to rely on statistical model calculations which are mostly performed in the scope of the Hauser-Feshbach statistical model. In order to reduce uncertainties and improve the predictive power of the statistical calculations, the models for nuclear-physics input parameter have to be improved. Therefore measurements in or close to the Gamow window - usually way below the Coulomb barrier - are necessary. In order to study the nuclear level density as well as the γ-ray strength function the ^{107}Ag(p,γ)^{108}Cd reaction was measured, for which the Gamow window is between 1.77 MeV and 3.83 MeV at T_9=3 [3].
By a comparison between experimentally obtained results to theory, different models can either be excluded or validated. Here, preliminary experimental results on the measurements of total and partial cross-section values of the ^{107}Ag(p,γ)^{108}Cd reaction as well as the comparison to statistical model calculations are presented.

⋆e-mail: fheim@ikp.uni-koeln.de

DOI: 10.1051/epjconf/201716501028

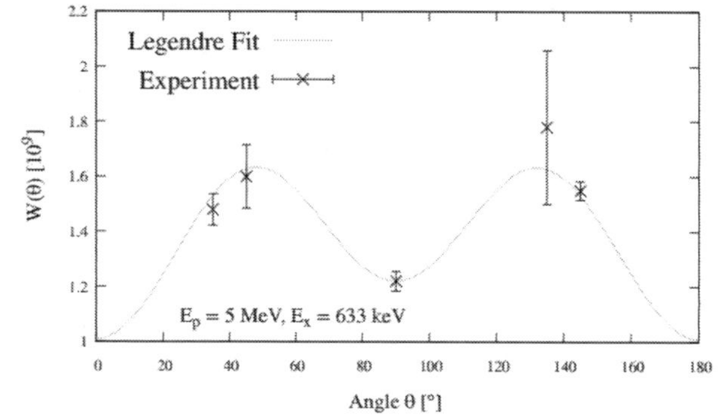

Figure 1. Angular distribution of the decay $2^+_1 \rightarrow g.s.$ in ^{108}Cd for an incident beam energy of 5 MeV. For the data points a sum of Legendre polynomials is fitted in order to obtain the total number of reactions.

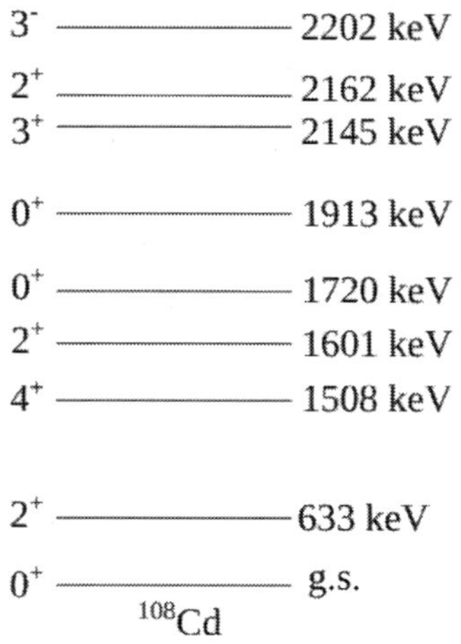

Figure 2. Levelscheme of ^{108}Cd.

2 Experimental method: The in-beam technique

The proton beam was delivered by the 10 MV FN tandem ion accelerator at the Institute for Nuclear Physics at the University of Cologne, Germany. The beam with an intensity of about 500 nA impinged on silver targets with a 99.5 % enrichment of ^{107}Ag. Target thicknesses from (2.70 to 3.19)$\times 10^{18}$ atoms/cm^2 were determined by means of a Rutherford Backscattering Spectrometry measurement. The ^{107}Ag(p,γ)^{108}Cd reaction was studied by means of in-beam γ-ray spectroscopy at four different proton beam energies between 3.5 MeV and 5.0 MeV using the γ-ray spectrometer HORUS [4], which consists of 14 high-purity germanium (HPGe) detectors. The detectors are mounted under five different angles with respect to the beam axis which allows the measurement of angular distributions. The total detection efficiency of HORUS amounts to about 2 % at an γ-ray energy of E_γ=1.3 MeV. The angular distributions are fitted on a sum of Legendre polynomials using the function and used for the determination of absolute cross-section values.

$$W(\Theta) = A_0 \left(1 + \sum_{k=2,4} \alpha_k P_k(cos\Theta) \right) \qquad (1)$$

Those angular distributions are obtained for each ground state transition at each beam energy and the sum of all A_0 coefficients represents the total number of reactions. The angular distribution for the $2^+_1 \rightarrow g.s.$ transition at a beam energy of E_p=5 MeV is shown in Fig. 1.

Total cross sections are obtained by summing up all ground state transitions for each beam energy and partial cross sections are given by the direct population of a respective level in ^{108}Cd. The Q value of this reaction amounts to $Q(p,\gamma) = 8134$ keV.

3 Experimental results

Not all of the seven observed ground-state transitions have the same intensity. The decay of the 2^+_1 state at E_γ=633 keV has by far the highest contribution to the total ground state population with about 80 %. The contribution from the direct de-excitation of the compound state into the ground state, i.e. the γ_0, has a negligible contribution with about one percent. The experimentally obtained

DOI: 10.1051/epjconf/201716501028

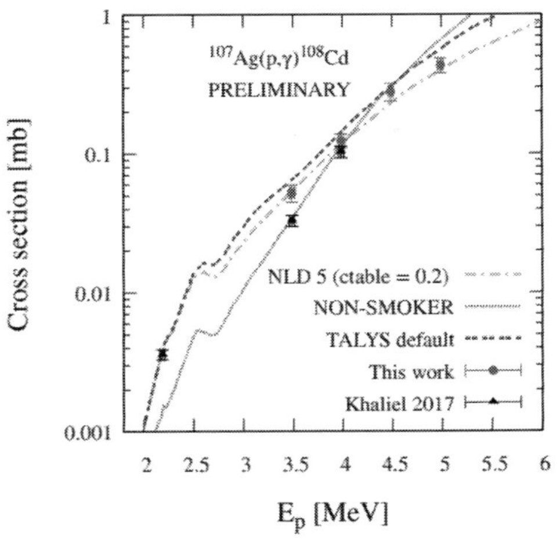

Figure 3. Total cross section results of the ^{107}Ag(p,γ)^{108}Cd reaction compared to statistical model calculations. The TALYS [5] default settings describe the data quite well, but overestimate the data slightly. The widely-used NON-SMOKER [9] code neither describes the energy trend nor the absolute values sufficiently well. Using the microscopic models for the γ-ray SF from Ref. [6] and the NLD from Ref. [7, 8] provides much better agreement with the experimental values. The microscopic NLD models have been slightly adjusted (see text for details) in order to obtain better agreement. For illustration only one adjusted NLD model is shown. The results from Khaliel *et al.* [10] do not completely agree with our results.

cross sections were compared to statistical model calculations using the TALYS code [5] in version 1.8. Various models have been used for the γ-ray strength function and nuclear level density. In detail, for the γ-ray strength function the TALYS γ-SF model 8 (Gogny D1M HFB+QRPA) [6] and for the nuclear level density (NLD) two microscopic models [7, 8] yielded excellent agreement with the experimental data. Both NLD models are based on a Hartree Fock Bogoliubov model. While the model from Ref. [7] is based on a zero-range Skyrme interaction the model from Ref [8] has the finite-range Gogny force implemented, which is more consistent with the γ-ray strength function model. As it is often used for microscopic models, called ρ_{HFM}, adjustment flexibility through a scaling function is provided within the TALYS code and applied for both NLD models for a better reproduction of the experimental values.

$$\rho(E_x, J, \pi) = e^{c \cdot \sqrt{E_x - \delta}} \cdot \rho_{HFM}(E_x - \delta, J, \pi) \tag{2}$$

By default $c = 0$ and the 'pairing shift' δ is obtained from the cumulative number of low-lying levels. The experimental results of the total cross-section measurement as well as the comparison to theoretical calculations are shown in Fig. 3. Furthermore, recent results from Khaliel *et al.* [10] are shown, whereas they are systematically lower than our values. This might be due to the fact that they only take two ground-state transitions into account and assume that there is no angular dependence. However, Fig. 1 clearly shows that the latter is inevitable for the determination of absolute cross sections. Note, that both adjusted NLD models are able to describe the number of low-lying levels sufficiently well. This is a crucial demand each global nuclear level density model has to fulfil.

Partial cross sections up to the decay to the 8th excited state at E_γ=2202 keV in ^{108}Cd have been observed in this experiment except for the 3^+ state at E_γ=2145 keV. Partial cross sections are predominantly sensitive to the γ-ray strength function and according to the Brink-Axel hypothesis the γ-ray transmission probability is independent from the excitation energy. Therefore, the partial cross sections hold important information about the γ-ray strength function at the measured energies. It was found that no theoretical γ-ray strength function, neither phenomenological nor the microscopic was able to correctly describe the experimentally determined partial cross sections. Therefore, the microscopic dipole-strength function based on the Gogny D1M HFB+QRPA model has been adjusted individually for a better reproduction of the partial cross-section values. The resulting theoretical calculations for the default dipole strength as well as for the adjusted strength are shown in Fig. 4.

In summary, total and partial cross sections of the ^{107}Ag(p,γ)^{108}Cd reaction were measured by means of in-beam γ-ray spectroscopy. Using the total cross-section results slight adjustments on two microscopic NLD models could be performed in order to ensure an excellent agreement between

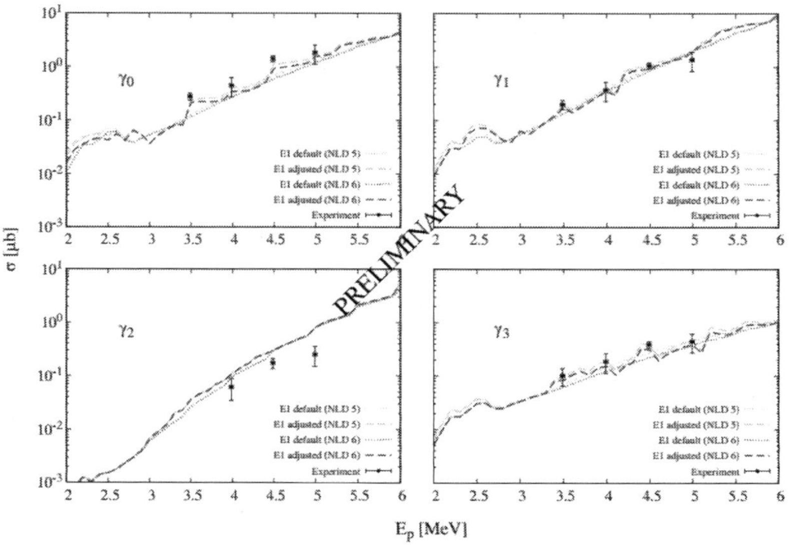

Figure 4. Experimentally determined partial cross sections compared to theoretical calculations. Both the default Gogny D1M HFB+QRPA γ-ray strength function as well as an adjusted one for a better reproduction of the experimental values has been used. Furthermore, both microscopic NLD models which have been adjusted via the total cross section measurement are used. Exemplary, partial cross sections for γ_0 to γ_3 transitions are shown.

experiment and theory. Furthermore, partial cross-section measurements were carried out and yielded an individual adjustment of the γ-ray strength function at γ-ray energies between 9.9 MeV and 13.1 MeV. Again, very good reproduction of the experimental results was achieved. As a next step, the total cross sections for beam energies of 2 MeV and 2.7 MeV, for which experiments have been already performed, will be evaluated in order to verify the current finding at lower energies.

Acknowledgement

This project was supported by the Deutsche Forschungsgemeinschaft under Contract No. ZI 510/8-1 and the ULDETIS project within the University of Cologne Excellence Initiative strategy.

References

[1] M. Arnould and S. Goriely, Phys. Rep. **384**, 1 (2003)
[2] T. Rauscher *et al.*, Rep. Prog. Phys. **76**, 066201 (2013)
[3] T. Rauscher, Phys. Rev. C **81**, 045807 (2010)
[4] L. Netterdon *et al.*, Nucl. Instr. Meth. A **758**, 69 (2014)
[5] A. Koning, S. Hilaire, M. Duijvestijn, International Conf. on Nuclear Data, Nice, France (2007)
[6] M. Martini, S. Hilaire, S. Goriely, A.J. Koning and S. Péru, Nuclear Data Sheets **118**, 273 (2014)
[7] S. Hilaire and S. Goriely Nucl. Phys., A **779**, 63 (2006)
[8] S. Hilaire, M. Girod, S. Goriely and A.J. Koning, Phys. Rev. C **86**, 064317 (2012)
[9] T. Rauscher and F.-K. Thielemann, Atomic Data and Nuclear Data Tables **79**, 47 (2001)
[10] A. Khaliel *et al.*, Phys. Rev. C **96**, 035806 (2017)

EPJ Web of Conferences **165**, 01029 (2017) DOI: 10.1051/epjconf/201716501029
NPA8 2017

Sub-barrier fusion cross section measurements with STELLA

M. Heine[1,★], *S.* Courtin[1,2], *G.* Fruet[1], *D.G.* Jenkins[3], *D.* Montanari[1,2], *P.* Adsley[3,4], *C.* Beck[1], *S.* Della Negra[4], *P.* Dené[1], *F.* Haas[1], *F.* Hammache[4], *G.* Heitz[1], *O.S.* Kirsebom[5], *M.* Krauth[1], *J.* Lesrel[4], *A.* Meyer[4], *L.* Morris[3], *P.H.* Regan[6,7], *M.* Richer[1], *M.* Rudigier[6], *N.* de Séréville[4], and *C.* Stodel[8]

[1] *IPHC-CNRS/Université de Strasbourg, 67037 Strasbourg, (France)*
[2] *USIAS/Université de Strasbourg, 67083 Strasbourg, (France)*
[3] *University of York, YO10 5DD York, (UK)*
[4] *IPN d'Orsay, UMR8608, CNRS/IN2P3, PSUD 11, F-91406 Orsay, (France)*
[5] *Åarhus University, DK-8000 Åarhus, (Denmark)*
[6] *University of Surrey, GU2 7XH Guildford, (UK)*
[7] *National Physical Laboratory, Teddington, TW11 0LD, (UK)*
[8] *GANIL, CEA/DSM-CNRS/IN2P3, 14076 Caen, (France)*

Abstract. The experimental setup STELLA (STELlar LAboratory) is designed for the measurement of deep sub-barrier light heavy ion fusion cross sections. For background suppression the γ-particle coincidence technique is used. In this project, $LaBr_3$ detectors from the UK FATIMA (FAst TIMing Array) collaboration are combined with annular silicon strip detectors customized at IPHC-CNRS, Strasbourg, and the setup is located at Andromède, IPN, Orsay. The commissioning of the experimental approach as well as a sub-barrier $^{12}C + ^{12}C \rightarrow {}^{24}Mg^*$ cross section measurement campaign are carried out.

1 Introduction

With the measurement of deep sub-barrier light heavy-ion fusion cross sections, both insights into nuclear cluster effects [1] as well as the S-factor at energies of astrophysical interest can be obtained. On the production route towards heavier elements during the carbon burning phase in massive stars, $^{12}C(^{12}C, {}^{24}Mg)$ was identified as a key reaction [2]. The relative energies of the fusion partners in the regime of astrophysical interest are as low as $E_{rel} = 2$ MeV. This is far below the Coulomb barrier of ^{12}C at $E_{rel} = 6.6$ MeV and the fusion cross sections for such measurements are expected to be in the order of nanobarn. Alongside to the use of high intensity beams up to 10 μA and compacted detector configurations ^{12}C fusion is detected *via* the coincident measurement of evaporation products and de-excitation gammas [3] from daughter nuclei of the ^{24}Mg compound.

In the $^{12}C + ^{12}C$ system the reaction channels, that include light charged particles, are

$$^{12}C + ^{12}C \rightarrow {}^{24}Mg^* \rightarrow {}^{23}Na + p, \quad (1)$$

$$^{24}Mg^* \rightarrow {}^{20}Ne + \alpha, \quad (2)$$

★e-mail: marcel.heine@iphc.cnrs.fr

with Q-values $Q_p = 2.24$ MeV and $Q_\alpha = 4.62$ MeV, respectively. Both reactions occur, in this energy domain, *via* the formation of an excited compound ^{24}Mg nucleus. It was shown that the de-excitation of (1) and (2) predominantly cascades through the first excited states of the daughter nuclei [4]. Alongside with the measurement of the corresponding particle energies the detection of the characteristic gamma lines at 440 keV and 1634 keV, respectively, is targeted in the first instance.

2 Measurement Station

In the first campaign at Andromède [5], a $^{12}C^{2,3+}$ beam with energies ranging from 11.1 MeV to 5.5 MeV is directed on (thin) fixed targets of natural ^{12}C with densities of a few tens of μg/cm^2. In the course of the measurements the beam intensity was gradually increased from 40 pnA to 800 pnA. The accuracy of the detection systems as well as the time alignment of the data acquisition are guaranteed and cross section measurements are carried out [6].

For the detection of de-excitation gammas, an array of 36 (1.5" x 2") $LaBr_3$ detectors from the UK FATIMA collaboration is developed with the focus on the maximization of the detection efficiency. By means of Geant4 simulations the geometrical acceptance was optimized to 21% of the 4Π solid angle with the full-energy peak detection efficiency of 8% (440 keV) and 2% (1634 keV) [7]. The simulations are furthermore utilized to reconstruct the decay pattern of the natural contaminant ^{138}La. In the comparison to experimental data a correction of an electronic walk of 10 keV per day is carried out. This ensures that the QDC channels of the targeted gamma energies don't drift and a similar correction is foreseen for the charged particle detectors.

The nominal positions for annular in-beam double sided silicon strip detectors are 3 cm upstream as well as 6 cm up- and downstream with respect to the target are equipped with annular detectors based on S1, S3 Micron ® chips. They have a strip pitch of 1505 μm (S1) and 886 μm (S3) and are protected by absorber foils against scattered beam as well as electrons produced in the target. Due to the compact placement the geometrical acceptance is 25% of the 4Π solid angle.

To withstand the heat input during bombardment with up to 10 μA beam, a rotating target mechanism is developed and commissioned. Amongst fixed target positions the target wheel hosts three slots for frames with a diameter of 4.6 cm that spin with around 1000 rpm.

For the absolute normalization, two monitor detectors are placed in extensions around 25 cm from the beam spot position. These barrier surface silicon detectors measure the well know Mott scattering rates from the ^{12}C+^{12}C reaction at 45°. In addition, the beam current is recorded by an integrator behind the experiment. From a combination of both measurements, the target thickness can be recognized as well directly from the experimental data.

3 First Results

During the commissioning, the ^{12}C+^{12}C reaction is measured to compare with sub-barrier data from the literature and to guarantee the accuracy of the analysis. The time synchronization of the time stamped trigger less data acquisition system is tested extracting time correlations between gammas (Compton events) as well as the gamma- and particle detectors (fusion reactions). In this first attempt the time coincidence window is set to 400 ns.

In the course of the experiment several calibration measurements with ^{152}Eu and ^{241}Am sources are carried out to guarantee the reliability of the energy gates for the gamma- and particle detectors, respectively. It turns out that a crude calibration at the beginning is sufficient for the gammas, while afterward data are calibrated in blocks of one hour based on the comparison to the simulated self activity pattern mentioned earlier.

DOI: 10.1051/epjconf/201716501029

The correlation of simultaneous gamma- and charged particle detector entries is shown in figure 1 for a beam energy of around E_{beam} = 10.4 MeV for a subset of data. The coincidence window of

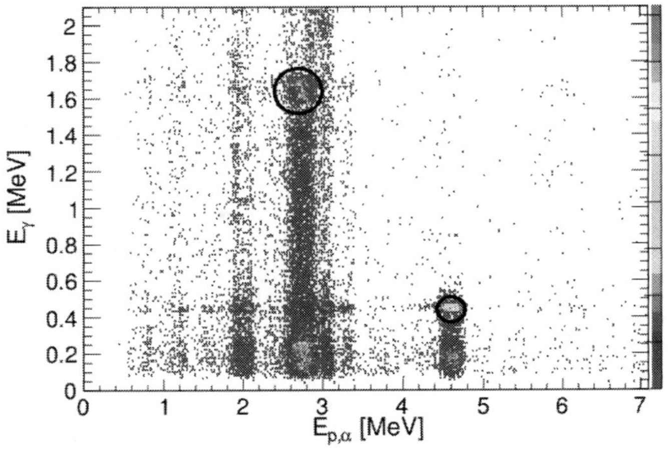

Figure 1. [Color online] Correlation of simultaneous gamma- and charged particle detector entries with a coincidence window of 400 ns. Fusion events from the ^{12}C+^{12}C reaction, populating the first excited state of the ^{23}Na or the ^{20}Ne residual nucleus, can be selected like indicated for the characteristic gammas at E_γ = 0.440 MeV or E_γ = 1.634 MeV resulting from their subsequent de-excitation.

400 ns used in this first analysis can still be minimized to reduce random coincidence background. For the calculation of fusion cross sections, the regions of interest around the full-energy depositions of the characteristic gammas from the ^{23}Na as well as ^{20}Ne daughter nuclei are highlighted with ellipses at the gamma energies of E_γ = 0.440 MeV or E_γ = 1.634 MeV, respectively.

It is also noteworthy to mention that the timing resolution of the detection system allows for a discrimination of protons and alpha particles based on the rise time of the electronic pulse. In this way, a background free identification of the respective reaction is realized and a comprehensive analysis for further measurements at different beam energy is on the way.

References

[1] D. Jenkins and S. Courtin, Jour. Phys. G **42**, 034010 (2015)
[2] A. Chieffi *et al.*, Astr. Phys. Jour. **502**, 737 (1998)
[3] C.L. Jiang *et al.*, Nucl. Instr. and Methods A **682**, 12 (2012)
[4] K.A. Erb *et al.*, Phys. Rev. C **22**, 22 (1980)
[5] S. Della Negra, Inn. Rev. p. 38 (2016)
[6] G. Fruet *et al.*, submitted to EPJ Web of Conferences (2017)
[7] M. Heine *et al.*, J. Phys. Conf. Ser **763**, 012005 (2016)

Isomer beam elastic scattering: ^{26m}Al(p, p) for astrophysics

D. Kahl[1,⋆], *H.* Shimizu[2], *H.* Yamaguchi[2], *K.* Abe[2], *O.* Beliuskina[2], *S. M.* Cha[3], *K. Y.* Chae[3], *A. A.* Chen[4], *Z.* Ge[5], *S.* Hayakawa[2], *N.* Imai[2], *N.* Iwasa[6], *A.* Kim[7], *D. H.* Kim[7], *M. J.* Kim[3], *S.* Kubono[5], *M. S.* Kwag[3], *J.* Liang[4], *J. Y.* Moon[8,9], *S.* Nishimura[5], *S.* Oka[10], *S. Y.* Park[6], *A.* Psaltis[4], *T.* Teranishi[10], *Y.* Ueno[10], and *L.* Yang[2]

[1]*School of Physics & Astronomy, the University of Edinburgh, Edinburgh, United Kingdom*
[2]*Center for Nuclear Study, the University of Tokyo, Wako, Japan*
[3]*Department of Physics, Sungkyunkwan University, Suwon, Korea*
[4]*Department of Physics & Astronomy, Hamilton, Ontario, Canada*
[5]*RIKEN Nishina Center, Wako, Japan*
[6]*Department of Physics, Tohoku University, Sendai, Japan*
[7]*Department of Physics, Ewha Womans University, Seoul, Korea*
[8]*Institute of Particle and Nuclear Studies, KEK, Japan*
[9]*Wako Nuclear Science Center, KEK, Wako, Japan*
[10]*Department of Physics, Kyushu University, Fukuoka, Japan*

Abstract. The advent of radioactive *ground-state* beams some three decades ago ultimately sparked a revolution in our understanding of nuclear physics. However, studies with radioactive *isomer* beams are sparse and have often required sophisticated apparatuses coupled with the technologies of ground-state beams due to typical mass differences on the order of hundreds of keV and vastly different lifetimes for isomers. We present an application of a isomeric beam of ^{26m}Al to one of the most famous observables in nuclear astrophysics: galactic ^{26}Al. The characteristic decay of ^{26}Al in the Galaxy was the first such specific radioactivity to be observed originating from outside the Earth some four decades ago. We present a newly-developed, novel technique to probe the structure of low-spin states in ^{27}Si. Using the Center for Nuclear Study low-energy radioisotope beam separator (CRIB), we report on the measurement of ^{26m}Al proton resonant elastic scattering conducted with a thick target in inverse kinematics. The preliminary results of this on-going study are presented.

1 Motivation

The observation of 1.809-MeV γ-rays associated with the decay of ^{26}Al across the Milky Way has attracted much attention in various subfields of nuclear astrophysics in the years since its discovery [1]. Observations point to massive stars as the main producers of the observed ^{26}Al [2], although it is unclear whether its production occurs mainly in the Wolf-Rayet phase or during subsequent supernova. The ground state of ^{26}Al, which we denote ^{26g}Al, has a spin-parity $J^\pi = 5^+$, a halflife of 0.72 Myr, and decays predominately through the first excited state in ^{26}Mg located at 1.809 MeV,

⋆e-mail: daid.kahl@ed.ac.uk

which then de-excites by emission of characteristic electromagnetic radiation. Nucleosynthesis of ^{26}Al is complicated by a low-laying isomeric state ^{26m}Al at 228 keV with $J^{\pi} = 0^{+}$ and a halflife of $T_{1/2} = 6.3$ s. The isomeric decay not only tends to bypass the production of the characteristic 1.809-MeV γ-ray, given its short lifetime it is unlikely to be transported intact from its site of production to a transparent region of space for observation. In an astrophysical plasma, the tail-end of the Planck distribution may be energetic enough to link the two species through thermally-induced transitions [3, 4]; these studies highlighted the complexity of these physics, yet there can be no doubt the states ^{26g,m}Al are almost certainly in statistical equilibrium at temperatures > 1 GK.

The ^{26m}Al(p, γ) stellar reaction rate is highly uncertain owing to sparse experimental information. At present, ^{26m}Al states in ^{27}Si have been studied via charged-particle spectroscopy [5] and in-beam γ spectroscopy [6], while mirror states in ^{27}Al were recently studied by a neutron-transfer reaction on ^{26m}Al [7]. No proton partial widths Γ_p are known, and only limited information is known about the nature of higher energy resonances which might be important for temperatures > 1 GK corresponding to those typically found in core collapse supernovae. We performed a measurement of ^{26m}Al(p, p) which should reveal any states with large proton partial widths and low ℓ transfer somewhat above the proton separation energy.

2 Experiment

We conducted a measurement of proton resonant elastic scattering with a mixed ^{26g,m}Al beam at the Center for Nuclear Study (CNS) low-energy radioactive ion beam (RIB) separator [8], called CRIB. To produce the beam, we used the inflight method and the ^{1}H(^{26}Mg, ^{26}Al)n reaction. The ^{26}Mg beam was accelerated to 6.65 MeV/u with typical intensities of 25–50 pnA. The beam then impinged on a Havar-windowed, 8 cm long gas cell [9] filled with H_2 gas and cooled to an effective temperature of 90 K with LN_2. To change the ^{26}Al isomeric purity, we produced the cocktail beam at different center-of-mass energies which turned out to be effective. We accomplished this with an energy-degrader foil upstream of the production target as well as by varying the H_2 gas pressure over 130–290 Torr (0.5 to 1.0 mg cm^{-2}). The cocktail beam was selected by its magnetic rigidity at the dispersive focal plane between two magnetic dipoles, and further purified by a Wien (velocity) filter before arriving at the experimental scattering chamber.

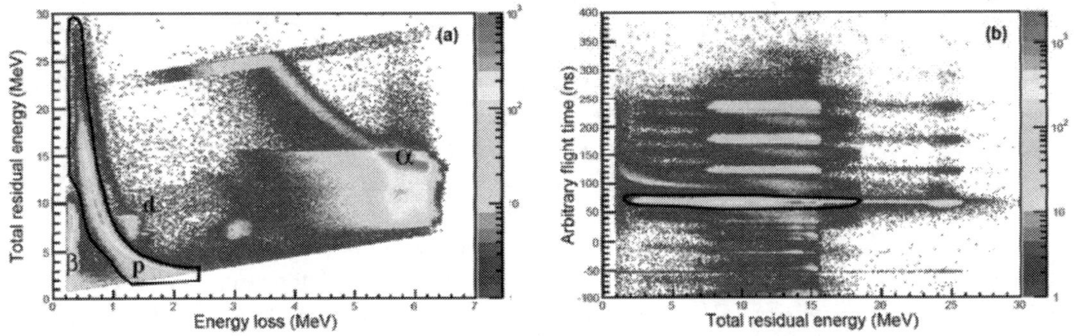

Figure 1. (Color online) Proton identification: (a) Energy loss versus residual energy from the Si telescope. (b) Residual energy from the Si telescope against time of flight between PPACa and the Si telescope. The depicted gates show the scattered protons.

The experimental setup consisted of two parallel plate avalanche counters (PPACs) [10] to track the beam, a target slider, two ΔE–E silicon telescopes to measure protons, and an array of ten NaI de-

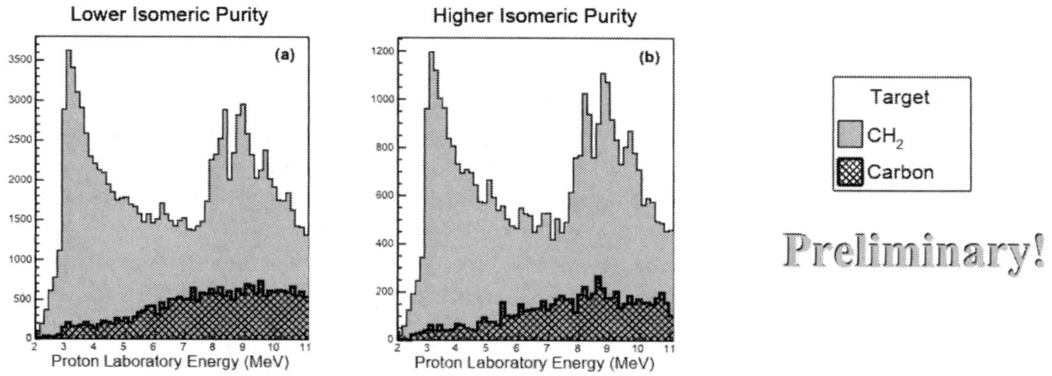

Figure 2. Residual *laboratory* energy histograms for protons scattered by ^{26}Al from two of the runs. Isomeric purity of the beam is lower in (a) than in (b).

tectors to measure γ-rays. The PPACs enabled us to track the beam ions event-by-event, to determine their trajectory and nuclear species. The ^{26}Al cocktail beam had an average intensity of 1.5×10^5 pps, 93% purity, and on-target energies of 68, 83, and 93 MeV; the main contaminants were the stable isotopes ^{23}Na and leaky primary beam ^{26}Mg [11]. The target slider held the secondary targets, which were a 7.5 mg cm^{-2} CH_2 foil as a proton target, a 10.6 mg cm^{-2} ^{nat}C foil for background subtraction, and 1 cm blocks of Al and plastic which are thick enough to completely stop the β^+-rays from the decay of ^{26m}Al. We made regular measurements of the isomeric purity by both directly measuring the positrons with the Si telescopes as well as measuring the annihilation γ-rays at 511 keV with the NaI array. During these decay measurements, we pulsed the primary beam in an on/off mode with a duty cycle of 24 s. The β^+ energy distribution and the derived halflife was completely consistent with ^{26m}Al, as shown in Fig. 3 of [11]. With the decay measurements we found the purity ^{26m}Al/^{26g,m}Al covered a range of roughly $50 \pm 20\%$ depending on the beam production conditions.

Protons elastically scattered from the CH_2 target were measured with two silicon telescopes placed at forward angles in the laboratory. Each silicon detector was 50×50 mm^2. We measured the position and energy loss of each proton with the first layer, which was nominally 75 μm thick and had 16 orthogonal strips on each side. The other telescope layers were 1.5 mm thick which we summed to get the proton residual energy. Protons were distinguished from other light ions by the ΔE–E method as shown in Fig. 2(a), and protons scattered by the ^{26}Al beam were further distinguished from other protons by their timing between PPACa and the Si telescope as shown in Fig. 2(b).

Selected proton spectra from the 0° telescope are shown in Fig. 2, where the spectra obtained with the C target are normalized to the number of incident ^{26}Al ions on CH_2 for comparison. These spectra are *preliminary* and taken in the *laboratory* frame. To obtain the center-of-mass energy of the protons from the laboratory energy for small θ, the kinematic compression is about a factor of four. No correction has been applied for the energy loss of protons in the target yet. Next, we will solve the kinematic equation with energy loss and angle for each event to construct the excitation function.

At present, despite the above limitations of the laboratory proton spectra, the basic features are quite informative. Firstly, we note that pure ^{26g}Al proton elastic scattering was previously measured up to 1.5 MeV in the center-of-mass frame, and no strong resonances were observed [12]. Thus, to obtain a given ^{26m}Al proton spectrum, we simply need to perform a background subtraction of the well-known Rutherford scattering cross-section scaled to the intensity of ^{26g}Al. The strong features that emerge in our spectra at proton energies higher than 8 MeV could be from either state of ^{26}Al, although these energies are likely to be too high to have an astrophysical impact; in the future, their origin may be clarified by a more careful analysis of their strength as a fuction of isomeric purity. Below proton energies of 3 MeV, the protons do not have enough energy to reach the second layer of the Si telescope. However, between these two energies, hints of peak-like structures may be present. Although we already know there are no strong proton resonances from ^{26g}Al in this energy region, further analysis of our proton scattering data as a function of isomeric purity would help confirm that any peaks arise from ^{26m}Al. In order to be resolved, these states must have large proton partial widths Γ_p, yet no strong states were observed in this region of the mirror nucleus via the (d, p) reaction [7], suggesting the states we observe may have $\ell > 0$. In the future, we will perform an R-Matrix fit on the proton scattering excitation function to extract the resonant properties of Γ_p, ℓ, and $E_{c.m.}$.

3 Summary

We reported the first experimental work to produce an RIB of ^{26m}Al and control its isomeric purity. Using this beam, we measured ^{26m}Al proton resonant elastic scattering for the first time. We observed hints of resonant-like structure with large Γ_p around 1–2 MeV in $E_{c.m.}$ in the ^{26m}Al+p system. Curiously, there are no strong ^{26g}Al proton resonances over the same energy region in ^{27}Si, possibly suggesting that ^{26m}Al is more efficiently destroyed by radiative proton capture than ^{26g}Al in high temperature astrophysical environments. The nuclear structure which gives rise to this behavior of unbound proton states in ^{27}Si will be interesting to investigate. As ^{26g,m}Al will be in thermal equilibrium for $T > 1$ GK, ^{26m}Al(p, γ) could be an efficient pathway to destroy ^{26g}Al in core collapse supernovae, which may impact the contribution of massive stars to galactic ^{26}Al. More definite conclusions will be possible in the near future after we complete our analysis.

Acknowledgments

This work was partly supported by grants from the STFC in the UK, the JSPS KAKENHI (Grants No. 25800125 and 16K05369) in Japan, NRF funded by the Korea government (MEST) (No. NRF-2014S1A2A2028636, NRF-2015R1D1A1A01056918, NRF-2016K1A3A7A09005579, and NRF-2016R1A5A1013277), and NSERC of Canada. We thank the CNS and RIKEN staff for operation of the ion source and accelerator during the machine time.

References

[1] W. A. Mahoney *et al.*, Astrophysical Journal **262**, 742 (1982).
[2] M. Limongi and A. Chieffi, Astrophysical Journal **647**, 483 (2006).
[3] R. C. Runkle, A. E. Champagne, and J. Engel, Astrophysical Journal **556**, 970 (2001).
[4] S. S. Gupta and B. S. Meyer, Physical Review C **64**, 025805 (2001).
[5] C. M. Deibel *et al.*, Physical Review C **80**, 035806 (2009).
[6] G. Lotay, P. J. Woods, D. Seweryniak *et al.*, Physical Review C **80**, 055802 (2009).
[7] S. Almaraz-Calderon, K. E. Rehm, N. Gerken *et al.*, Physical Review Letters **119**, 072701 (2017).
[8] Y. Yanagisawa *et al.*, Nuclear Instruments and Methods in Physics Research A **539**, 74 (2005).
[9] H. Yamaguchi *et al.*, Nuclear Instruments and Methods in Physics Research A **589**, 150 (2008).
[10] H. Kumagai *et al.*, Nuclear Instruments and Methods in Physics Research A **470**, 562 (2001)
[11] D. Kahl, H. Yamaguchi, H. Shimizu *et al.*, Nuovo Cimento C **39**, 362 (2016).
[12] S. T. Pittman *et al.*, Physical Review C **85**, 065804 (2012).

EPJ Web of Conferences **165**, 01031 (2017)
NPA8 2017
DOI: 10.1051/epjconf/201716501031

Improved experimental determination of the branching ratio for β-delayed α decay of ^{16}N

Oliver S. Kirsebom[1,*], *Irene M.* Alonso[2], *Tom* Berry[3], *Maria J. G.* Borge[4], *Esben R.* Christensen[1], *Daniel* Doherty[5], *Paloma D.* Fernández[6], *Miguel M.* Flores[4], *Hans O. U.* Fynbo[1], *Mariano C.* Gallardo[7], *Nicolas* Hubbard[5], *Jesper H.* Jensen[1], *Håkan* Johansson[6], *Björn* Jonson[6], *Razvan* Lica[4], *Morten V.* Lund[1], *Angel P.* Martínez[2], *Michael K.* Munch[1], *Karsten* Riisager[1], *Olivier* Sorlin[4,8], *Jacobus A.* Swartz[1], *Olof* Tengblad[2], and *Victoria* Vedia[7]

[1]*Department of Physics and Astronomy, Aarhus University, DK-8000 Aarhus, Denmark*
[2]*Instituto de Estructura de la Materia, CSIC, E-28006 Madrid, Spain*
[3]*Department of Physics, University of Surrey, Guildford, GU2 7XH, UK*
[4]*CERN, 1211 Geneva 23, Switzerland*
[5]*Department of Physics, University of York, York YO10 5DD, UK*
[6]*Fundamental Physics, Chalmers University of Technology, S-41296 Göteborg, Sweden*
[7]*Grupo de Física Nuclear, Universidad Complutense de Madrid, E-28040 Madrid, Spain*
[8]*GANIL, CEA/DSM-CNRS/IN2P3, Bvd Henri Becquerel, 14076 Caen, France*

Abstract. We report on the preliminary results of an experimental study of the β decay of ^{16}N, aiming to determine the branching ratio of the $\beta\alpha$ channel with a precision of $\leq 5\%$.

1 Introduction

During hydrostatic helium burning carbon is converted to oxygen via the reaction ^{12}C$(\alpha, \gamma)^{16}$O. The temperature at which this takes place is $T \sim 10^8$ K, implying that reactions preferentially occur close to a center-of-mass energy of $E_{\rm cm} \sim 0.3$ MeV. At such low energies the cross section is too small to be measured directly in laboratory experiments and must be determined by extrapolating the data obtained at higher energies ($E_{\rm cm} \gtrsim 1.0$ MeV). Typically, R-matrix theory [1] is used to parametrize the, a priori unknown, energy dependence of the cross section in terms of the properties of the levels in ^{16}O that are involved in the reaction. Since the same levels can be observed in other reactions many of the relevant parameters can be accurately determined or at least constrained by indirect techniques. Following this approach the cross section at 0.3 MeV has been determined with an estimated precision of $\sim 20\%$ while a precision of at least $\sim 10\%$ is desired [2]. Among the indirect techniques used the β-delayed α decay of ^{16}N, shown in Fig. 1, has proven useful to determine how strongly the 7.12 MeV level in ^{16}O couples to the α + ^{12}C channel, and hence constrain the level's significant contribution ($\sim 54\%$) to the capture cross section at 0.3 MeV. This requires precise measurements of the branching ratio to the 7.12 MeV level, the branching ratio for α emission, and the shape of the α spectrum. Here, we bring the preliminary results of a new experimental study of the β decay of ^{16}N recently performed at the ISOLDE facility [3] with the aim of obtaining precise values for the branching ratios.

* Acknowledges support from the Villum Foundation; e-mail: oliskir@phys.au.dk

DOI: 10.1051/epjconf/201716501031

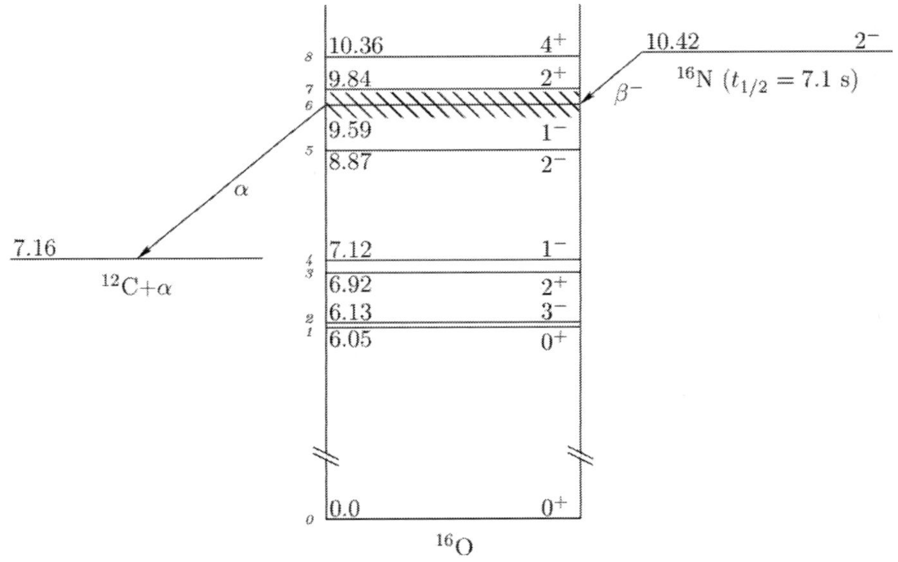

Figure 1. ^{16}N decay scheme [4]. Energies are in MeV relative to the ^{16}O ground state. The levels in the ^{16}O are indexed 0–8. The main decay branches go to the ground state (28%) and the the 6.13 MeV level (66%) with smaller branches going to the levels at 7.12 MeV (4.8%) and 8.87 MeV (1.1%) while even smaller branches ($< 10^{-3}$) go to the other levels including an α branch of $\sim 10^{-5}$ indicated by the two arrows.

2 Existing knowledge about the β decay of ^{16}N

We proceed by giving a brief review of the existing empirical knowledge about the β decay of ^{16}N [4]. The main decay branches go to the ground state (28%) and the the 3^- level at 6.13 MeV (66%) with smaller branches going to the 1^- level at 7.12 MeV (4.8%) and the 2^- level at 8.87 MeV (1.1%) while even smaller branches ($< 10^{-3}$) go to the other levels, including a β-delayed α-decay branch of $\sim 10^{-5}$ to the ground state of ^{12}C. It is remarkable that much of the existing data on the β decay of ^{16}N stems from γ-ray studies performed in the 1950s and 1960s using NaI detectors with poor energy resolution. For example, the branching ratio to the 7.12 MeV level is known with a precision of $\sim$ 8% based on three independent measurements of the intensity ratio of the 6.13 and 7.12 MeV γ-rays performed in the 1950s [5–7]. With modern HPGe detectors it should be possible to obtain significantly improved values, not only for the branching ratio to the 7.12 MeV level, but also for the branching ratios to some of the other levels.[1] On the other hand, the ratio of the two main branches, which account for 94% of the total intensity, has been determined very precisely ($\sim$ 1%) by careful measurements of the shape of the β spectrum.

The α-decay branching ratio was first determined to have a value of $1.20(5) \times 10^{-5}$ by Kaufmann *et al.* [9]. More recently, Zhao *et al.* have obtained the value $1.3(3) \times 10^{-5}$ [10], while Refsgaard *et al.* find $1.49(5) \times 10^{-5}$ with a possible systematic uncertainty of -0.10×10^{-5} [11]. There is evidently a significant discrepancy between the values of Kaufmann *et al.* and Refsgaard *et al.*, while the value of Zhao *et al.* has sufficiently large error bars to be consistent with either of the two other values. A new measurement of α-decay branching ratio is needed to resolve the discrepancy.

[1] Tang *et al.* [8] have recently reported a new value for the braching ratio to the 7.12 MeV level with an uncertainty of 4%, but the details of the measurement have not yet been made available.

3 New experiment and preliminary results

In order to close some of the gaps in our knowledge about the β decay of ^{16}N, we have studied the decay at ISOLDE [3]. In this study, performed in May 2016, a 30 keV beam of ^{14}N^{16}N$^+$ molecular ions was delivered to the ISOLDE Decay Station (IDS) [12] at an average rate of $\sim 2 \times 10^4$ s^{-1} for a total of 32 hours. The ions were stopped in a thin (30 μg/cm^2) carbon foil surrounded by five double-sided silicon strip detectors (DSSD) and four high-purity germanium (HPGe) clovers, allowing for the simultaneous detection of charged particles and γ rays, while auxiliary detectors were used to check that the beam was being fully transmitted to the center of the setup and fully stopped in the foil. Three of the DSSDs were sufficiently thin (40–60 μm) to allow the α spectrum to be clearly separated from the β background, as shown in Fig. 2. The other two DSSDs were much thicker (300 μm and 1 mm)

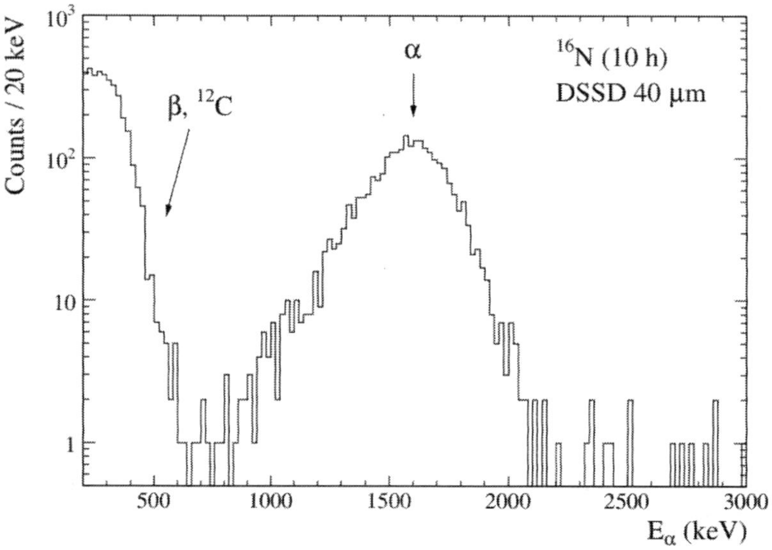

Figure 2. Energy spectrum measured in the 40-μm thick DSSD. The α particles emitted in the decay of ^{16}N have energies between 0.8 MeV and 2.2 MeV and are seen to be well separated from the β particles and the ^{12}C recoils. The data shown here represents 1/3 of the total acquired data. The remaining data have yet to be analyzed.

and served primarily to detect the β particles. The γ-ray spectrum measured in the HPGe clovers (Fig. 3) contains several γ-rays from the decay of ^{16}N with no evidence of other radioactive isotopes. In order to convert the observed γ-ray yields to intensity ratios it is necessary to correct for the energy dependent detection efficiency of the HPGe array. An absolutely calibrated ^{152}Eu source was used to determine the detection efficiency at low energies ($E_\gamma < 1.5$ MeV), while $\beta\gamma$ and $\gamma\gamma$ coincidences will be used to extend the calibration up to 7 MeV.

4 Summary and outlook

The β decay of ^{16}N has been studied in an experiment at the ISOLDE Decay Station in which both charged particles and γ rays were detected. It is expected that the data will constrain the branching ratio for β-delayed α emission and the branching ratio to the 7.12 MeV level with a precision of 5% or better. The data analysis is nearing completion and the results will soon be published, including an assessment of the impact on the inferred cross section of the ^{12}C$(\alpha,\gamma)^{16}$O reaction at $E_{cm} = 0.3$ MeV.

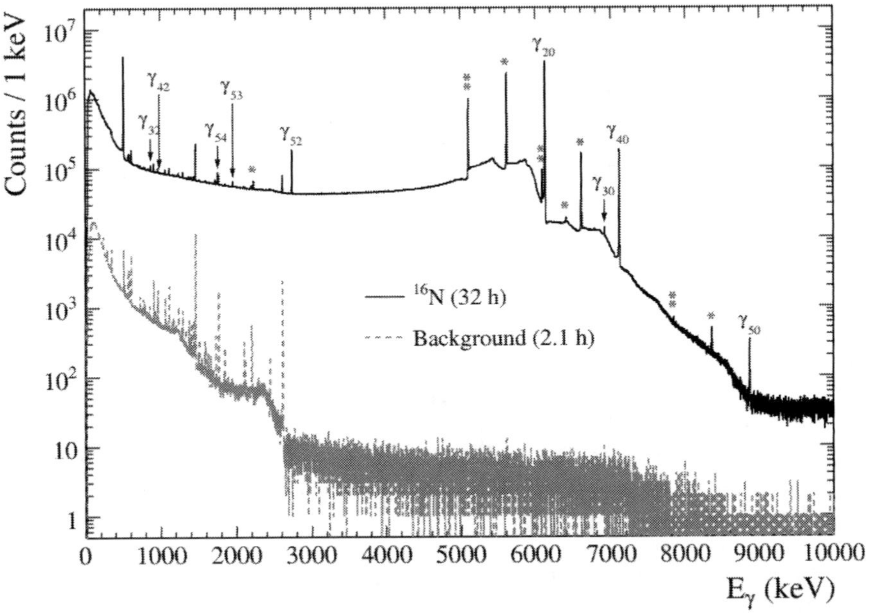

Figure 3. Add-back γ-ray spectrum obtained in the present experiment. ^{16}N: solid (black) line. Background: dashed (red) line. Photopeaks due to transitions $i \rightarrow j$ in ^{16}O are indicated by the symbol γ_{ij} while first- and second-escape peaks are indicated by a single asterisk ($*$) and a double asterisk ($**$), respectively.

References

[1] A.M. Lane, R.G. Thomas, Reviews of Modern Physics **30**, 257 (1958)

[2] R.J. deBoer, J. Görres, M. Wiescher, R.E. Azuma, A. Best, C.R. Brune, C.E. Fields, S. Jones, M. Pignatari, D. Sayre et al., Rev. Mod. Phys. **89**, 035007 (2017)

[3] R. Catherall, W. Andreazza, M. Breitenfeldt, A. Dorsival, G.J. Focker, T.P. Gharsa, T.J. Giles, J.L. Grenard, F. Locci, P. Martins et al., Journal of Physics G: Nuclear and Particle Physics **44**, 094002 (2017)

[4] D. Tilley, H. Weller, C. Cheves, Nuclear Physics A **564**, 1 (1993)

[5] C.H. Millar, G.A. Bartholomew, B.B. Kinsey, Phys. Rev. **81**, 150 (1951)

[6] B.J. Toppel, Phys. Rev. **103**, 141 (1956)

[7] D.E. Alburger, A. Gallmann, D.H. Wilkinson, Phys. Rev. **116**, 939 (1959)

[8] X.D. Tang, K.E. Rehm, I. Ahmad, C.R. Brune, A. Champagne, J.P. Greene, A. Hecht, D.J. Henderson, R.V.F. Janssens, C.L. Jiang et al., Phys. Rev. C **81**, 045809 (2010)

[9] W. Kaufmann, H. Wäffler, Nuclear Physics **24**, 62 (1961)

[10] Z. Zhao, R.H. France, K.S. Lai, M. Gai, E.L. Wilds, R.A. Kryger, J.A. Winger, K.B. Beard, Phys. Rev. C **48**, 429 (1993)

[11] J. Refsgaard, O. Kirsebom, E. Dijck, H. Fynbo, M. Lund, M. Portela, R. Raabe, G. Randisi, F. Renzi, S. Sambi et al., Physics Letters B **752**, 296 (2016)

[12] H. Fynbo, O.S. Kirseboom, O. Tengblad, Journal of Physics G: Nuclear and Particle Physics **44**, 044005 (2017)

DOI: 10.1051/epjconf/201716501032

The Trojan Horse Method for nuclear astrophysics and its recent applications

L. Lamia[1,2,⋆], C. Spitaleri[1,2], M. Mazzocco[3,4], A. Boiano[6], C. Boiano[13], C. Broggini[4], A. Caciolli[3,4], R. Depalo[3,4], A. Di Pietro[2], P. Figuera[2], F. Galtarossa[15], G.L. Guardo[1,2], M. Gulino[1,10], S. Hayakawa[7], S. Kubono[8,14], M. La Cognata[1], M. La Commara[6,9], G. La Rana[6,9], M. Lattuada[1,2], R. Menegazzo[4], A. Pakou[5], C. Parascandolo[6], D. Piatti[3,4], D. Pierroutsakou[6], R.G. Pizzone[1], S.M.R. Puglia[1,2], S. Romano[1,2], G.G. Rapisarda[1], A.M. Sanchez-Benitez[17], M.L. Sergi[1], O. Sgouros[5], H. Silva[18], F. Soramel[3,4], V. Soukeras[5], E. Strano[3,4], D. Torresi[1], O. Trippella[19,20], A. Tumino[1,10], H. Yamaguchi[7], F.L. Villante[11,12], and G.L. Zhang[16]

[1] *Laboratori Nazionali del Sud di Catania, LNS-INFN, Italy*
[2] *Dipartimento di Fisica e Astronomia, Università di Catania, Italy*
[3] *Dipartimento di Fisica e Astronomia, Università di Padova, Italy*
[4] *INFN-Sezione di Padova, Italy*
[5] *Department of Physics and HINP, University of Ioannina, Greece*
[6] *INFN-Sezione di Napoli, Italy*
[7] *Center for Nuclear Study, University of Tokyo, Japan*
[8] *Riken Nishina Center, Wako, Saitama, Japan*
[9] *Dipartimento di Scienze Fisiche, Università di Napoli, Italy*
[10] *Facoltà di Ingegneria e Architettura, Università degli Studi di Enna Kore, Italy*
[11] *Laboratori Nazionali del Gran Sasso LNGS-INFN, Assergi (AQ), Italy*
[12] *Dipartimento di Scienze Fisiche e Chimiche, Università de L'Aquila, Italy*
[13] *INFN-Sezione di Milano, Italy*
[14] *Institute of Modern Physics, Chinese Academy of Sciences, China*
[15] *Laboratori Nazionali di Legnaro, LNL-INFN, Italy*
[16] *School of Physics and Nuclear Energy Engineering, Bihang University*
[17] *Universitad de Huelva, Spain*
[18] *Universitad de Lisboa, Purtugal*
[19] *Dipartimento di Fisica e Geologia, Università di Perugia, Italy*
[20] *INFN-Sezione di Perugia, Italy*

Abstract. The Trojan Horse Method (THM) has been applied extensively for the last 25 years to measure nuclear reaction cross sections of interest for astrophysics. Although it has been mainly applied for charged particle-induced reactions, recently it has been found to have also a relevant role for neutron-induced reactions. Here, some advantages of THM will be discussed and the preliminary results of the cosmological relevant ^{7}Be(n,α)^{4}He cross section measurement are discussed.

⋆e-mail: llamia@lns.infn.it

1 Introduction

Experimental nuclear astrophysics aims at measuring nuclear reaction cross sections of interest for both primordial and stellar nucleosynthesis in the corresponding Gamow windows. Often *extrapolations* from high energy data down to Gamow energies are used, although they could suffer of uncertainties related, for instance, to the presence of unknown low-lying (or even subthreshold) resonances or to the *electron screening* phenomenon [1, 2]. This effect, due to the electron (atomic) cloud surrounding the positively charged nucleus, alters the cross-section at low-energies, thus partially shielding the pure Coulomb repulsion between the charged-interacting nuclei. Electron screening understanding is far to be completely achieved since the current available theoretical dynamics models (i.e. adiabatic approximation) largely underestimate the electron screening potential values with respect to those measured in terrestrial laboratories.

2 The Trojan Horse Method

The Trojan Horse Method (THM) [3–6] allows one to extract the bare-nucleus cross-section of a charged-particle induced reaction a+x→c+C at astrophysical energies free of Coulomb suppression, by properly selecting the quasi-free (QF) contribution of an appropriate reaction a+A→c+C+s, performed at energies well above the Coulomb barrier, where the nucleus A has a dominant x⊕s cluster configuration. The a+x→c+C reaction studied with THM is not affected by tunneling, since the $a + x$ interaction occurs in the pure nuclear field without the influence of the Coulomb barrier. This allows to the experimentalist to measure the corresponding S(E)-factor even at energies at which direct measurements can access only via the extrapolations. However, it must be stressed here that the TH S(E)-factor determination requires a normalization procedure to the available direct measurements performed at higher energies, thus rendering the THM a complementary experimental technique for nuclear astrophysics aimed at reaching the ultra-low energy region of interest for astrophysical applications. THM has been used in studying several problems, ranging from BBN [7], light element burning reactions [8–12], CNO reactions [13–15], explosive nucleosynthesis involving RIB [16, 17] and removing/producing neutron reactions [18, 19].

2.1 Some details on THM data analysis

The THM can be regarded as an extension to the ultra low-energies of QF reactions, widely used in the past for evaluating nuclear structure with particular regard to the cluster configurations, making it possible to apply the method to nuclear astrophysics [4, 20, 21]. By performing a devoted experiment for studying the three-body reaction a+A→c+C+s, it is possible to connect the three-body cross section with the one of interest for astrophysics through the relation [5, 6]

$$\frac{d^3\sigma}{dE_c C\Omega_c C\Omega_C} \propto KF \cdot \mid \Phi(\vec{p_s}) \mid^2 \cdot \left(\frac{d\sigma}{d\Omega}\right)\Bigg|_{a-x}^{HOES} \tag{1}$$

where KF represents the kinematical factor, $\mid \Phi(\vec{p_s}) \mid^2$ is the square of the momentum distribution for the $x - s$ relative motion inside the TH-nucleus A, and $\frac{d\sigma}{d\Omega}\Big|_{a-x}^{HOES}$ the half-off energy shell cross section. This last quantity represents the "bare-nucleus" cross section of interest for astrophysics, that needs to be corrected for the penetrability through the Coulomb barrier and normalized to the available high-energy direct data. More advanced approaches have been introduced in later years, allowing us to justify and generalize Eq. 1, releasing some approximations contained in it (see, for instance, [5, 6, 15, 19]). In order to assess the method, some of the most important sources of uncertainties in a typical THM experiment and/or data analysis need to be underlined:

- **Energy resolution effects.** THM data are expressed in terms of the relative energy between two-out-of three c and C detected particles, being their detection sufficient for reconstructing completely the full kinematic of a three-body exit channel reaction by applying energy and momentum conservation rules. Energy resolution effects on the relative E_{cC} energy can be evaluated via the standard errors propagation theory, taking into account both energy and angular resolution due to the adopted experimental setup. For instance, in the case of the ^{11}B+p reaction discussed in [22, 23], this has been evaluated to be ~40 keV;
- **Determination of the experimental momentum distribution.** The experimental determination of the momentum distribution for the intercluster $x - s$ motion is one of the most important steps of a typical THM analysis, this reflecting the presence of the quasi-free reaction mechanism. Its behavior is then evaluated in terms of both Plane Wave Impulse Approximation (PWIA) and Distorted Wave Born Approximation (DWBA). By referring to the recent ^{10}B+p case [24, 25], its has been found that the two approaches, PW and DWBA, nicely agree within lower momentum of the exiting neutron, being the same result confirmed in a variety of THM experiments discussed in the literature (see [13] for instance);
- **Correction for barrier penetration effects.** THM cross section data are not affectced by barrier penetration effects. For charged particle-induced reactions, one has to take into account Coulomb as well as centrifugal effects while in the case of neutron-induced reactions only the latter need to be considered. Thus, THM data need to be corrected for an analytical function describing the penetration through the barrier, as given in several text books, i.e. [1]. For such purposes, ones has to fix a cut-off radius in terms of the standard formula $r=1.2*(A_1^{1/3}+A_2^{1/3})$ fm [1], even if such choice introduces systematic uncertainties on the final results. This further source of uncertainty has been investigated in THM papers such as [23], for which an overall uncertainty of ~14% on the final evaluation of the zero-energy S(E)-factor has been found;
- **Influence of the momentum distribution in THM data.** The role of the momentum distribution and, in particular, of the FWHM of the experimental momentum distribution on THM data has been investigated in [27], where we have evaluated the role of the d-state when describing the deuteron ground-state wave function. In particular, both the s and d state wave functions have been calculated by using the exact form and the asymptotic one, their shape being the one shown in the left panel of Fig.1. In order to evaluate the impact on THM data, we have introduced the d state contribution in the THM analysis of [23] and we found a maximum variation of about 0.5% on the absolute value of the S(0) for the ^{11}B+p reaction.

3 The recent ^{7}Be(n,α)^{4}He THM investigation

To shed light on the cosmological lithium-problem of interest in the Big Bang Nucleosynthesis (BBN), nuclear physic solutions have been extensively investigated [7, 28]. Among the nuclear reactions producing or destroying lithium, the ones involving the radioactive ^{7}Be isotope ($t_{1/2}$=53.22±0.06 d) and neutrons have been matter of recent studies aimed at investigating the (n,p) and (n,α) reaction channels at Gamow energies (about 100 keV) [29, 30]. To complement such information, the Trojan Horse Method (THM) appears to be suitable since it allows us to study neutron-induced reactions with unstable beams overcoming all the experimental difficulties related to the direct measurements of such

DOI: 10.1051/epjconf/201716501032

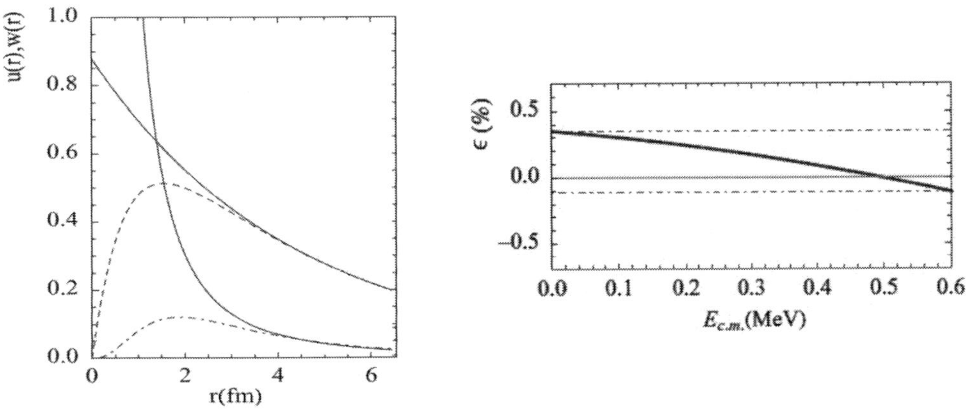

Figure 1. (Left panel). Deuteron ground state wave function: the exact (dashed) and asymptotic (full line) solution for the *s*-state ($u(r)$, blue color) and the *d*-state ($w(r)$, purple color). **(Right panel).** Influence of the *d*-state component of the deuteron wave function on the ^{11}B-p THM analysis, leading to a maximum variation of 0.5% [27].

processes [5, 6]. To this aim, a devoted experimental research program started in 2015 in order to study the ^{7}Be-n induced reactions by means of THM with two independent measurements performed at the Radioactive Ion Beam in-flight facility EXOTIC [31, 32] INFN-LNL (Legnaro, Italy) and at at Center-for-Nuclear-Study Radioactive Isotope Beam (CRIB) (Riken, Japan). The preliminary results of the CRIB experiment are discussed in this volume (see contribution from S. Hayakawa in this volume).

In general, the ^{7}Be(n,α)^{4}He reaction (Q=18.99 MeV) has been studied by means of the THM applied to the quasi-free reaction ^{2}H(^{7}Be, α^4He)p (Q=16.765 MeV), by using a 20.4 MeV ^{7}Be beam impinging on a 400 μg/cm^2 thick CD_2 target. By referring to the polar diagram of Fig.2 (left panel), deuteron has been used as TH-nucleus, undergoing its break-up in neutron (participant) and proton (spectator). Additionally, only quasi-free (QF) events are the ones of interest for THM analysis, corresponding to the kinematical conditions for which, in the exit channel, proton maintains the same momentum distribution it had inside the deuteron before the occurrence of break-up [5, 6]. For the purposes of the experiment, only the two emerging alpha particles have been detected while the kinematical quantities of the undetected proton have been reconstructed by means of momentum-energy conservation laws. The BELICOS (BEryllium and LIthium in the COSmos) experiment has been performed at the facility EXOTIC, where the production of the unstable ^{7}Be beam has been already accomplished in the past. In the present case, the ^{7}Be beam has been produced by means of a 33 MeV ^{7}Li beam interacting with a 1 bar H_2 cryogenic gas target. At the end of the beam line, an intensity of 5-8$\cdot 10^5$ pps and a purity of about 99% were measured. The adopted experimental setup for the present ^{7}Be+^{2}H experiment is sketched in Fig.2 (right panel). It is part of the EXPADES array described in [33]. The detectors have been located around the so-called QF angular pairs, i.e. the angular pairs at which alpha particles are emitted in correspondence of low-momenta of the undetected proton, thus completely covering the kinematic region at which the contribution of the QF reaction mechanism is expected to be dominant. A symmetrical configuration of the detection system has been chosen to double statistics. The alpha particles emitted in the angular range 27°±8° have been detected by means of a ΔE-E telescope made up of an ionization chamber (IC), (ΔE stage), and two 300 μm silicon detectors acting as E stage (T2

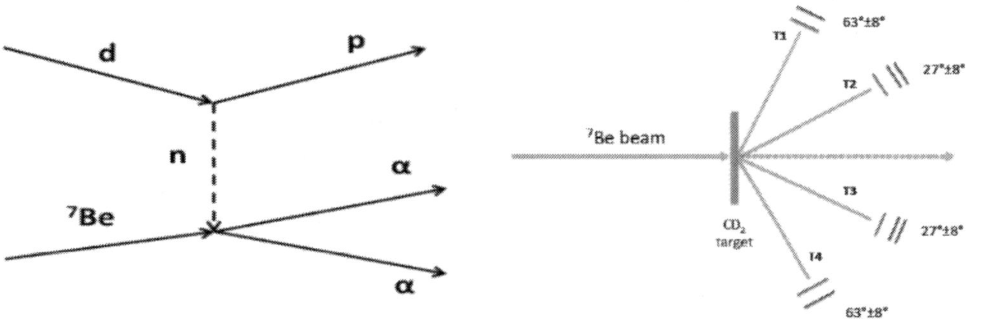

Figure 2. (Left panel). The polar diagram describing the quasi-free 7Be+d interaction. The lower pole describes the ^{7}Be-n interaction of interest.**(Right panel).** The experimental setup adopted for the BELICOS experiment.

and T3, in Fig.2). The IC have had an active depth of 61.5 mm and have been filled with 100 mbar isobutane gas. Entrance and exit windows were made up by 1.5 μm thick mylar foils with an effective area of 60x60 mm^2 to match the E silicon-stage. The further stage of the analysis foresees the selection of the events corresponding to the three-body reaction channel ^{2}H(^{7}Be, α^4He)p. The coincidence events between T2-T4 and T1-T3 have been studied. By using the standard ΔE-E technique to select the Z=2 loci in the telescopes, the alpha-alpha events of interest have been reconstructed once the energy loss in the CD_2 target as well as in the IC has been properly evaluated. To assess the proper selection of the exit channel the experimental Q-value spectrum has been deduced for the selected events (Fig.3, left panel). The experimental peak is centered at an energy of about 16.76 MeV, in agreement with the theoretical one of 16.765 MeV. A Gaussian fit of such a peak leads to a FWHM of about 2 MeV, reflecting the experimental FWHM of the ^{7}Be beam (FWHM$\approx$1 MeV), energy loss effects in the CD_2 target ($\approx$0.7 MeV) and angular resolution ($\pm$0.4°) effects. In order to select the QF-reaction mechanism, on which the full THM data analysis is based, the trend of the momentum distribution for the p-n intercluster motion inside deuteron has been studied and the preliminary result is shown in the right panel of Fig.3 (black dots). The red-curve of Fig.3 represents the theorectial momentum distribution for the p-n intercluster motion usually described in terms of the squared modulus of the Hulthén wave function in mometum space. The agreement between our data and the theoretical distribution is a strong evidence of the presence of the QF-reaction mechanism thus allowing us to further proceed in the extraction of the ^{7}Be(n,α)^{4}He cross section. The data analysis is still on going.

References

[1] Rolfs, C. & Rodney W. 1988, *Cauldrons in the Cosmos* (The Univ. of Chicago, Chicago)
[2] Spitaleri C., Bertulani C.A., Fortunato L. et al. 2016, Phys. Lett. B, **755**, 275-278
[3] G. Baur 1986, Phys. Lett., **178**, 135
[4] Spitaleri C. et al., Problems of Fundamental Modern Physics, edited by R. Cherubini, P. Dalpiaz, and B. Minetti (World Scientific, Singapore, 1991), p. 211.
[5] Tribble R.E., Bertulani C.A., La Cognata M. et al. 2014, Rep. Prog. Phys., **77**, 106901
[6] Spitaleri C., La Cognata M., Lamia L. et al. 2016, Eur. Phys. J. A, **52**, 77
[7] Pizzone R.G., R. Spartà, Bertulani C.A. et al. 2014, The Astrophys. Journ., **786**, 112

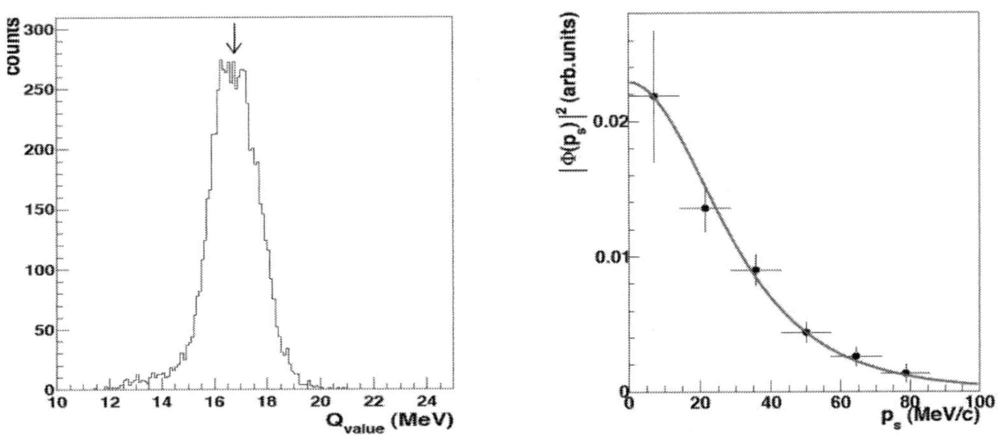

Figure 3. (Left panel). The experimental Q-value spectrum for the selected $^2H(^7Be,\alpha^4He)p$ events. The arrow marks the position of the theoretical value. No other competing channels are detectable. **(Right panel).** The preliminary experimental momentum distribution (black dots) compared with the theoretical one (red line) give in terms of the modulus square of the Hulthén wave function in momentum space.

[8] Tumino A., Spitaleri C., Mukhamedzhanov A. et al. 2011, Phys.Lett. B, **705**, 546
[9] Pizzone R.G., Spitaleri C., Bertulani C.A. et al. 2013, Phys. Rev. C, **87**, 025805
[10] Lamia L., Spitaleri C., Tognelli E. et al. 2015, The Astrophys. Journ., **811**, 99
[11] Chengbo Li, Qungang Wen, Yuanyong Fu et al. 2015, Phys. Rev. C, **92**, 025805
[12] Grineviciute J., Lamia L., Mukhamedzhanov A. et al. 2015, Phys. Rev. C, **91**, 014601
[13] Sergi M.L., Spitaleri C., La Cognata M. et al. 2015, Phys.Rev.C, **91**, 065803
[14] Palmerini S., Sergi M.L., La Cognata M. et al. 2013, The Astrophysical Journal, **764**, 128
[15] La Cognata M., Spitaleri C., Trippella O. et al. 2013, The Astr. Phys. Journ., **777**, 143
[16] Cherubini S., Gulino M., Spitaleri C. et al. 2015, Phys., Rev. C, **92**, 015805
[17] Pizzone R.G., Roederer B., McCleskey M. et al. 2016, EpJ A, **52**, 24
[18] Gulino M., Spitaleri C., Tang X.D. et al. 2013, Phys. Rev. C, **87**, 012801
[19] Guardo G.L., Spitaleri C., Lamia L. et al., 2017, Phys. Rev. C, **95**, 025807
[20] Zadro M., Miljanic D., Spitaleri C. et al. 1989, Phys. Rev. C, **40**, 181
[21] Calvi G., Lattuada M., Miljanic D. et al., 1990, Phys. Rev. C, **41**, 1848
[22] Spitaleri C., Lamia L., Tumino A. et al. 2004, Phys.Rev.C, **69** 055806
[23] Lamia L., Spitaleri C., Carlin N. et al. 2012, J. Phys. G, **39**, 015106
[24] Spitaleri C., Lamia L., Puglia S.M.R. et al. 2014 *Phys.Rev.C* **90** 035801
[25] Spitaleri C., Puglia S.M.R., La Cognata M. et al. 2017 *Phys.Rev.C* **95** 035801
[26] Angulo, C. et al. 1999, Nucl. Phys. A, **656**, 3
[27] Lamia L., La Cognata M., Spitaleri C. et al. 2012, Phys. Rev. C, **85**, 025805
[28] Broggini C., Canton L., Fiorentini G. et al. 2012, J. Cosmol. Astropart. Phys., **6**, 30
[29] Barbagallo M., Musumarra A., Cosentino L. et al. 2016, Phys. Rev. Lett., **117**, 152701
[30] Kawabata T., Fujikawa Y., Furuno T. et al. Phys. Rev. Lett. 2017, **118**, 052701
[31] Maidikov V.Z. et al. 2004, Nucl. Phys. A, 746, 389c

[32] Farinon F., Glodariu T., Mazzocco M. et al. 2008, Nucl. Instr. and Meth. B, **266**, 4097
[33] Pierroutsakou D., Boiano A., Boiano C. et al. 2016, Nucl. Instr. and Meth. A, **834**, 46

EPJ Web of Conferences **165**, 01033 (2017) DOI: 10.1051/epjconf/201716501033
NPA8 2017

Constraining astrophysical reaction rates: using the storage rings at FAIR/GSI

Christoph Langer[1,⋆], *Jan* Glorius[2], *Zuzana* Slavkovská[1], *Sergey* Litvinov[2], *Yuri A.* Litvinov[2], and *René* Reifarth[1]

[1]*Goethe University Frankfurt, Max-von-Laue Strasse 1, Frankfurt a. M., Germany*
[2]*GSI Darmstadt, Planckstrasse 1, 64291 Darmstadt, Germany*

Abstract. Ion optical calculations for a storage ring at the present GSI facility for direct proton-induced reactions relevant for different astrophysical processes are presented. As an example case, the ^{59}Cu(p,γ) and ^{59}Cu(p,α) reactions are shown. The branching of these two reactions is important in X-ray burst scenarios, since it determines the breakout out of the major ^{56}Ni waiting point.

1 Introduction

One of the biggest challenges in Nuclear Astrophysics is the extrapolation of charged-particle induced cross sections, typically measured at high energies, to the astrophysically relevant low energies. This process relies heavily on predictions from theory and often results in large uncertainties and even conflicting statements. As such, direct measurements at the low stellar energies would be ideal.

However, measurements of charged-particle induced reactions, like (p,γ), (α,γ), (α,p), typically suffer from extremely low cross sections at the interesting stellar energies. One way to overcome this problem experimentally is to increase the number of beam particles. This is of course limited by each facility and especially for radioactive ions, it is mostly impossible.

If radioactive ions are involved in the reaction, almost always experiments in inverse kinematics need to be performed. However, the production rates of radioactive ions at currently operational radioactive beam facilities are often still too low to perform direct experiments in inverse kinematics with the standard geometry of a fixed target and the beam impinging on it once (*single pass experiment*). This approach has its own advantages; however, the unreacted beam after the target is typically lost for further reactions.

In this respect, astrophysical reaction measurements employing storage rings (*multi pass*) enable unprecedented studies of reactions, which are not accessible so far. High luminosities on the order of 10^{25} cm^{-2} s^{-1} can be achieved at the relevant astrophysical energies (see Fig. 1, in which different Gamow window ranges for some astrophysical processes are shown). This leads to acceptable rates even for relatively low cross sections. These luminosities are achieved by revolving the unreacted beam at a high frequency in a storage ring while constantly readjusting the momentum spread of the beam (*e.g.* induced while traversing the target) by using an electron cooler. This device efficiently cools the beam to a given energy spread (which can be very well fine-tuned).

⋆e-mail: langer@iap.uni-frankfurt.de

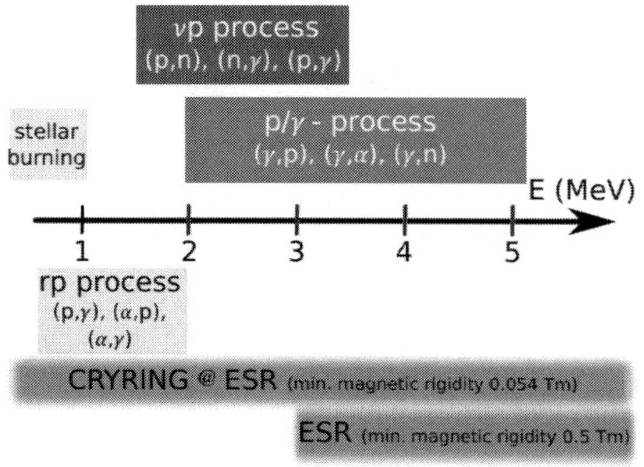

Figure 1. Typical Gamow window energy ranges for (p, γ) reactions for different astrophysical processes. Other reactions involved in the particular process are also shown. The expected experimental coverage of the future CRYRING@ESR [7] and the existing ESR storage ring at GSI [1] can be seen.

Following the reaction in a gas-jet target, the product has a different magnetic rigidity than the beam according to $B\rho \sim A/Z$. As a consequence, the trajectories in a dipole are differing. A standard position-sensitive particle detector, like a Si detector (DSSSD), is then inserted to detect the products in the dipole without intercepting the circulating unreacted beam. Fig. 2 shows the kinematical relationship between the laboratory energy (typically denoted in MeV/A) and the center-of-mass system, corresponding to the Gamow window. As can be seen: α-induced reactions require much lower beam energies, which inherently complicates the experiment.

The success of this method using the existing ESR storage ring at GSI [1] has been shown in [2] and in a recent experiment employing the ^{124}Xe(p, γ) reaction [3]. Both reactions belong to the astrophysical γ-process and the measurements allow to constrain input parameters for Hauser-Feshbach rate calculations.

In the following chapter, ion-optics calculations are presented for a typical reaction important for Nuclear Astrophysics showing possibilities and current limitations.

2 Kinematics and ion-optics calculations

The rapid proton-capture (rp) process drives the energy generation of type I X-ray bursts [4]. An important branching point is identified at the radioactive ^{59}Cu isotope. It has been shown recently, that the (p, γ) - (p, α) branching on ^{59}Cu determines the breakout out of the important ^{56}Ni waiting point [5, 6], see left part of Fig. 3. None of the reactions involved (*i.e.* ^{59}Cu(p, γ) and ^{59}Cu(p, α)) have been measured so far, especially not in the astrophysically relevant energy region.

This is an interesting candidate reaction to be studied using the storage rings at FAIR/GSI. The isotope ^{59}Cu can be sufficiently produced using the fragmentation method at GSI. Its half-life is long enough to be stored easily for some time in the ring for preparation of the beam and carrying out the measurement. A dense liquid hydrogen target ($\approx 10^{14}$ particles/cm^2) is used for the reaction surrounded by X-ray detectors for normalization of the cross section [2]. A standard double-sided silicon strip detector is placed into the ultra high vacuum of the storage ring in the dipole for detection of the reaction residuals, which move on a different trajectory than the unreacted beam. In a recent experiment, the feasibility of this method has been shown successfully [3].

The maximum deflection angle of the ^{60}Zn product after the ^{59}Cu(p, γ) reaction is 0.1 degrees, as can be seen in the right part of Fig. 3. In the case of ^{56}Ni following the ^{59}Cu(p, α) reaction, the maximum angle is 3.7 degrees. Typically, in the storage rings at GSI, the particles drift behind the target for several meters before they enter into the first quadrupoles. The large difference in

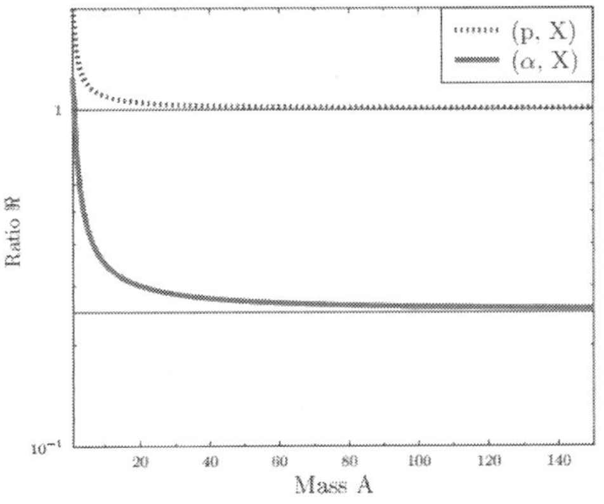

Figure 2. The kinematical ratio $\Re$ defined here as E^*_{lab}/E_{CM}, in which E^*_{lab} is given in the typical units of MeV/A used for the beam energy. If *e.g.* a (p,γ) reaction with $A = 59$ needs to be studied in inverse kinematics and the Gamow window centers at 4 MeV, a beam energy of $\Re(A = 59) \times 4 = 4.1$ MeV/A is required. On the other hand, for the same mass but an (α,γ) reaction, a beam energy of 1.3 MeV/A is required (since $\Re \approx 0.3$), which is experimentally much more challenging.

the laboratory scattering angle between the two reactions results in significant differences in the x-y particle distribution just before the first quadrupoles, as can be seen in the left part of Fig. 4. Also indicated is the beam pipe aperture, which significantly cuts into the acceptance actually required for the (p,α) reaction case. It is subsequently clear, that a measurement of a (p,α) reaction is much more challenging and requires a different approach compared to the (p,γ) case. For an induced (p,γ) reaction, the heavy residual nucleus moves almost on the same trajectory as the unreacted beam before the dipole. Because of the different A/Z-ratio, the magnetic rigidity is different for the unreacted beam and the residual. Consequently, the trajectories inside and after the dipole are differing and this allows to disentangle both: whereas the unreacted beam is recycled in the ring, the residual is detected with almost 100% detection efficiency in the DSSSD.

As can be concluded from Fig. 4, ^{59}Cu(p,γ) can be measured without any losses due to acceptance. A DSSSD with standard size of (5 x 5) cm^2 covers the complete distribution of the (p,γ) reaction residuals in the dipole at a distance to the unreacted revolving beam of $\approx$8 cm. The ^{59}Cu(p,α) residuals will be cut almost completely by the acceptance of the ESR and, thus, are not measurable using

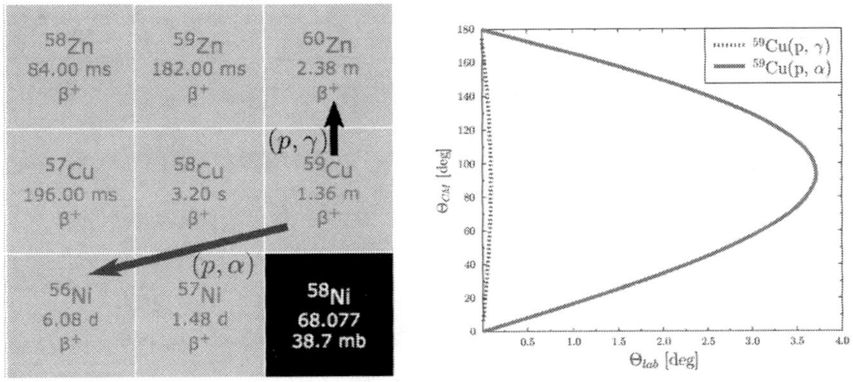

Figure 3. Left: Part of the rp process around the important ^{56}Ni waiting point. The ^{59}Cu(p,γ) reaction leads to ^{60}Zn, and, as such, successfully breaking out of the ^{56}Ni waiting point. The competing reaction ^{59}Cu(p,α), instead feeds back into ^{56}Ni. Right: Transformation of the center-of-mass scattering angle of the heavy residuals into the laboratory azimuthal angle Θ_{lab} for the two example reactions considered here. The CM energy of both reactions is set to 4 MeV to be still measurable in the ESR (see Fig. 2).

 DOI: 10.1051/epjconf/201716501033

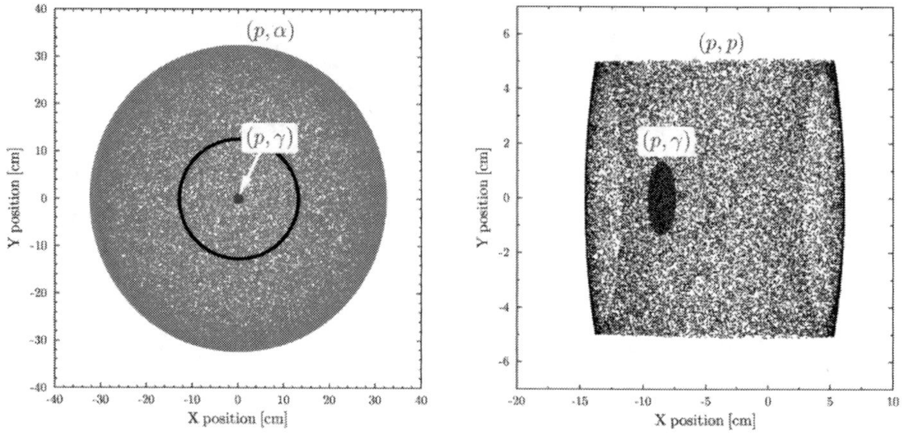

Figure 4. Left: The x-y distribution of heavy residuals following the ^{59}Cu(p,α) and ^{59}Cu(p,γ) reaction before entering into the first quadrupole in the ESR after the target (drift path is roughly 5 meters). MOCADI has been used to simulate the beam transport [8]. The black circle indicates the typical ESR beam line aperture. Right: The x-y distribution in the dipole (at 53.3 degrees, which is the position used in [3]) of the ^{59}Cu(p,γ) residuals. They can be sufficiently measured with a (5 x 5) cm^2 DSSSD detector. In black, also the elastic scattering contribution from ^{59}Cu(p,p) is shown. The MOCADI simulations include realistic beam apertures; for the reaction, no internal decay patterns of the single residuals are incorporated.

this technique and measurement position. According to Rutherford scattering, at low center-of-mass energies the elastic scattering contribution is expected to be high. The distribution of the (p,p) events can also be seen in the right part of Fig. 4. However, as has been shown before, depending on the cross section of the ^{59}Cu(p,γ) reaction, it should be possible to disentangle both distributions [2, 3]. Another idea could be to block the elastic scattering contribution already before entering into the dipole, since its spread is much wider than in the (p,γ) case.

References

[1] B. Franzke, Nuclear Instruments and Methods in Physics Research Section B: Beam Interactions with Materials and Atoms **24**, 18 (1987)

[2] B. Mei, T. Aumann, S. Bishop, K. Blaum, K. Boretzky, F. Bosch, C. Brandau, H. Bräuning, T. Davinson, I. Dillmann et al., Phys. Rev. C **92**, 035803 (2015)

[3] Experiment E108b at GSI in 2016

[4] H. Schatz, A. Aprahamian, J. Görres, M. Wiescher, T. Rauscher, J. Rembges, F.K. Thielemann, B. Pfeiffer, P. Möller, K.L. Kratz et al., Physics Reports **294**, 167 (1998)

[5] R.H. Cyburt, A.M. Amthor, A. Heger, E. Johnson, L. Keek, Z. Meisel, H. Schatz, K. Smith, The Astrophysical Journal **830**, 55 (2016)

[6] C. Langer, N. Klapper, C. Köppchen, S. Dababneh, R. Reifarth, submitted to Jordanian Journal of Physics (2017)

[7] M. Lestinsky, V. Andrianov, B. Aurand, V. Bagnoud, D. Bernhardt, H. Beyer, S. Bishop, K. Blaum, A. Bleile, A. Borovik et al., The European Physical Journal Special Topics **225**, 797 (2016)

[8] H. Weick et al., https://web-docs.gsi.de/~weick/mocadi/index.html, 2017

EPJ Web of Conferences **165**, 01034 (2017) DOI: 10.1051/epjconf/201716501034
NPA8 2017

A fast and complete GEANT4 and ROOT Object-Oriented Toolkit: GROOT

D. Lattuada[1,*], *D. L.*Balabanski[1], *S.* Chesnevskaya[1], *M.* Costa[2], *V.* Crucillà[2], *G. L.* Guardo[1], *M.* La Cognata[2], *C.* Matei[1], *R. G.* Pizzone[2], *S.* Romano[2,3], *C.* Spitaleri[2], *A.* Tumino[2,4], and *Y.* Xu[1]

[1] *ELI-NP / IFIN-HH, Magurele, Romania*
[2] *INFN, Laboratori Nazionali del Sud, Catania, Italy*
[3] *Dipartimento di Fisica e Astronomia, Università degli Studi, Catania, Italy*
[4] *Facoltà di Ingegneria e Architettura, Università Kore, Enna, Italy*

Abstract. Present and future gamma-beam facilities represent a great opportunity to validate and evaluate the cross-sections of many photonuclear reactions at near-threshold energies. Monte Carlo (MC) simulations are very important to evaluate the reaction rates and to maximize the detection efficiency but, unfortunately, they can be very cpu-time-consuming and in some cases very hard to reproduce, especially when exploring near-threshold cross-section. We developed a software that makes use of the validated tracking GEANT4 libraries and the n-body event generator of ROOT in order to provide a fast, realiable and complete MC tool to be used for nuclear physics experiments. This tool is indeed intended to be used for photonuclear reactions at γ-beam facilities with ELISSA (ELI Silicon Strip Array), a new detector array under development at the Extreme Light Infrastructure - Nuclear Physics (ELI-NP). We discuss the results of MC simulations performed to evaluate the effects of the electromagnetic induced background, of the straggling due to the target thickness and of the resolution of the silicon detectors.

1 Introduction

Nuclear reactions near threshold energies are of great interest in nuclear astrophysics. In the last few decades, direct and indirect methods have been shedding light on the uncertainties on nuclear cross-sections at very low energies[1]. Nuclear physics experiments have a long history record and Monte Carlo (MC) simulations have always played a crucial role in their preparation, in data analysis and for proper understanding the experimental results. At ELI-NP[2] in Bucharest, together with the Laboratori Nazionali del Sud (LNS) of INFN in Catania, we are developing a MC tool to be used in nuclear physics experiments focused on complex detectors and nuclear reactions with low cross-sections, using open-source scientific libraries with a strong record of validation. The code was developed for being used in nuclear experiments with silicon detectors involving photonuclear reactions induced by the intense gamma beam produced by the Gamma Beam System (GBS) at ELI-NP. The idea of this toolkit initially came from the current lack of validated photonuclear models working at low energies in GEANT4[3] and the need of a GEANT4-based code to provide detailed information on electromagnetic (and hadronic) background in nuclear physics experiments. The open-source C++ GROOT

*e-mail: dario.lattuada@eli-np.ro

 DOI: 10.1051/epjconf/201716501034

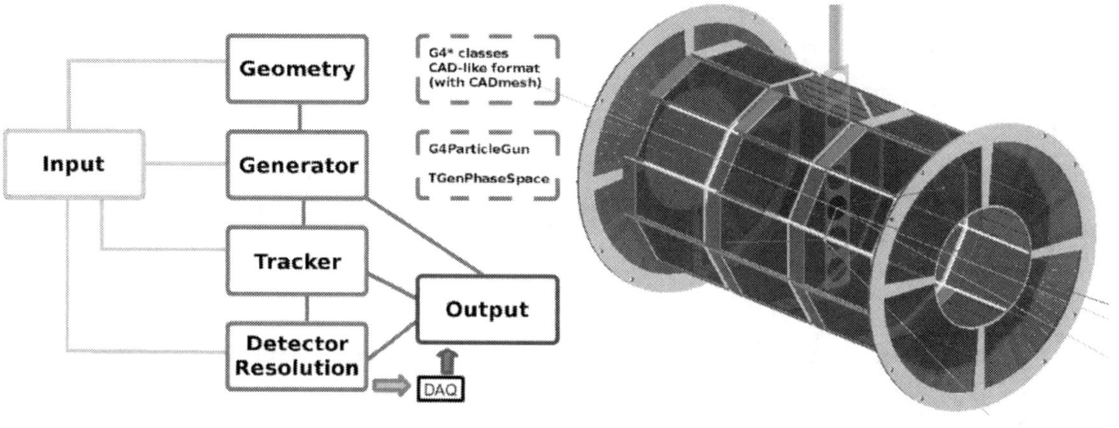

Figure 1. *Left*: A scheme of the MC code. *Right*: The geometry of ELISSA imported with CADMesh in GROOT (GEANT4 visualization interface). Gamma (*green*) and electron (*red*) tracks are shown.

(GEANT4 and ROOT Object-Oriented Toolkit) code we are currenlty developing mainly relies on ROOT[4], GEANT4 (10.02.p01) and the Standard Template Library (STL) of C++. Also, for the geometry we implemented the possibility to use CAD-like format as input, by means of CADMesh[5], a CAD Interface for GEANT4. The goal is to provide the users with a complete, realiable, fast and user-friendly framework to be used in the preparation, running and analysis phases of any nuclear physics experiments.

2 The Code

Since GROOT relies on GEANT4 for the most, it also follows its basic concepts and philosophy. The structure of the GROOT code (shown in Fig.1) follows a typical GEANT4 example. The program has both a classic event generator based on GEANT4 libraries and an external event generator (based on CERN ROOT libraries). Since photonuclear reactions near threshold are not properly handled by GEANT4 to date [6], the ROOT classes *TGenPhaseSpace* and *TLorentzVector* are used to provide a reliable and fast algorythm generating the ejectiles, starting from the input phase-space distribution. The geometry of the detector can be defined as a standard GEANT4 *DetectorConstruction* file by using G4 classes or imported in *stl* format by means of an external tool and its classes. The tracking of particles entirely relies on GEANT4 and its classes (default is *G4EmStandardPhysics_option3*). A (temporarily external) simple program is used to simulate the detector's angular and energy resolution. This will soon be incorporated into the main program. The output can be written to disk in ASCII, in ROOT format or both, with different levels of verbosity.

3 Results

The tool has been and will be used for past and future experiments. We present here some preliminary results of experiments with nuclear reactions of astrophysical interest with nearly monochromatic gamma beams with different profiles. We performed an experiment on ^{7}Li(γ,t)^{4}He at 5-10MeV energies at HIγS (Duke University) in 2017 and this is a good candidate for zero-day experiment at ELI-NP. Also, the ^{24}Mg(γ,α)^{20}Ne reaction is an important reaction for the silicon burning in pre-supernova

DOI: 10.1051/epjconf/201716501034

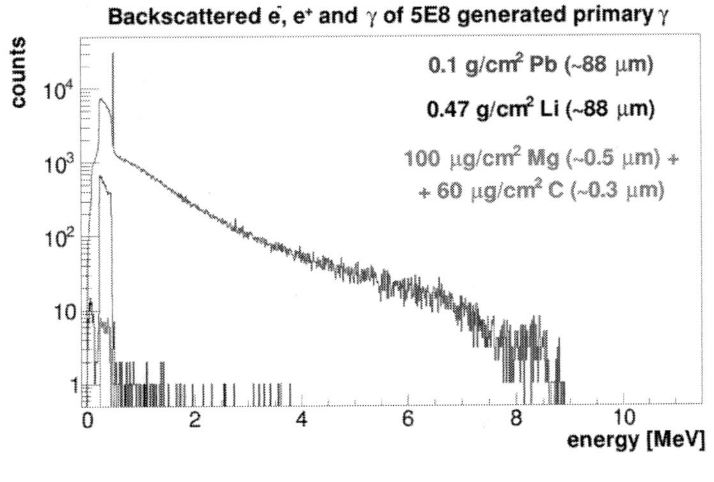

Figure 2. Backscattered electrons, positrons and gamma off a 11.4 MeV gamma beam wit ELI-NP GBS-like profile on different targets.

phase. Working with electromagnetic beams and silicon detectors the induced electromagnetic background can be an issue. We show the results of MC simulations with GROOT in Figure 2. Here we show the simulated expected background on ELISSA[7] due to a 11.4 MeV gamma beam with ELI-NP profile (2.5mm beam spot and 0.5% bandwith, $5e^8$ events are generated, corresponding just to few seconds of beam time) impinging on targets with different thickness and composition. Unsurprisingly enough, the rate of e.m. background (mainly made of e^-, but also e^+ and γ) strongly depends on the target thickness and the atomic number of the element. It is clear that it is preferable to work with light elements and thin targets to avoid high rates that could negatively affect the threshold selection and damage your detectors eventually. For a realistic 250μg/cm^2 ^{24}Mg target with a thin carbon backing, the expected e.m. background on ELISSA with the proposed GBS of ELI-NP is very low (10-100Hz, mostly at energies below the threshold). This does not fully apply for a thick target of lithium since the Compton scattering is significantly higher. Finally, the background is definitely huge for a thick lead target, as expected. Both the Compton scattering and the e^-e^+ couple creation are important and they could represent an obstacle for proper particle detection, threshold selection, detector damage and also for the dead time.

We also show some simulation results for the HIγS experiment, where a 5-10MeV gamma beam with impinged on a lithium target to study the photodissociation of ^{7}Li(γ,t)^{4}He. In Figure 3 the energies and angles of tritons and alphas reaching the detector after losing energy inside the target are plotted on top (ELISSA setup is shown in Figure 1 for reference, only the X3 "rings" are considered). On bottom the same is plotted after applying a Gaussian smearing with an energy-dependent sigma in order to simulate the detector energy resolution, as reported from Micron and calculated in [8, 9]. The first proposed target was a 250μg/cm^2 lithium hydride while the second one was a slightly thinner gold-coated pure lithium target. The difference on the energy loss in the targets is evident. The choice of the target is thus strongly related to the energy spectra of the ejectiles. Estabilishing wether one is preferable over the other depends on the interplay between many factors, such as the precision of the measurement, the electronic noise and, last but not least, the target feasibility and robustness. Also, it should be noted that it might be usefeul to rotate the target in order to exploit the whole detector instead of dealing with a "blind" ring in the central ring at 90° with respect to the beam direction.

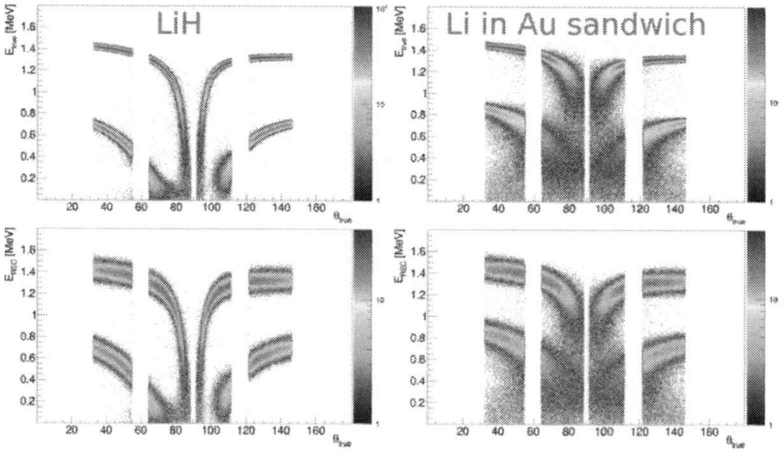

Figure 3. *Top*: For both LiH (*left*) and gold-coated lithium (*right*) targets, the energy of the simulated alpha particles from $^7Li(\gamma,t)$ reaction reaching the detector is plotted versus their angle of arrival. *Bottom*: the same but including the energy resolution effect.

In conclusion, even though a proper validation is going to be performed very soon, the reasonable results we show here are realiable since they are obtained with the use of widely validated libraries and classes from GEANT4 and ROOT. The code, still under development at ELI-NP and LNS, is applying to become the reference tool to be used for the photonuclear physics with silicon detectors at ELI-NP.

We acknowledge the support from the Extreme Light Infrastructure Nuclear Physics (ELI-NP) Phase II, a project cofinanced by the Romanian Government and the European Union through the European Regional Development Fund - the Competitiveness Operational Programme (1/07.07.2016, COP, ID 1334).

References

[1] O. Tesileanu et al., 2016 Charged particle detection at ELI-NP, Rom. Rep. Phys. 68 S699

[2] D. Filipescu et al., Eur. Phys. J., A51, 185 (2015)

[3] GEANT4: A Simulation toolkit - GEANT4 Collaboration (Agostinelli, S. et al.), Nucl. Inst. & Meth. A506 (2003) 250-303 SLAC-PUB-9350, FERMILAB-PUB-03-339.

[4] ROOT - An Object Oriented Data Analysis Framework - Rene Brun and Fons Rademakers, Proceedings AIHENP'96 Workshop, Lausanne, Sep. 1996, Nucl. Inst. & Meth. in Phys. Res. A 389 (1997) 81-86. See also http://root.cern.ch/.

[5] A CAD Interface for GEANT4, Poole, C. M. and Cornelius, I. and Trapp, J. V. and Langton, C. M., Australasian Physical & Engineering Science in Medicine, September 2012,

[6] http://GEANT4.cern.ch/support/proc_mod_catalog/processes/hadronic/G4PhotoNuclearProcess.html

[7] M. La Cognata et al., 2017 Journal of Instrumentation 12 C03079

[8] G. L. Guardo et al., this volume.

[9] S. Chesnevskaya et al., this volume.

EPJ Web of Conferences **165**, 01035 (2017)
NPA8 2017
DOI: 10.1051/epjconf/201716501035

Catalysis of Nuclear Reactions by Electrons

Matej Lipoglavšek*

Jožef Stefan Institute, Jamova cesta 39, 1000 Ljubljana, Slovenia

Abstract. Electron screening enhances nuclear reaction cross sections at low energies. We studied the nuclear reaction $^{1}H(^{19}F,\alpha\gamma)^{16}O$ in inverse kinematics in different solid hydrogen targets. Measured resonance strengths differed by up to a factor of 10 in different targets. We also studied the $^{2}H(p,\gamma)^{3}He$ fusion reaction and observed electrons emitted as reaction products instead of γ rays. In this case electron screening greatly enhances internal conversion probability.

1 Introduction

The screening of nuclear charges by electrons enhances nuclear reaction cross sections at low energies [1]. The effect is important in astrophysical scenarios, where almost all nuclear reactions happen at low energies and nuclei are not alone in the stellar plasma. The predicted effect was confirmed experimentally, but its measured magnitude was very often way above predictions [2-11]. Although the theory predicts an independence of electron screening on the nuclear surroundings, measurements of the same reactions in different environments gave different magnitudes of electron screening. This called for a new look at the electron screening effect. Our group proposed a new model of electron screening based on the hydrogen molecular ion, where the electron resides primarily between the two protons [11]. We also predicted a dependence of electron screening on the position of hydrogen in the crystal lattice. To test our model we observed 5.6 MeV electrons in the fusion reaction of protons and deuterons [12]. This proves that atomic electrons actively participate in the nuclear reaction and not just merely decrease the Coulomb barrier from an atomic shell.

2 Experiments

To measure the largest electron screening effect we perform nuclear reaction experiments in inverse kinematics. Beams of hydrogen to fluorine ions are accelerated by the 2 MV Tandetron accelerator at Jožef Stefan Institute. We try to use the same hydrogen targets for all experiments. This gives us a unique opportunity to compare electron screening in the same targets for different beams. The targets used are the following. For a standard hydrogen target we use polyimide (Kapton, $C_{22}H_{10}N_2O_5$) due to its relative insensitivity to radiation damage. The second target is titanium hydride pressed from TiH powder into a 1 mm deep hole in a Cu backing which helps to cool the target after heating with the beam.

* Corresponding author: matej.lipoglavsek@ijs.si

The third sample is a 250 μm thick graphite foil obtained from Chempur. The fourth target are 250 μm thick palladium foils from Goodfellow, loaded with hydrogen from gas phase. The final target is a tungsten sample, produced by Plansee. The graphite and tungsten samples were implanted with 5 keV hydrogen ions from the ion gun with beam intensity of 0.8 mA/cm^2. In the fluorine experiment we also used an amorphous carbon target produced by plasma discharge in methane gas. This target quickly loses hydrogen upon ion bombardment and was unavailable for other experiments. All targets represent thick hydrogen targets. The Kapton, graphite, amorphous carbon, Pd and TiH targets have uniform hydrogen distribution, while the W target exhibited a surface hydrogen peak and a uniform distribution in the bulk (see below). The beam dose is deduced from the measurement of electric charge deposited inside an electrically insulated target chamber. During the experiment the hydrogen concentration in the sample is controlled by repeatedly measuring at the same beam energy after each measurement at a different energy [11]. The results from the experiment with the fluorine beam are described below. Gamma rays were measured with a Ge detector placed at 135° with respect to the beam direction at a distance of 5.7 cm from the target.

3 Results

The nuclear reaction $^{1}H(^{19}F,\alpha\gamma)^{16}O$ was studied by measuring the yields of the 6129 keV oxygen γ ray around the c.m.s. energy E_r=323.4(1) keV resonance [13]. The measured yields are shown in Figs. 1 and 2.

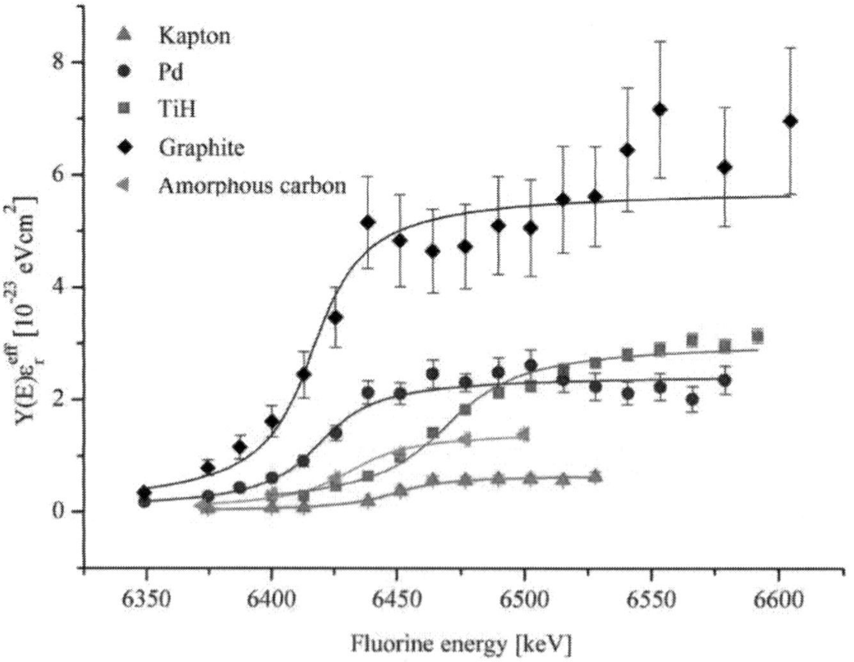

Fig. 1. Gamma-ray yields for the $^{1}H(^{19}F,\alpha\gamma)^{16}O$ reaction multiplied by the effective stopping power as a function of fluorine beam energy for amorphous carbon, graphite, TiH, Pd and Kapton targets near the resonance energy of 323 keV in c.m.s. (or 6.469 MeV in laboratory system) [14]. The solid lines represent fits with eq. 1. The error bars are roughly proportional to the amount of hydrogen in the target.

The data for Kapton, graphite, amorphous carbon, Pd and TiH targets were fitted with a function describing thick target resonance yields Y [14,15]

$$Y = \frac{\lambda^2 \cdot \omega\gamma}{2\pi \cdot \varepsilon_r}\left[\arctan\left(\frac{E-E_r}{\Gamma/2}\right)+\frac{\pi}{2}\right], \quad (1)$$

where λ is the de Broglie wavelength of the beam, ε_r its effective stopping power in the target calculated with the SRIM code [16] and Γ the resonance width. From the fits we deduced resonance strengths ωγ and for the first time observed a dependence of resonance strength on the type of hydrogen target. Moreover, the deduced differences are not small, the resonance strengths in graphite and Kapton differ by about a factor of 10.

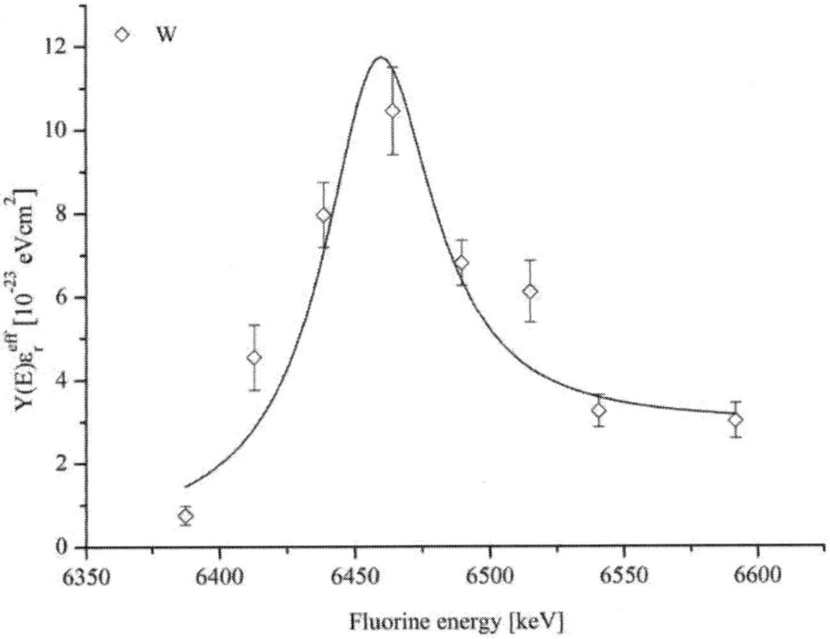

Fig. 2. Gamma-ray yields for the $^1H(^{19}F,\alpha\gamma)^{16}O$ reaction multiplied by the effective stopping power as a function of fluorine beam energy for the tungsten target near the resonance energy of 323 keV in c.m.s. [14]. The solid line represents a fit with a combination of thin and thick target yields.

The data for the W target were fitted with a combination of thin and thick target yield curve with an assumption that the resonance strengths are the same on the surface and in the bulk. The surface peak is not due to impurities on the surface, but rather due to a larger availability of hydrogen trapping sites on the surface than in the bulk. Some fitted parameters are listed in Table 1 for each target. In addition, Table 1 lists the difference between the measured resonance energy in the normal kinematics reaction [13] and our inverse kinematics measurements. Electron screening potentials were deduced from enhancement factor

$$f(E) = \frac{\sigma_s}{\sigma_b} = \frac{\omega\gamma_s}{\omega\gamma_b} = \frac{e^{-2\pi\eta(E+Ue)}}{e^{-2\pi\eta(E)}}, \quad (2)$$

where σ_s and σ_b represent the screened and bare cross sections. Similarly $\omega\gamma_s$ and $\omega\gamma_b$ are screened and bare resonance strengths. The Sommerfeld parameter $\eta = Z_1 Z_2 e^2/(4\pi\varepsilon_0\hbar c)\cdot(\mu c^2/2E)^{1/2}$. Here Z_1 and Z_2 are the charge numbers of the interacting nuclei and μ their reduced mass. The electron screening potentials U_e were deduced by

DOI: 10.1051/epjconf/201716501035

assuming that $\omega\gamma_b$ is the resonance strength measured for Kapton. This means that our results represent only lower limits on $\omega\gamma_s$.

Table 1. Electron screening potentials U_e, resonance energies E_r and resonance energy differences Δ for targets studied in the fluorine reaction.

Target	U_e [keV]	E_r [keV]	Δ [keV]
Kapton	/	322.4±0.2	1.0±0.2
Amorphous Carbon	36±6	321.6±0.2	1.8±0.2
TiH	73±6	323.4±0.1	0.0±0.1
Pd	63±6	321.0±0.4	2.4±0.4
W	74±15	322.9±0.3	0.5±0.3
Graphite	115±8	320.8±0.7	2.6±0.7

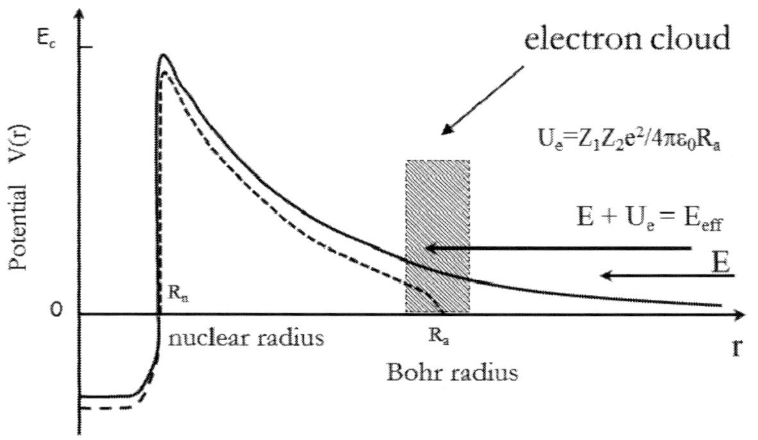

Fig. 3. A schematic picture of static electron screening with electrons on a shell at Bohr radius. A change of nuclear binding energy is indicated.

First of all one notices that the resonance energy shifts have no connection to the measured electron screening potentials, meaning that the two phenomena have a different origin, despite the fact that they have been previously treated together [17]. Secondly, all resonance shifts are towards lower beam energies compared to the normal kinematics reaction. If the resonance shifts were due to impurities on target surfaces, one would expect shifts to higher energies. This means that the shifts are most likely real and not due to some experimental difficulty. All resonance shifts are compatible with the U_e=2.19 keV maximum prediction for the electron screening potential in the adiabatic limit. It is, therefore, possible that atomic electrons lower the binding energy of the proton captured in ^{20}Ne. This can be inferred from the static electron screening picture of ref. [1], since the Coulomb potential of the electrons has to extend inside the nucleus (see Fig. 3). To confirm this explanation one would need to calculate the equilibrium electron density at the place of the proton in the solid target, where the reaction occurs. However, this is not straight forward, since proton positions are not precisely known, especially in implanted and radiation damaged targets.

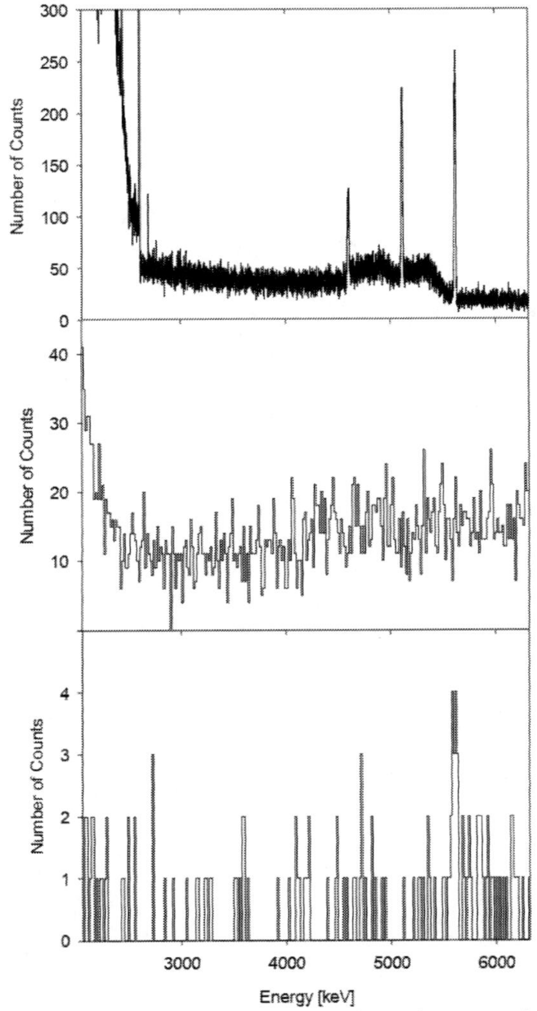

Fig. 4 Upper panel: the spectrum from the germanium detector. The three distinct peaks are the full absorption, single escape and double escape γ rays from the $^{2}H(p,\gamma)^{3}He$ reaction. Natural background γ radiation can be seen below the 2615 keV ^{208}Pb peak. Middle panel: the spectrum of the CsI(Tl) detector in the same energy range. Bottom panel: the spectrum of cumulative energy deposited in the CsI(Tl) and silicon detectors, when the two signals arrived within 40 ns of each other. The only statistically significant peak in the bottom spectrum is the one at 5.6 MeV.

4 Internal Conversion with Electron Screening

To test our new model of electron screening based on hydrogen molecular ion, we studied electron screening in the fusion reaction of protons and deuterons. Normally at high energies this reaction produces a ^{3}He nucleus and a γ ray with an energy almost equal to the reaction Q-value of 5.5 MeV. Such γ rays are seen in the top panel of Fig. 4. However, in the case of muon catalysed fusion of protons and deuterons, a muon flies away with 5.5 MeV instead of a γ ray [18]. In this case the internal conversion probability increases due to the proximity of the muon to the nucleus. The same could happen when electron screening forces an electron closer to the nucleus. The atomic electron then catalyses the nuclear reaction and is ejected with a high energy.

We performed another experiment with the 2 MV Tandetron accelerator, which produced protons with the energy of 260 keV, the minimum achievable energy. Deuterons were implanted into graphite targets in the same way as protons in the above described fluorine experiment. The detector setup consisted of a germanium detector for γ-ray detection placed 57 mm from the target at an angle of 135° with respect to the beam direction. The electrons were separated from γ rays using the ΔE-E technique. For the ΔE detector we used a 0.5 mm thick silicon detector placed 36 mm from the target at an angle of 135° with respect to the beam direction. Behind it, 53 mm from the target, we placed a 10 mm thick CsI(Tl) scintillation detector. Further experimental details are given in ref. [12]. The main experimental result is shown in Fig. 4. The upper panel of the figure shows a part of the γ-ray spectrum from the germanium detector. The three distinct peaks clearly seen are the full absorption (5626 keV), single escape (5626–511 keV) and double escape (5626–1022 keV) peaks resulting from the ^{2}H(p,γ)^{3}He reaction. The full absorption energy of 5626 keV is the sum of the reaction Q value (5493 keV) and beam energy in the center-of-mass system, together with the Doppler correction. The middle panel of Fig. 4 shows the spectrum from the CsI(Tl) detector in the same energy range. Due to the low detection efficiency, no γ ray lines can be seen in the spectrum. The only feature of this spectrum is that the continuous background is lower between about 2.5 and 4 MeV than at higher energy. This is due to the presence of high-energy cosmic muon signals. However, when we demanded simultaneous detection of signals from the CsI(Tl) and silicon detectors, a line became visible in the spectrum. This is demonstrated in the bottom panel of Fig. 4, which shows a cumulative spectrum of the sum of measured energies in the silicon and CsI(Tl) detectors. Only coincidence events where the two signals arrived within 40 ns of each other were taken into account. In addition, the energy detected in the silicon detector was required to lie between 120 and 240 keV, corresponding to the energy deposition of 5.6 MeV electrons in a 0.5 mm thick layer of silicon. The bottom spectrum also shows a peak at about 5.6 MeV. We interpreted this peak as due to conversion electrons emitted after the fusion of protons and deuterons.

References

1. H. J. Assenbaum, K. Langanke and C. Rolfs, Z. Phys. A 327 (1987) 461.
2. L. Lamia et al., Astron. Astrophys. 541 (2012) **A**158.
3. A. Huke et al.,Phys.Rev. **C** 78, (2008) 015803.
4. K. Czerski et al., J. Phys. **G** 35 (2008) 014012.
5. F. Raiola et al.,Eur.Phys.J. **A** 19 (2004) 283; F. Raiola et al., Phys Lett **B** 547 (2002) 193.
6. J. Kasagi, Prog. Theor. Phys. Suppl. 154 (2004) 365.
7. J. Cruz et al., Phys. Lett. **B** 624, (2005) 181; J. Cruz et al., J. Phys. **G** 35 (2008) 014004.
8. F. Raiola et al., J. Phys. **G** 31 (2005) 1141.
9. M. Lipoglavsek et al., Eur. Phys. J. **A** 44 (2010) 71.
10. J. Vesic et al., Eur. Phys. J. **A** 50 (2014) 153.
11. A. Cvetinovic et al., Phys. Rev. **C** 92 (2015) 065801.
12. M. Lipoglavsek et al., Phys. Lett. **B** 773 (2017) 553.
13. K. Spyrou et al., Eur. Phys. J. **A** 7 (2000) 79.
14. A. Cvetinović, PhD Thesis, University of Ljubljana (2015), unpublished.
15. C. Iliadis, Nuclear Physics of Stars, Wiley, Hoboken, NJ, USA (2007).
16. J.F. Ziegler et al., The Stopping and Range of Ions in Matter (Lulu Press Co., Morrisville, NC, 2008); www.srim.org.
17. K. Kettner et al., J. Phys. **G** 32 (2006) 489.
18. L. Alvarez et al., Phys.Rev. 105 (1957) 1127.

EPJ Web of Conferences **165**, 01036 (2017) DOI: 10.1051/epjconf/201716501036
NPA8 2017

The role of ^{13}C excited states in α+^{9}Be reaction and scattering cross sections

Ivano Lombardo[1,3,⋆], *Daniele* Dell'Aquila[1,2], and *Mariano* Vigilante[1,2]

[1] *Dip. di Fisica "E. Pancini", Univ. di Napoli Federico II, via Cintia, Napoli, Italy*
[2] *Istituto Nazionale di Fisica Nucleare, Sezione di Catania, via S. Sofia, Catania, Italy*
[3] *Istituto Nazionale di Fisica Nucleare, Sezione di Napoli, via Cintia, Napoli, Italy*

Abstract. The spectroscopy of ^{13}C excited states above the alpha emission threshold has been investigated by studying data on ^{9}Be(α,α)^{9}Be and ^{9}Be(α,n)^{12}C collisions at bombarding energies in the range E_α ≈2-10 MeV. A multichannel R-matrix fit allowed us to better constrain the spectroscopy of ^{13}C states above the α threshold (10.65 MeV). Preliminary results on this analysis will be discussed.

1 Introduction

The study of the structure of light isotopes allows us to understand the effects of clusterization in light non-self-conjugated nuclei [1–10]. The presence of rotational bands built on molecular states has been suggested in several papers [11–13]. Furthermore, in recent times, some theoretical works [12, 14] predicted the possible existence of states corresponding to the coupling of valence neutrons to the ^{12}C Hoyle state. The analysis of resonant elastic scattering with a gas target in inverse kinematics [2, 15], of α-transfer reactions with magnetic spectrometers [16, 17], and of cluster correlations with high segmentation hodoscopes [18–22], represent powerful tools to shed light on these aspects. In this framework, we performed a comprehensive R-matrix fit of α+^{9}Be elastic (α_0) and inelastic (α_1 and α_2) scattering data in the energy range E≃ 3.5 – 10 MeV at several angles [23, 24]. To carefully determine the partial decay widths of states above the α decay threshold we included in the fit procedure also ^{9}Be$(\alpha, n_0)^{12}$C$_{gs}$ and ^{9}Be$(\alpha, n_1)^{12}$C$_{4.44}$ cross section data taken from [25, 26]. With the exception of ^{9}Be$(\alpha, n_2)^{12}$C$_{7.65}$ (that is open above E_α =2.8 MeV) and radiative capture reactions, these reaction channels are the only ones open at low energies. This analysis allows us to improve the (poorly known) spectroscopy of excited states in ^{13}C in the E$_x$ ≃12-17 MeV region [27]. Furthermore, a better knowledge of high-energy resonance parameters (especially for broad states) can improve low-energy extrapolations of the ^{9}Be$(\alpha, n)^{12}$C reaction S-factor, that plays a key role in the description of ^{12}C nucleosynthesis during a supernova explosions via the triple process ^{4}He(αn,γ)^{9}Be followed by ^{9}Be$(\alpha, n)^{12}$C [25]. This reaction has also a strong importance in applied physics, being at the basis of standard neutron sources [28]. Preliminary results of these studies will be discussed.

⋆e-mail: ivano.lombardo@ct.infn.it

2 Selection of experimental data

The main part of data that we included in the *R*-matrix analysis were obtained with an experiment done at the TTT3 tandem accelerator in Napoli [29]. Details concerning the experimental configuration are reported in Ref. [23]. Our data set was complemented with low energy data of Ref. [30] and by including also neutron data from Ref. [25] and [26].

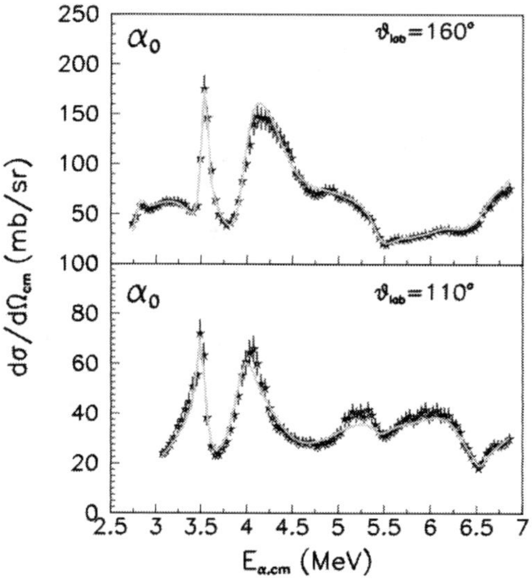

Figure 1. α+^{9}Be elastic scattering differential cross section at 160° and 110°. Open stars: experimental data. Green line: preliminary *R*-matrix best fit.

3 *R*-matrix fit of data: preliminary results

Experimental data described in Section 2 were fitted simultaneously with the *R*-matrix code AZURE2 [31, 32]. Details of the adopted procedure will be given in a forthcoming paper. We used the table of states reported in Ref. [27], complemented by more recent results of Ref. [2] as starting parameters for the states included in the *R*-matrix analysis. A few J^π values that were tentatively reported in the literature were varied to better describe all the details of excitations functions. The overall data reproduction is quite satisfactory, as reported for example, in Figure 1 for the elastic scattering differential cross section at 160° and 110°.

Some preliminary considerations can be drawn from the *R*-matrix analysis reported here. The small peak at E_{cm} ≃2.76 MeV can be attributed to a $9/2^-$ state at 13.41 MeV, in agreement with the tentative value given by Ref. [33]. In this way we can reproduce simultaneously well all the excitation functions studied here. Analogously, the main peak, seen at E_{cm} ≃3.5 MeV, is attributed to the presence of a $5/2^-$ state at 14.13 MeV, having a total width of about 94 keV and a vanishingly small neutron partial width. This J^π value is in contrast with the $9/2^-$ tentative assignment made in Ref. [12] to fulfill the E_x versus $J(J+1)$ relationship typical of rotational bands. Further and much more extended details on the present data analysis will be reported in a future work.

4 Conclusions

In this proceeding we report some preliminary results on the investigation of ^{13}C spectroscopy at excitation energies larger than the α emission threshold ($E_x > 10.648$ MeV). We performed a simultaneous R-matrix fit of various reaction channels related to α and neutron emission. Thank to this analysis, it was possible to remove some ambiguities in J^π assignments for some states that are candidates for a molecular structure configuration, e.g. for the 14.13 MeV state.

References

[1] W.V. Oertzen, M. Freer, Y. Kanada-En'yo, Phys. Rep. **432**, 43 (2006)
[2] M. Freer et al., Phys. Rev. C **84**, 034317 (2011)
[3] M. Freer et al., Phys. Rev. C **90**, 054324 (2014)
[4] I. Lombardo, D. Dell'Aquila, M. Vigilante, J. Phys. Conf. Ser. **730**, 012016 (2016)
[5] C. Spitaleri et al., Phys. Rev. **C95**, 035801 (2017)
[6] G. Cardella et al., Nucl. Instr. and Meth. A **799**, 64 (2015)
[7] D. Dell'Aquila et al., J. Phys. Conf. Ser. **876**, 012006 (2017)
[8] D. Dell'Aquila et al., Phys. Rev. Lett. **119**, 132501 (2017)
[9] R. Smith et al., Phys. Rev. Lett. **119**, 132502 (2017)
[10] I. Lombardo et al., Phys. Lett. B **748**, 178 (2015)
[11] N. Furutachi, M. Kimura, Phys. Rev. C **83**, 021303(R) (2011)
[12] M. Milin, W. von Oertzen, Eur. Phys. J. A **14**, 295 (2002)
[13] Y. Chiba, M. Kimura, J. Phys. Conf. Ser. **569**, 012047 (2014)
[14] T. Yamada, Y. Funaki, Phys. Rev. C **92**, 034326 (2015)
[15] H. Yamaguchi et al., Phys. Rev. C **87**, 034303 (2013)
[16] C. Wheldon et al., Phys. Rev. C **86**, 044328 (2012)
[17] M. Rodrigues, AIP Conf. Proc. **1351**, 125 (2011)
[18] D. Dell'Aquila et al., Phys. Rev. C **93**, 024611 (2016)
[19] D. Dell'Aquila et al., EPJ Web of Conf. **117**, 06011 (2016)
[20] D. Dell'Aquila, Il Nuov. Cim. C **39**, 272 (2016)
[21] E.V. Pagano et al., EPJ Web Conf. **117**, 10008 (2016)
[22] D. Dell'Aquila et al., Nucl. Instrum. Meth. Phys. Res. A **877**, 227 (2018)
[23] I. Lombardo et al., Nucl. Instrum. Meth. Phys. Res. **B 302**, 19 (2013)
[24] I. Lombardo et al., J. Phys. Conf. Ser. **569**, 012068 (2014)
[25] M. Avila et al., Phys. Rev. C **53**, 2486 (1996)
[26] L. van der Zwan, K. Geiger, Nucl. Phys. A **152**, 481 (1970)
[27] F. Ajzenberg-Selove, Nucl. Phys. A **523**, 1 (1991)
[28] S. Notarrigo, F. Porto, A. Rubbino, S. Sambataro, Nucl. Phys. **A125**, 28 (1969)
[29] I. Lombardo et al., J. Phys. G **40**, 125102 (2013)
[30] Z. Saleh et al., Ann. der Phys. **7**, 76 (1974)
[31] R. Azuma et al., Phys. Rev. C **81**, 045805 (2010)
[32] M. Wiescher et al., Phys. Rev. C **95**, 044617 (2017)
[33] J. Goss et al., Phys. Rev. C **7**, 1837 (1973)

EPJ Web of Conferences **165**, 01037 (2017) DOI: 10.1051/epjconf/201716501037
NPA8 2017

Breakup of ^{8}B on ^{58}Ni at energies around the Coulomb barrier and the astrophysical $S_{17}(0)$ factor revisited

J.C. Morales-Rivera[1,2], *T.L.* Belyaeva[1], *P.* Amador-Valenzuela[2], *E.F.* Aguilera[2], *E.* Martinez-Quiroz[2], and *J.J.* Kolata[3]

[1] *Universidad Autónoma del Estado de México, C. P. 50000, Toluca, México*
[2] *Depto. de Aceleradores, Instituto Nacional de Investigaciones Nucleares, A. P. 18-1027, C. P. 11801 México, D. F., México*
[3] *Physics Department, University of Notre Dame, Notre Dame, Indiana, 46556, USA*

Abstract. Calculations of breakup and direct proton transfer for the ^{8}B+^{58}Ni system at energies around the Coulomb barrier ($E_{\mathrm{B,lab}}$=22.95 MeV) were performed by the continuum-discretized coupled channels (CDCC) method and the coupled-reaction-channels (CRC) method, respectively. For the ^{7}Be +^{58}Ni interaction, we used a semimicroscopic optical model potential (OMP) that combines microscopic calculations of the mean-field double folding potential and a phenomenological construction of the dynamical polarization potential (DPP). The ^{7}Be angular distribution at E_{lab} =25.75 MeV from the ^{8}B breakup on ^{58}Ni was calculated and the spectroscopic factor for $^8\mathrm{B} \rightarrow {}^7\mathrm{Be}+p$ vertex, $S_{\mathrm{expt}} = 1.10 \pm 0.05$, was deduced. The astrophysical $S_{17}(0)$ factor was calculated equal to 20.7 ±1.1 eV·b, being in good agreement with the previously reported values.

1 Introduction

The study of nuclear reactions with the proton-halo exotic ^{8}B nuclei is of great interest for nuclear astrophysics in view of the problem of stellar nucleosynthesis and the production of high-energy neutrinos in the Sun. In particular, the breakup of ^{8}B in the field of heavy targets can provide information about an inverse process, the proton radiative capture by ^{7}Be, which occurs in the Sun at energies about 20 keV. The ^{8}B+^{58}Ni system has been extensively studied both experimentally and theoretically by different research groups around the world. We studied the breakup of ^{8}B in the field of ^{58}Ni with the realistic ^{7}Be core-target potential calculated in the semi-microscopic OMP. The analysis of the breakup, transfer and elastic-scattering cross sections allowed us to obtain the experimental spectroscopic factor S_{expt} and extract the astrophysical $S_{17}(0)$ factor by using the ANC method. A comparison was made with calculations performed by using the Woods-Saxon potentials previously reported [1].

2 Elastic scattering calculations

For the ^{7}Be+^{58}Ni interaction an optical potential of the form $U = V_F + V_P + iW + V_C$ was used, where V_F is a double folded potential, $iW = i(W_V f(x_V) + W_D f(x_D))$, V_C represent the absorption and Coulomb potential, respectively and V_P is the DPP, implemented by S. A. Goncharov [2]:

DOI: 10.1051/epjconf/201716501037

$$V_P = \alpha(E)W_V(E)f(x_V) + \beta(E)\cdot 4W_D(E)\frac{df(x_D)}{dx_D}, \tag{1}$$

where the Woods-Saxon form factor $f(x_{V,D})$ was used. To calculate V_F, we used for the projectile an empirical density model that was constructed on the basis of the global parametrization, described in [3]; for the target, we considered an appropriately normalized empirical charge density in a three-parameter modified Fermi form with values taken from [4]. The radial part of the potential were calculated by using the CDM3Y6-Paris nucleon-nucleon effective interaction [5]. To find the OMP parameters for the $p+^7$Be and $p+^{58}$Ni interactions, the systematics proposed in [6] was used.

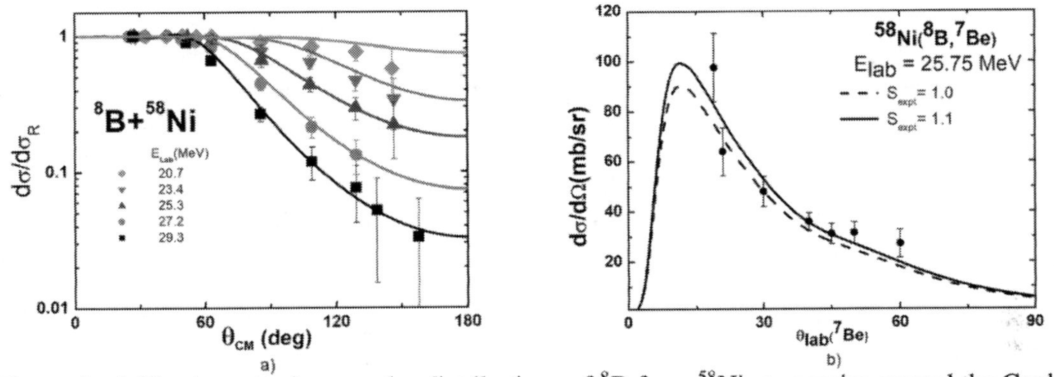

Figure 1: a) Elastic scattering angular distributions of ^{8}B from ^{58}Ni at energies around the Coulomb barrier calculated by the CDCC in comparison with the experimental data [8]. b) The differential breakup cross section of the ^{8}B+^{58}Ni reaction at $E_{\rm lab}$ =25.75 MeV. CDCC calculations performed with two values of the spectroscopic factors are shown in comparison with the experimental data of [9].

The elastic scattering angular distributions were calculated for the ^{8}B+^{58}Ni system at laboratory energies $E_{\rm lab}$ = 20.7, 23.4, 25.3, 27.2 and 29.3 MeV using the FRESCO code [7]. Figure 1 (a) shows the calculations in comparison with the data reported in [8]. Our results agree well with the data, particularly at energies above the Coulomb barrier. At energies below the barrier, the calculations slightly differ from the experimental data. Table 1 shows the DPP parameters used to fit the ^{7}Be+^{58}Ni system data of [8]. For the energies 20.7 and 23.4 MeV, values of $r_{V,D}$ =1.2 and $a_{V,D}$ =0.4 fm were used, while for the rest of the energies, the values $r_{V,D}$ =1.28 and $a_{V,D}$ =0.45 fm showed the best fit to the data. The values W_V =90 and W_D =5 MeV were kept constant for all energies.

Table 1: DPP parameters used for the ^{7}Be+^{58}Ni system.

E_{lab} $[MeV]$	α	β	E_{lab} $[MeV]$	α	β	E_{lab} $[MeV]$	α	β
20.7	-1.25	15.0	25.3	-0.28	0	27.2	-0.08	0
23.4	0	0	25.75	-0.23	0	29.3	0	0

3 Breakup and transfer analysis

We assume a cluster structure of ^{8}B= $p \oplus {}^7$Be. The valence proton has an orbital angular momentum l, thus having total angular momentum relative to the core $\mathbf{J} = \mathbf{l} + \mathbf{s}$. In the case of breakup above

 DOI: 10.1051/epjconf/201716501037

Table 2: Breakup and reaction cross sections for the ^{8}B+^{58}Ni system.

$E_{\rm lab}$ (MeV)	$\sigma_{\rm bu\ th}$ (mb)	$\sigma_{\rm R\ th}$ (mb)	$\sigma^a_{\rm R\ exp}$ (mb)
23.4	194.56	382.26	365±50
25.3	198.92	606.92	515±50
27.2	204.80	797.80	827±45
29.3	209.24	978.24	1007±40

^{a} Experimental data taken from ref. [8]

the Coulomb barrier, the excited states of proton in the continuum were represented by 167 bins with orbital angular momenta l =0-4 up to energies of 6 MeV. Figure 1 (b) shows the CDCC calculations of the ^{8}B breakup differential cross section for the ^{8}B+^{58}Ni reaction with spectroscopic factors $S_{\rm expt}$ = 1.0 and $S_{\rm expt}$ = 1.1, respectively, corresponding to the ^{8}B→^{7}Be+p vertex . The results are compared with the data of [9].

We calculated the direct proton transfer in the ^{58}Ni(^{8}B,^{7}Be)^{59}Cu reaction, which can contribute to the ^{58}Ni(^{8}B,^{7}Be) reaction cross section. Excited states of ^{59}Cu up to E_x = 3.580 MeV were taken into consideration. The calculation showed that proton stripping provides less then 3% of the total ^{7}Be emission cross sections.

Table 2 shows the breakup and reaction cross sections calculated for the ^{8}B+^{58}Ni system at energies around the Coulomb barrier in comparison with the data taken from Ref. [8]. The reaction cross sections were obtained by fitting the elastic scattering angular distributions using CDCC calculations. An accepted value of the spectroscopic factor for the ^{8}B→^{7}Be+p vertex, $S_{\rm expt}$=1.10±0.05, allowed us to estimate the ANC, C^2 = 0.54 ±0.03 fm^{-1}, and the astrophysical $S_{17}(0)$ factor to be equal to 20.7 ±1.1 eV·b, which are in good accordance with the previously published results [1].

4 Conclusions

We have performed CDCC calculations of the elastic scattering, breakup, direct proton transfer, and reaction cross sections for the ^{8}B+^{58}Ni system at energies around the Coulomb barrier. All cross sections were calculated by using the ^{7}Be+^{58}Ni semi-microscopical optical model potential containing the folding and DPP parts. The direct proton transfer contribution to the reaction cross section is about 3%. The astrophysical $S_{17}(0)$ factor equal to 20.7 ±1.1 eV·b was calculated using the ANC method, being in good agreement with previously reported values.

References

[1] T.L. Belyaeva *et at.*, Phys. Rev. C **80**, 064617 (2009).
[2] S.A. Goncharov and A. Izadpanah, Phys. At. Nucl. **70**, 18 (2007).
[3] J.H. Sorensen and A. Winther, Nucl. Phys. **A**550, 329 (1992).
[4] C. W. De Jager, H.De Vries and C. De Vries, At .Data Nucl. Data Tables **14**, 479 (1974).
[5] G. R. Satchler and W. G. Love, Phys. Rep. **55**, 183 (1973).
[6] F. D. Becchetti Jr. and G. W. Greenlees, Phys. Rev. **182**, 1190 (1969).
[7] I. J. Thompson, FRESCO user's manual and code (available from author).
[8] E. F. Aguilera *et al.*, Phys. Rev. C **79**, 021601(R) (2009).
[9] J. J. Kolata et al., Phys. Rev. C **63**, 024616 (2001).

EPJ Web of Conferences **165**, 01038 (2017)   DOI: 10.1051/epjconf/201716501038
NPA8 2017

Study of the ^{2}H(p, γ)^{3}He reaction in the Big Bang Nucleosynthesis energy range at LUNA

Viviana Mossa (for the LUNA collaboration)[1]

[1] *Università degli Studi di Bari and INFN sez. Bari*

Abstract. Deuterium is the first nucleus produced in the Universe, whose accumulation marks the beginning of the so called Big Bang Nucleosynthesis (BBN). Its primordial abundance is very sensitive to some cosmological parameters like the baryon density and the number of the neutrino families. Presently the main obstacle to an accurate theoretical deuterium abundance evaluation is due to the poor knowledge of the ^{2}H(p, γ)^{3}He cross section at BBN energies. The aim of the present work is to describe the experimental approach proposed by the LUNA collaboration, whose goal is to measure, with unprecedented precision, the total and the differential cross section of the reaction in the $30 < E_{c.m.} [keV] < 300$ energy range.

1 Introduction

The Big Bang Nucleosynthesis describes the production of light nuclides in the first minutes of cosmic time through the reaction chain shown in figure 1. The nucleosynthesis network begins with the formation of deuterium in the process p(n, γ)^{2}H. Nearly all the free neutrons end up bound in the most stable light element ^{4}He, while ^{2}H, ^{3}He, ^{7}Li and ^{6}Li nuclei form in residual quantities. The synthesis of heavier nuclei is hindered by the absence of stable nuclei with A = 8.
The primordial abundance of light isotopes depends on the competition between the expansion rate of the Universe and the yields of the reactions shown in figure 1. In the standard cosmology the expansion rate of the Universe is governed by the Friedmann equation $H^2 = \frac{8\pi}{3}G\rho$, where H is the Hubble parameter, G the gravitational constant and ρ the energy density, dominated, in the early Universe, by the "radiation", namely any massless or extremely relativistic particle. Being photons and neutrinos the only known relativistic particles at the BBN epoch, the radiation density can be expressed as $\rho = \rho_\gamma\left[1 + \frac{7}{8}\left(\frac{4}{11}\right)^{4/3} N_{eff}\right]$, where ρ_γ is the photon density and N_{eff} the contribution of other relativistic species ($N_{eff} = 3.046$ if only the three known neutrino families are taken into account). Consequently the primordial abundance of light isotopes depends on the baryon density Ω_b, usually expressed normalized to the baryon-to-photon ratio η and on the number of neutrino families or any other relativistic species N_{eff}. BBN theory can thus provide a powerful tool to constrain these cosmological parameters.

2 Physics motivation for a new ^{2}H(p, γ)^{3}He measurement at BBN energies

Recent observations of Cosmic Microwave Background (CMB) anisotropies by the Plank satellite [1] constrain the cosmological parameters with such a high level of accuracy that the primordial deu-

DOI: 10.1051/epjconf/201716501038

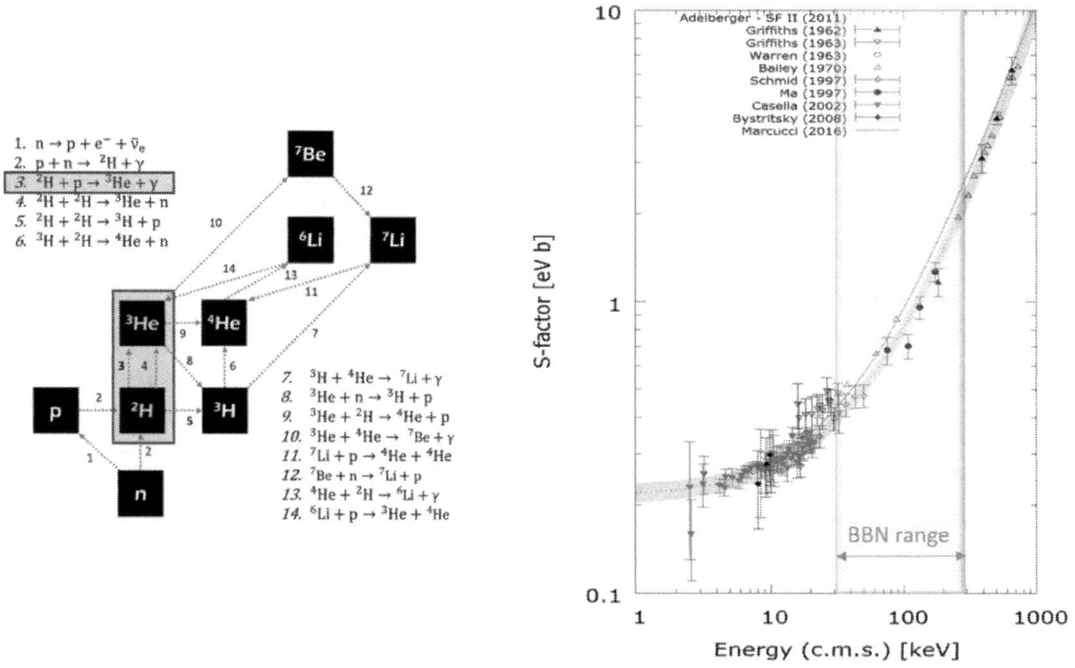

Figure 1. On the left the leading processes of Big Bang Nucleosynthesis. On the right $^2H(p,\gamma)^3He$ S-factor as a function of center of mass energy. The blue band represents the 68% lower and upper bound of the adopted quadratic best fit of Adelberger et al.

terium abundance can be inferred with remarkable precision. In particular, assuming the Plank baryonic density in a standard cosmological scenario and using the standard BBN dynamics, it is possible to compute the abundance of primordial deuterium $(D/H)_{BBN} = (2.65 \pm 0.07) \times 10^{-5}$ [2].
The result of BBN theory has to be compared with the direct observations of Damped Lyman-Alpha (DLA) systems at high redshift according to which the deuterium abundance is smaller than the indirect and model-dependent cosmological determination with a higher accuracy $(D/H)_{obs} = (2.547 \pm 0.033) \times 10^{-5}$ [3].
These two deuterium abundance determinations, while broadly consistent, are off by about one standard deviation. This is mainly due to the uncertainty with which the cross sections of the nuclear processes involved in deuterium creation and destruction are known. Presently the main source of uncertainty is the radiative capture process $^2H(p,\gamma)^3He$ converting deuterium into helium, because of the poor knowledge of its S-factor at BBN energies. In fact, as shown in figure 1 (right side), while the low energy limit of this reaction cross section is well known thanks to the data coming from the LUNA measurement performed within the solar Gamow peak [4], in the BBN energy range only a single data set is currently available with a systematic error of 9%.
The astrophysical S-factor of the $^2H(p,\gamma)^3He$ reaction is of high interest also for theoretical nuclear physics, in particular for what concern "ab-initio" calculations [5], providing an accurate prediction for the total and the differential cross section at low energies. Conversely, as highlighted by the plot in figure 1, in the BBN energy range, the theoretical result is systematically higher than data [6] by 20-30%. This huge difference between theory and data makes the situation undefined, letting some author to adopt the theoretical curve and others the S-factor value obtained from the direct measurements.

3 The ^{2}H(p, γ)^{3}He reaction at BBN energies at LUNA

The measurement of the ^{2}H(p, γ)^{3}He cross section is ongoing at the Laboratory for Underground Nuclear Astrophysics (LUNA) taking advantage of the low background of the underground Gran Sasso Laboratories and of the experience accumulated in more than twenty years of scientific activity on precision measurements.

The new ^{2}H(p, γ)^{3}He measurement at LUNA will exploit a 400 kV electrostatic accelerator [7] able to provide intense current of proton up to 500 μA with a precise absolute energy ($\pm$ 0.3 keV), low spread (< 0.1keV) and long-term stability (5 eV/h) and a windowless deuterium gas target, consisting of three-stage pumping system able to increase the gas pressure from the accelerator high vacuum to the mbar level in the target chamber. Since no real window is present, the energy degradation of the proton beam is negligible and the final ion energy is precisely determined. The beam current is measured by a constant-gradient calorimeter characterized by two sides, a hot one heated to 70°C by thermoresistors and a cold one cooled to -5°C by a refrigerating system [8]. When the ion beam hits the hot side, it contributes to its heating and reduces correspondingly the electric power needed to keep the temperature gradient constant. The current impinging on the target is thus related to the power difference between beam-on and beam-off conditions.

The experimental procedure consists of two main phases characterized by two different set-ups. The former foresees a deuterium gas target 10 cm long at 0.3 mbar of pressure and a cylindrical BGO detector having a length of 28 cm with a radial thickness of 7 cm. The crystal is optically divided into six sectors, each covering an azimuthal angle of 60 degrees and the chamber and the calorimeter are hosted inside the BGO hole. Thanks to the high detection efficiency (about 60%) in the range of interest for the ^{2}H(p, γ)^{3}He reaction (5.5 MeV), this set-up will provide a cross section measurement down to very low energies. The efficiency calibration can be obtained at a few per cent level using Monte Carlo simulations of the set-up, tuned with radioactive sources (^{137}Cs, ^{60}Co and ^{88}Y) at low energies and with the well-known resonant reaction ^{14}N(p, γ)^{15}O at E_r = 259 keV, emitting γ rays in the p+d energy range. In order to reduce systematic errors, the spatial position of the resonance has been identified inside the chamber using a well collimated NaI detector in close geometry with the target: keeping the scintillator fixed in the measurement position, the proton energy has been increased and decreased in 2-3 keV steps until the achievement of the maximum counting rate corresponding to the resonance peak.

The latter phase, instead, will cover the medium-high energies (70 < $E_{c.m.}$[keV] < 260) using a High Purity Germanium detector (HPGe) in a close geometry with the interaction chamber. This set-up allows to study the angular distribution of the emitted $\gamma-$rays, exploiting the high energy resolution of the HPGe detector and the Doppler effect affecting the photons emitted along the beam line. Using an extended deuterium gas target (33 cm long) at 0.3 mbar of pressure, in fact, the emitted photons are Doppler broadened and the peak width will be dominated by kinematics. Since the angular distribution of the emitted gammas will determine the peak shape, starting from the $\gamma-$ray energy spectrum, it is possible to extract the differential cross section of the process. This measurement requires the detection efficiency to be known with high accuracy along all the target chamber. The procedure adopted for measuring it couples the main HPGe detector (Ge1) to an auxiliary one (Ge2) and exploits the coincidence between two γ-rays emitted in cascade. The main decay channel of the E_r = 259 keV resonance (BR=57.8 %) is particularly useful to determine the germanium efficiency, because it is composed by 2 photons with energy E_{γ_1} = 1384 keV (close to the energy of ^{60}Co $\gamma-$rays) and E_{γ_2} = 6172 keV (close to the energy of photons produced by the ^{2}H(p, γ)^{3}He reaction). Whenever the auxiliary counter detects an event γ_1, it enables the other counter that can thus detect the corresponding cascade emitted photon γ_2. The detection efficiency of Ge1 can be evaluated by the ratio between the number of events that has triggered the acquisition and the number of events actually acquired by the

 DOI: 10.1051/epjconf/201716501038

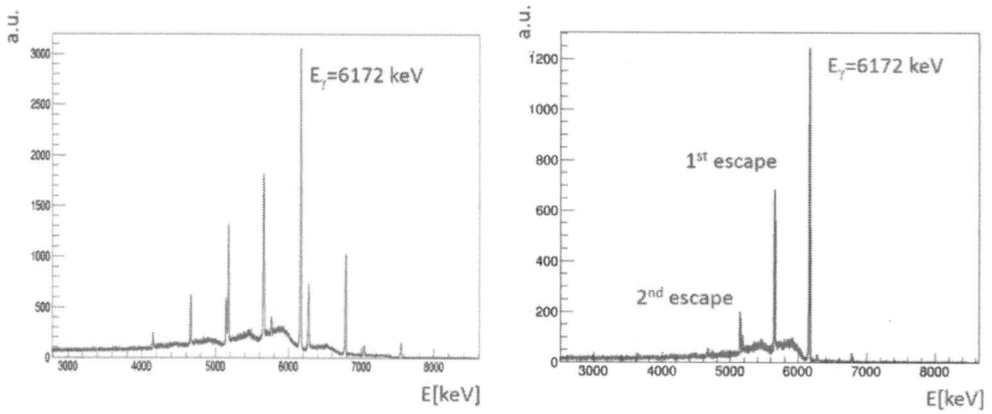

Figure 2. $^{14}N(p, \gamma)^{15}O$ resonance spectra: on the left the one acquired in inclusive configuration, on the right the one acquired in coincidence configuration.

detector itself. The effect of the coincidence acquisition is shown in figure 2, where the spectrum of the $^{14}N(p, \gamma)^{15}O$ resonance acquired by Ge1 in inclusive configuration is shown on the left. Selecting the resonance emitted photon at 1.3 MeV with Ge2, the resulting coincidence spectrum of Ge1 (on the right in fig. 2) shows the corresponding cascade photon at 6.1 MeV, with its first and second escape peaks.

4 Conclusion

At the moment the BGO phase is completed, while the HPGe data taking and the data analysis of both phases are still on going. The low background and the high quality of the LUNA facility at LNGS will allow the $^{2}H(p, \gamma)^{3}He$ cross section to be measured, filling the lack of experimental data in the BBN energy range with a higher precision with respect to the existing one and verifying theoretical nuclear studies. This study is of primary importance also to derive the baryon density of Universe with high accuracy, similar to the one obtained with CMB experiments. Moreover, it allows to constrain the existence of the so called "dark radiation", composed by undiscovered relativistic species permeating the universe, such as sterile neutrinos.

References

[1] Plank collaboration, A& A **594 A13**, 63 (2016)
[2] Di Valentino E. *et al.*, Phys. Rev. D **90**, 023543 (2014)
[3] Cooke R. *et al.*, Astrophys. J. **781**, 31 (2016)
[4] Casella C. *et al.*, Nucl. Phys. A **706**, 203 (2002)
[5] Marcucci L.E. *et al.*, Phys. Rev. Lett. **116**, 10250 (2016)
[6] Ma L. *et al.*, Phys. Rev. C **55**, 2 (1997)
[7] Formicola A. *et al.*, Nucl. Inst. Meth. **A 507**, 609(2003)
[8] Casella C. *et al.*, Nucl. Inst. Meth. **A 489**, 160 (2002)

EPJ Web of Conferences **165**, 01039 (2017)
NPA8 2017
DOI: 10.1051/epjconf/201716501039

A mechanism to account for well known peculiarities of low mass AGB star nucleosynthesis

Sara Palmerini[1,2,⋆], *Oscar* Trippella[1,2], *Diego* Vescovi[1,2], and *Maurizio* Busso[1,2]

[1] *Dipartimento di Fisica e Geologia, Università degli Studi di Perugia, Italia*
[2] *INFN, Sezione di Perugia, Italia*

Abstract. We present here the application of a model for a mass circulation mechanism induced by the stellar magnetic field to study peculiar aspects of AGB star nucleosynthesis. The mixing scheme is based on a previously suggested magnetic-buoyancy process [1, 2] and here shown to account adequately for the formation of the ^{13}C neutron source for s-processes. In particular our analysis results are focused on addressing the constrains to AGB nucleosynthesis coming from the isotopic composition of presolar grains recovered in meteorites. It turns out that *n*-captures driven by the magnetically-induced mixing can account for the isotopic abundance ratios of *s*-elements recorded.

1 Introduction

The Asymptotic Giant Branch (AGB) phase is a late stage of evolution of low mass stars ($M \leq 6M_\odot$) that is well known to be site for the production of nuclei heavier than Fe via the slow neutron capture process (*s*−process). Two reactions, the $^{13}C(\alpha,n)^{16}O$ and the $^{22}Ne(\alpha,n)^{25}Mg$, are the sources of neutrons for nucleosynthesis of AGB stars. Since the $^{22}Ne(\alpha,n)^{25}Mg$ can not be efficiently burnt in stars with mass smaller than $3M_\odot$ because of the low temperatures, the *s*−elements from Sr to Pb observed in low mass AGB stars have then to be produced trough the capture of neutrons provided by the ^{13}C source. The crucial point is to understand how a congruous amount of ^{13}C can form in the He-rich region of these stars. The mechanism suggest years ago by [3] and widely accepted is that protons are injected from the envelope into the He-rich layers at the moment of the third dredge-up (hereafter TDU), namely when at the end of a thermal instability due to a He-burning episode the convection penetrates into the radiative stellar region and brings to the stellar surface fresh synthesized materials. Afterwards, when the H-shell burning restarts, protons penetrated below the edge of the convection yield to the formation of ^{13}C in the He-rich region via the $^{12}C(p,\gamma)^{13}N(\beta^+)^{13}C$ reaction chain, later followed by the activation of the $^{13}C(\alpha,n)^{16}O$ reaction. Notice that the relation between the quantity of mixed protons and the number of nuclei synthesized by *s*-processing is not linear, indeed if too much protons are injected in the zone of interest ^{14}N forms rather than ^{13}C and it will hamper the production of heavy nuclei because of its particularly large neutron capture cross section.

Until recently, despite decades of studies and the observational evidence that *s*-process elements are produced by low mass stars ascending the second giant branch, the formation of the so called

⋆e-mail: sara.palmerini@pg.infn.it

DOI: 10.1051/epjconf/201716501039

$^{13}C - pocket$ has not been explained by self-consistent models, forcing researchers to extensive parameterizations. In the last years mass circulation mechanism induced by stellar magnetic field have been studies by several authors (eg. [4, 5]). In particular [1] demonstrated that in red giant stars magnetized parcels of matter might arrange themselves in buoyant structures that expand from radiative layers to convective envelopes. [2] shown this process to be a reliable trigger for the proton injection in the He-rich region during the TDU, then for the $^{13}C - pocket$ formation. Even if in this scenario the number of mixed protons is smaller than what usually claimed in literature [6], the model well reproduces the s-element abundances measured in young open clusters and on C-rich AGB stars. This is so because the s/C ratio in the material dredged-up from the He-shell to the envelope is significantly larger than in parametric models (see e.g., [7] and [8]) thanks to s different slope of the proton injection, which allows the formation of a ^{13}C reservoir of several $10^{-3}M_{\odot}$ but a smaller production of ^{14}N.

In this paper we compare results of our model for n-captures driven by magnetically-induced mixing with records of isotopic abundance ratios of s-elements in presolar SiC grains, which offer precise constraints on AGB nucleosynthesis and on the $^{13}C - pocket$ in particular.

2 Nucleosynthesis models, grains and results

Following the approach outlined by [2, 9] and [10], we computed the proton penetration into the He-rich layers, the consequent formation of the ^{13}C reservoir and the s-element nucleosynthesis for a set of selected AGB stellar models with mass from 1.5 to $3M_{\odot}$ and metallicity from 1/3 to 1 solar. With this small database of stellar nucleosynthesis, we tried to reproduce the relative distributions of s-process isotopes and elements, as measured in SiC grains. We used for comparisons the sample of grains analyzed by [11] and the data about Mainstream SiC grains present in the on-line Presolar Grain Database of the Washington University in St. Louis ([12] and subsequent updates).

The comparisons between our model predictions for isotopic and elemental ratios, pulse after pulse, and the measured data are shown in Figure 1. The model lines are characterized by dots that represent the various TPs. We show results for the isotopic ratios of Zr and Ba as the most rappresentative ones among light s- (*ls*) and heavy s- (*hs*) elements, respectively. Despite the large error bars of grain data, Figure 1 presents a satisfactory agreement between models and measurements: model curves fit the SiC isotopic abundances with the same accuracy of the best results achieved by other authors and available in the literature [11, 13–15]. In general, we account for the measurements of Ba isotopes at the same level of accuracy of the quoted authors ([13, 14]). Perhaps the discrepancy between data and calculations in the lowermost part of the measured points in panel C, which is not well fitted (neither by us, nor by previous works) might hints to a problem in the nuclear parameters controlling the nucleosynthesis of s-process isotopes. Moreover, the presented AGB models reach the low values of ^{91}Zr/^{94}Zr and the high ones of ^{96}Zr/^{94}Zr and ^{92}Zr/^{94}Zr isotopic ratios better than before, although not all the points are reproduced. This is so because neutron captures at low neutron density are favored by the extension and the shape of the "magnetically" formed $^{13}C - pocket$, which permit also a better reproduction of Sr isotopic ratios, as already illustrated in panel F and in [2]. Since a crucial constraint on the $^{13}C - pocket$ (its distribution and form) are the relative trends of *ls* and *hs* nuclei [11], we plot Zr versus Ba isotopic ratios in panel D and E, where our model curves account for almost the whole area covered by the data points. This good agreement for Zr and Ba is the most important result in this work, because it proves that magnetically-induced mixing can yield for the formation of a $^{13}C - pocket$, whose characteristics are fixed by the stellar physics and cannot be further fine tuned, that fulfill the AGB nucleosynthesis constraints of SiC grain composition. This result comes with an extra property: the same model of magnetic induce mixing when applied to the radiative layers above the H-burning shell reproduces also the isotopic abundances of nuclei affected

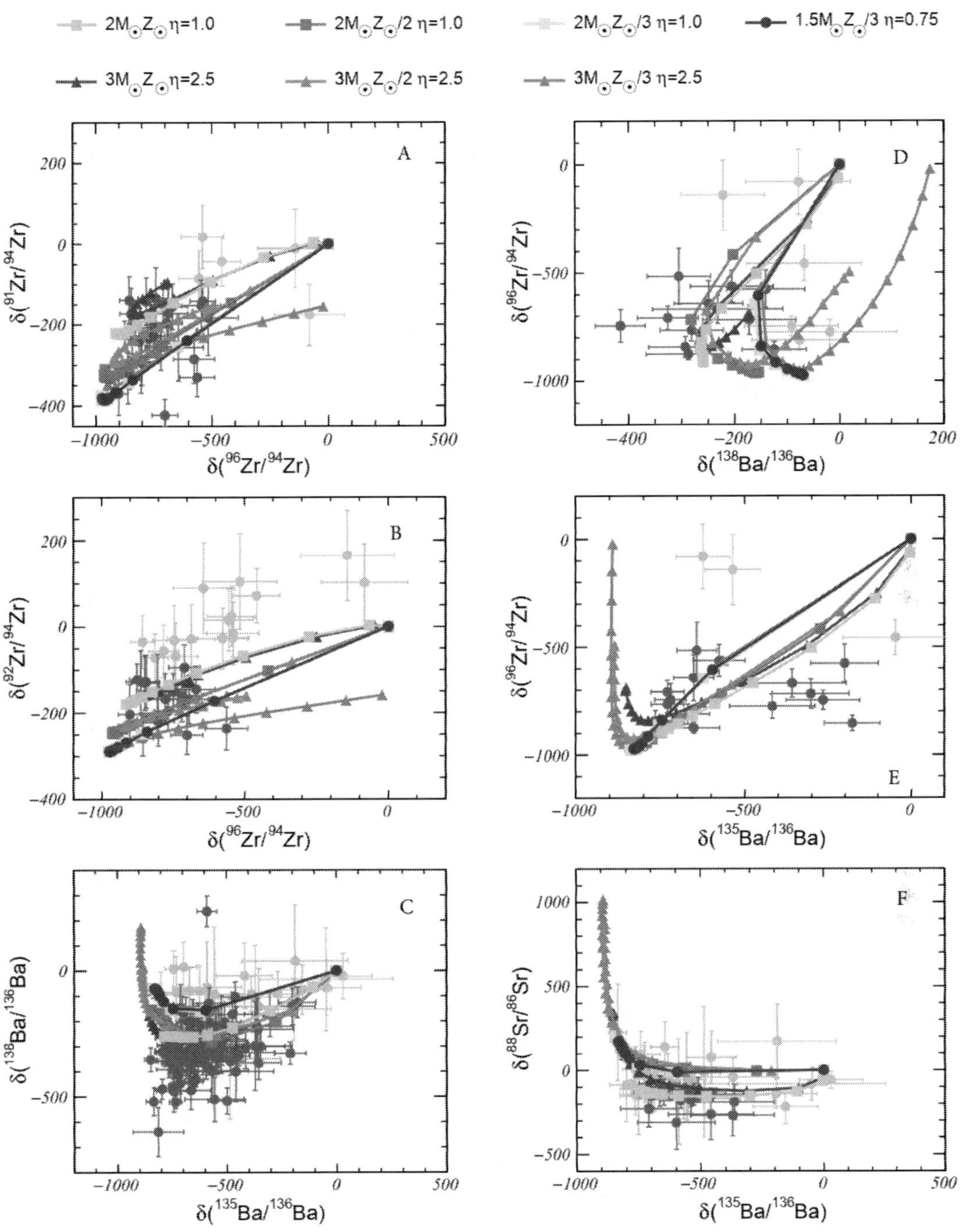

Figure 1. Comparison between our model predictions and presolar-grain relative isotopic abundances. The plot reports δ (per mill) values for Zr, Ba and Sr isotopic ratios. Data points (full dots) refer to measurements from [11] and [12]. We report in a darker color the grains having error bars smaller than the 70% of the measurements themselves. The curves show the evolution in time of the abundances in the stellar envelopes; the dots along the lines represent the various TPs. Model calculations are for AGB stars with mass from 1.5 to $3M_\odot$ and metallicity from 1/3 to 1 solar.

 DOI: 10.1051/epjconf/201716501039

by proton captures, typical of evolved low mass stars [9], therefore ours sees to be the unique approach for the moment to offer a rather general tool for treating deep mixing in evolved red giants.

References

[1] Nucci, M. C. & Busso, M., ApJ **787**, 141 (2014)
[2] Trippella, O., Busso, M., Palmerini, S. et al., ApJ **818**, 125 (2016)
[3] Straniero, O., Gallino, R., Busso, M., et al., ApJ **440**, L85 (1995)
[4] Denissenkov, P. A., & Merryfield, W. J., ApJ **727**, L8 (2011)
[5] Busso, M., Wasserburg, G. J., Nollett, K. M., & Calandra, A., ApJ **671**, 802 (2007)
[6] Maiorca, E., Magrini, L., Busso, M., et al., ApJ **747**, 53 (2012)
[7] Busso, M., Gallino, R., & Wasserburg, G. J., ARA&A **37**, 239 (1999)
[8] Bisterzo, S., Travaglio, C., Gallino, R., et al., ApJ **787**, 10 (2014)
[9] Palmerini,S., Trippella, O. & Busso, M., MNRAS **467**, 1193 (2017)
[10] Palmerini,S., Trippella, O. , Busso, M., et sl. GeoCoA in press (2017)
[11] Liu N., Savina M. R. & Gallino R., et al. , ApJ **803**, 12 (2015)
[12] Hynes, K.M. & Gyngard, F., LPSC XL, Abstract 1198, (2009)
[13] Lugaro M., Tagliente G., Karakas A. I., et al., ApJ **780**, 95 (2014)
[14] Liu, N., Gallino, R., Bisterzo, S., et al. ApJ **786**, 66 (2014).
[15] Liu, N., Gallino, R., Bisterzo, S., et al. ApJ **788**, 66 (2014).

Direct cross section measurement for the ^{18}O(p, γ)^{19}F reaction at astrophysical energies at LUNA

F. R. Pantaleo[1,2,⋆] and *A.* Best[3,4], *G.* Imbriani[3,4], *R.* Perrino[2,⋆⋆]
for the LUNA Collaboration

[1] *Universita' degli Studi di Bari, Dipartimento Interateneo di Fisica "M. Merlin", Bari, Italy*
[2] *Istituto Nazionale di Fisica Nucleare (INFN), Sezione di Bari, Italy*
[3] *Universita' degli Studi di Napoli, Dipartimento di Fisica "E. Pancini", Napoli, Italy*
[4] *Istituto Nazionale di Fisica Nucleare (INFN), Sezione di Napoli, Italy*

Abstract. $^{18}O(p,\gamma)^{19}F$ plays an important role in the AGB star scenarios. The low energy cross section could be influenced by a hypothetical low energy resonance at 95 keV and by the tails of the higher energy broad states. The 95 keV resonance lies in the energy window corresponding to the relevant stellar temperature range of 40-50 MK.
Measurements of the direct cross section were performed at the Laboratory for Underground Nuclear Astrophysics (LUNA), including the unobserved low energy resonance, the higher energy resonances and the non-resonant component, taking advantage of the extremely low environmental background. Here we report on the experimental setup and the status of the analysis.

1 Introduction

The $^{18}O(p,\gamma)^{19}F$ reaction ($Q \simeq 7.99$ MeV) represents the bridge between the CNO cycle and other cycles involved in the production of heavier nuclei, which are active during shell H burning.
A $^{18}O/^{16}O$ isotopic ratio equal to $2.09^{+0.13}_{-0.12} \cdot 10^{-3}$ is present in matter within our solar system [1]. Investigations of oxygen isotopic ratios through pre-solar grains, gathered over the years from primitive meteorites and interplanetary dust particles, found a value $\leq 1.5 \cdot 10^{-3}$ [2], reflecting a substantial ^{18}O depletion compared to the solar system value.

It has been suggested that the low-mass asymptotic giant branch (AGB) stars are an ^{18}O depletion site [3] due to *cool bottom processing* (CBP) [2], in which the $^{18}O + p$ reaction occurs.
Within the CBP temperature regime (40 MK), the $^{18}O(p,\gamma)^{19}F$ reaction rate may be influenced by an unobserved low-energy resonance at $E_{res} = 95 \pm 3$ keV, that corresponds to the E_X=8084 keV±3 keV [4] level in the ^{19}F nucleus. The 95 keV resonance strength is disputed by [5] and [6]. An experimental upper limit of the 95 keV resonance strength [5] could have a negligible impact on the $^{18}O(p,\gamma)^{19}F$ reaction rate over the entire temperature range (see Fig. 1(a)), otherwise a larger resonance strength estimated by [6] could have a dominant contribution to the reaction rate over the AGB stars temperature range (see Fig. 1(b)).

⋆e-mail: francesca.pantaleo@ba.infn.it
⋆⋆permanent address: INFN Lecce, Italy

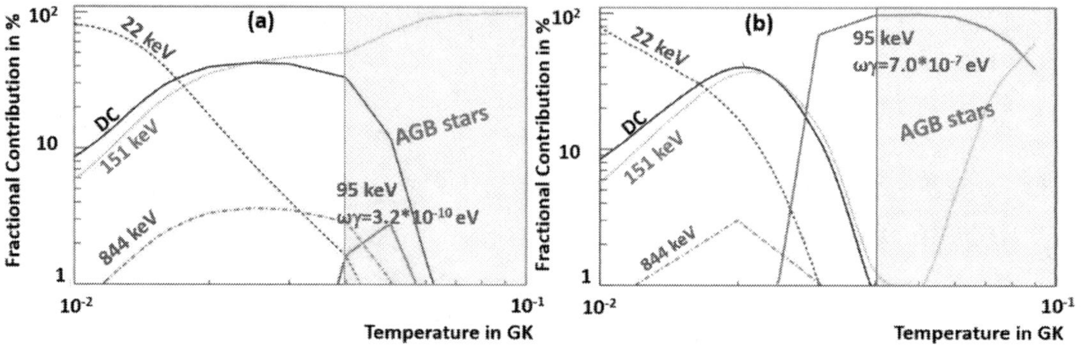

Figure 1. Reaction rates contributions in the AGB range of stars, (a) follows as [5] and (b) follows as [6].

A direct observation of the disputed resonance is desirable, and measurements of the non-resonant cross section at lower energies can support the extrapolation of the non-resonant contribution towards lower energies of astrophysical interest.

Further measurements regarding the off-resonant and on-resonant higher energy range, were done in order to improve our knowledge of levels in the ^{19}F compound nucleus that are relevant to nuclear astrophysics and to complete available data known in literature [7], [8], [9].

2 Experimental setup

The $^{18}O(p,\gamma)^{19}F$ reaction was investigated using the accelerator at the LUNA2 facility [10], which provides very accurate and reproducible ion beams of protons in the energy range of 50 keV $\leq E_p^{lab} \leq$ 400 keV with a maximum current (on-target) approximately equal to 300 μA. The facility has two beamlines, one dedicated to solid target and the other one to gas target experiments. The solid target beamline has been used for studying the reaction of interest for the present work.

The Ta_2O_5 targets were prepared by anodic oxidation [11] of thin tantalum disks (0.3 mm x 40 mm diameter) with an isotopic enrichment 99% in ^{18}O. These targets have to meet a number of specific requirements: thickness must be uniform and such to sustain the high beam currents over extended times; known and constant stoichiometry, to allow for accurate beam energy loss calculation; known and possibly enriched isotopic composition, to allow for measurable yields.

A high efficiency BGO summing crystal detector was used for total cross section measurement. The experimental setup also hosts a High Purity Germanium (HPGe) detector, positioned at 55° with respect to the beam direction, which provides a much lower detection efficiency compared to the BGO detector but a greatly improved energy resolution in order to measure branchings and higher energy resonances.
The BGO and the HPGe setup are shielded by 10 cm and 15 cm of lead in order to reduce the environmental background contributions, respectively.

3 Measurements

The experimental campaign was split in two phases. The first phase is characterized by a high efficiency BGO summing crystal detector. During the second phase a high energy resolution HPGe

 DOI: 10.1051/epjconf/201716501040

detector was used. Both approaches used a solid target setup and covered the common range of data taking (140-400 keV). The first phase extended the range of measurements to energies down to 89 keV, to probe the existence of the hypothetical low energy resonance using the high efficiency BGO summing crystal detector.
Five resonances were measured within 89-400 keV. For the first time a measurement was done in order to estimate the resonance strength for the low-energy resonance at $E_{res} = 95 \pm 3$ keV , and in addition measurements were done to reduce the error on the 151 keV, 216 keV, 274 keV and 334 keV resonance strengths. Two experimental spectra acquired at 95 keV and at the energy of the strongest resonance of $^{18}O(p,\gamma)^{19}F$ reaction 151 keV, are shown in Figures 2.(a) and (b). A signal was observed at 95 keV with the BGO setup, (see Fig. 2(c)). Clear individual transitions for the 151 keV resonance demonstrate resolution of the HPGe detector (see Figs. 2.(b) and (d)).

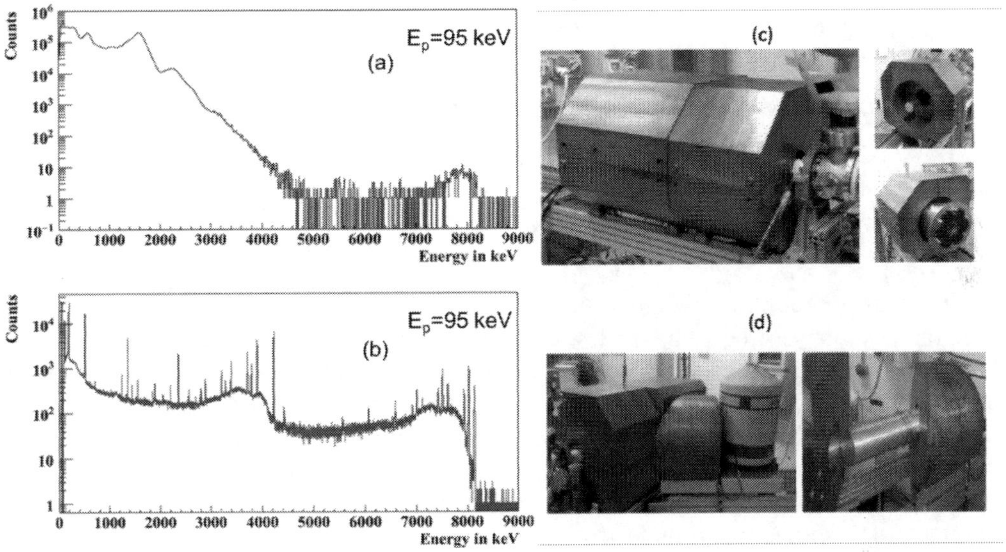

Figure 2. Experimental spectra are shown: (a) acquired at 95 keV, the energy of the low energy resonance, (b) acquired at 151 keV, the energy of the strongest resonance of $^{18}O(p,\gamma)^{19}F$ reaction. Detector setups are shown: (c) BGO and (d) HPGe.

The excitation function including all BGO measurements is shown in Fig. 3, except data lower than about 100 keV because the analysis of the low energy part is ongoing.

Several measurements of environmental backgrounds were done during the two phases, and beam induced background measurements were performed only during the first phase in order to understand the influence of the different contaminants. With Q values of 8.0 MeV, the signals are in regions at which the background rate is greatly reduced thanks to the underground location of LUNA.
It is dominated by radiative capture of thermal neutrons in the detector materials, but it is only on the order of 10^{-4} counts per MeV per second.

4 Analysis and Conclusions

The data analysis for the two experimental phases is ongoing and will be finalized soon.
The first focus is the low-energy resonance, and the estimates of the strengths regarding the other

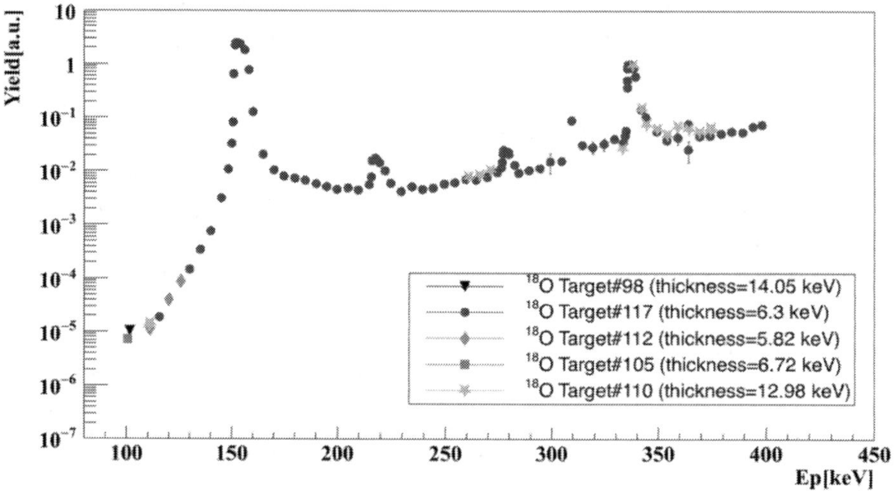

Figure 3. Excitation function from the BGO measurements.

resonances are ongoing, with the BGO data sets.

A preliminary analysis regarding the low-energy resonance with tail effects coming from the 151 keV resonance and the direct capture contributions is ongoing. Data analysis from complete HPGe data sets is also ongoing. In particular the determination of the γ-ray branchings and resonance strengths regarding on-resonance energy component are almost finalized. The γ-ray branchings regarding the 216 keV proton energy resonance were calculated for the first time at LUNA. For the remaining resonances at 151 keV, 274 keV, 334 keV proton energies, the γ-ray branchings obtained at LUNA were compared with previous published data [7].

References

[1] P. C. Scott et al., Astron. Astrophys. **456**, 675 (2006)
[2] S. Palmerini et al., Astrophys. J. **729**, 3 (2011)
[3] L. R. Nittler et al., Astrophys. J. **682**, 1450 (2008)
[4] D. R. Tilley et al., Nucl. Phys. A **595**, 1 (1995)
[5] M. Q. Buckner et al., Phys. Rev. C **86**, 065804 (2012)
[6] H. T. Fortune et al., Phys. Rev. C **88**, 015801 (2013)
[7] M. Wiescher et al., Nucl. Phys. **A349** 165-216 (1980)
[8] R. B. Vogelaar et al., Physical Review C **42**, 753 (1990)
[9] C. Iliadis et al., Nucl. Phys. **A841** 251 (2010)
[10] A. Formicola et al, Nucl. Instrum. Methods **A507**, 609-616 (2003).
[11] A. Caciolli et al. Eur. Phys. J. A **48** 144 (2012)

EPJ Web of Conferences **165**, 01041 (2017) DOI: 10.1051/epjconf/201716501041
NPA8 2017

The RIB in-flight facility EXOTIC at INFN-LNL

C. Parascandolo[1,★], *A.* Boiano[1], *C.* Boiano[2], *M.* La Commara[3,1], *G.* La Rana[3,1], *M.* Mazzocco[4,5], *D.* Pierroutsakou[1], *C.* Signorini[4,5], *F.* Soramel[4,5], and *E.* Strano[4,5]

[1] *INFN - Napoli, Via Cintia, I-80126 Napoli, Italy*
[2] *INFN - Milano, Via Celoria 16, I-20133 Milano, Italy*
[3] *Dipartimento di Fisica, Università di Napoli, Via Cintia, I-80126 Napoli, Italy*
[4] *Dipartimento di Fisica e Astronomia, Università di Padova, Via Marzolo 8, I-35131 Padova, Italy*
[5] *INFN - Padova, Via Marzolo 8, I-35131 Padova, Italy*

Abstract. The EXOTIC facility, installed at the INFN-Laboratori Nazionali di Legnaro (LNL), is devoted to the in-flight production of light Radioactive Ion Beams (RIBs) in the energy range between 3-5 MeV/nucleon. The research field accessible using the EXOTIC RIBs is the investigation of reaction mechanisms and nuclear structure, resonant scattering measurements and measurements of nuclear reaction cross sections of astrophysical interest.

1 Introduction

During the last decades there has been a growing interest in the possibility of using Radioactive Ion Beams (RIBs) for a variety of fundamental studies in physics. Experiments with RIBs allow to explore the properties of isotopes with a proton-to-neutron ratio very different from the stable ones, measure reaction cross sections of astrophysical interest occurring in explosive environments, constrain the isospin-dependent nucleon-nucleon interaction in neutron-rich nuclei and in neutron stars, synthesize superheavy elements and test physics beyond the standard model.

2 EXOTIC facility and Experimental Set-up

The EXOTIC facility [1–4] is dedicated to the in-flight production of low-energy light RIBs, by inverse kinematics nuclear reactions using the intense heavy-ion beams from the INFN-LNL Tandem XTU accelerator hitting a light gas target (H_2, D_2, ^{3}He and ^{4}He). The main features of the facility, shown in Fig. 1, are a large RIB acceptance of the optics elements and a large capability to suppress all the undesired scattered beams. It consists of: i) a production gas target that is a 5 cm-long cylindrical cell walled with windows made of 2.2 μm thick Havar foil, operating at room or at liquid N_2 temperatures at pressures up to 1 atm; ii) a beam selection and transport system consisting of: a triplet of large diameter quadrupole lenses, a 30° bending magnet, a Wien filter and a second triplet of quadrupole lenses placed before the reaction chamber. The Wien filter eliminates to a very large extent the tails of the primary beam that pass through the system with the same magnetic rigidity.

★e-mail: concetta.parascandolo@na.infn.it

 DOI: 10.1051/epjconf/201716501041

In order to stop the RIB contaminants, different slit sets are installed along the beamline and can be adjusted to the envelope of the produced RIB. The facility has a length of 8.34 m and presents the following charateristics: $\Delta E/E = \pm 10\%$, $\Delta p/p = \pm 5\%$, $\Delta\theta = \pm 50$ mrad, $\Delta\phi = \pm 65$ mrad, $\Delta\Omega \approx 10$ msr, $B\rho = 0.98$ Tm. So far, RIBs of ^{7}Be, ^{8}B, ^{17}F, ^{15}O, ^{8}Li, ^{10}C and ^{11}C at 3-5 MeV/nucleon have been delivered with intensities about 10^6, 10^3, 10^5, $4\text{x}10^4$, 10^5, $5\text{x}10^3$ and $2\text{x}10^5$ pps, respectively, and with 98-99% purity (except for ^{8}B that has a lower purity).

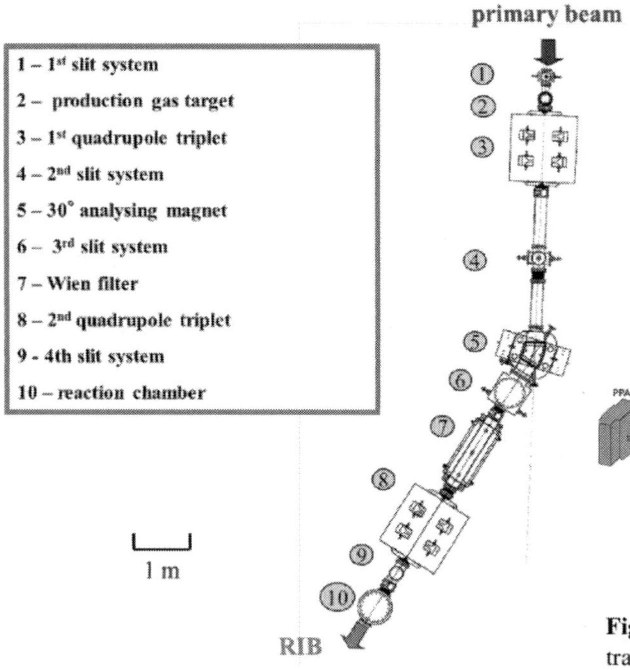

Figure 1. Layout of the EXOTIC facility.

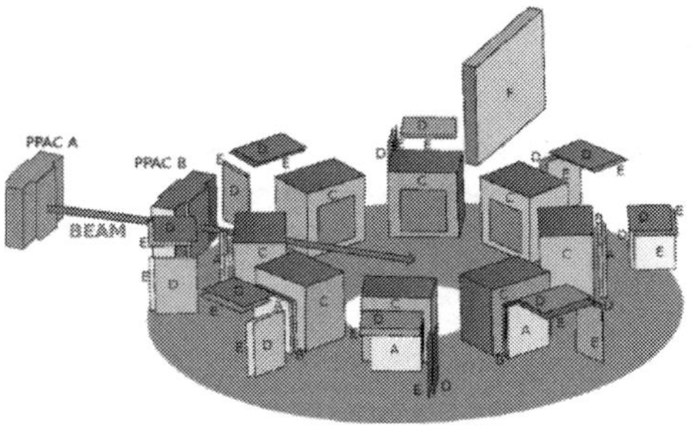

Figure 2. The position sensitive PPACs for the RIB tracking system and the detection system, EXPADES, placed in the reaction chamber (element 10 of Fig. 1) at the end of the EXOTIC facility.

The experimental set-up [5] of the EXOTIC facility, shown in Fig. 2 and situated at the end of the beamline, consists of: a) the RIB tracking system and b) the detection system, EXPADES, a new charged-particle telescope array. EXPADES is flexible to suit many experiments and it can be coupled with γ-ray and neutron arrays.

The Parallel Plate Avalanche Counters (PPACs) of the tracking system are position-sensitive and fast detectors, which can sustain counting rates up to $\sim 10^6$ Hz. They are placed 909 mm (PPAC A) and 365 mm (PPAC B) upstream the reaction target, at the EXOTIC focal plane. The PPACs are filled with C_4H_{10} gas at a working pressure of 10-20 mbar. Each PPAC has an active area of 62x62 mm^2 and is sealed with 1.5 μm-thick mylar windows. The detector has a central cathode and two anodes, placed symmetrically at 2.4 mm from the cathode. The position information of a particle crossing the PPAC is extracted from the anode signals by using a delay-line readout. The 1 mm resolution of the two PPACs allows us to reconstruct the position of the event on the reaction target with a 2.3 mm position resolution. The cathode signal is used as a reference time for Time of Flight measurements (giving a START signal with 0.9 ns FWHM time resolution) and for trigger purposes.

EXPADES is an array of eight telescopes arranged in a cylindrical configuration around the reaction target (Fig. 2). The telescope structure is flexible and is composed of two Double Side Silicon

Strip Detectors (DSSSDs) and/or Ionization Chamber (IC), depending on the experimental requests. We use 40/60 μm-thick DSSSDs for the ΔE stage (B in Fig.2) and 300 μm-thick DSSSDs for the E_{res} one (A in Fig.2). Each DSSSD has 64-mm long 32x32 strips, with 2 mm pitch size and 40 μm inter-strip separation, defining thus a $\sim$ 2x2 mm^2 pixel structure. For experiments requiring the detection of more energetic particles than those stopped in the E_{res}, 1 mm-thick DSSSDs can substitute the 300 μm-thick ones or can be used in addition to the previous stages. ASIC-based electronics was employed for the treatment of the E_{res} signals, with a high granularity and a very low cost. For the signal readout of the ΔE stage a compact low-noise electronics with very good energy and timing characteristics was developed by our collaboration. A valid alternative to allow the ΔE-E_{res} identification of reaction products with range in silicon shorter than 40/60 μm is the use of a transverse field IC (C in Fig. 2) that can give the possibility to tune the effective thickness of the ΔE layer by changing the gas pressure. The IC has 65x65 mm^2 entrance and exit windows (1.5 μm-thick mylar) and can be filled with CF_4 gas at pressures up to 100 mbar. The low-noise charge-sensitive preamplifiers for the ΔE DSSSDs (D in Fig. 2), those of the ICs (not present in Fig. 2) as well as the electronic boards (E in Fig. 2) for the E_{res} DSSSDs are placed under vacuum close to the detectors. In this way, we manage to keep as low as possible the DSSSDs electronic thresholds, typically 300-500 keV.

3 Experimental program and perspectives

The experimental program at EXOTIC is varied and aims at different topics.

Firstly, we focused our attention on the study of reaction mechanisms induced by light radioactive (exotic) nuclei impinging on medium- and heavy-mass targets at energies near to the Coulomb barrier [6–10]. Indeed, the peculiar features of exotic nuclei influence the reaction mechanisms giving a picture that is rather different from that of well bound species [11]. In these measurements the charged products emitted in direct nuclear reactions and the light charged particles emitted in fusion-evaporation reactions should be charge and mass identified.

Furthemore, we started to study α clustering phenomena in light exotic nuclei [12], employing the Thick Target Inverse Kinematics (TTIK) scattering technique [13], with the RIB impinging on a ^{4}He gas target. The pressure of the gas is tuned such that the RIB completely stops in the gas while the energetic recoiling light target nuclei, due to their low rate of energy loss, can traverse the gas and be detected. The TTIK method is useful for measurements with low-intensity RIBs since it allows to measure the elastic scattering excitation function and the reduced α-width directly, over a wide energy range by using a single beam energy, since the gas acts as a target but also as a degrader. In particular, we searched for α-cluster states in ^{19}Ne above its α-decay threshold measuring, for the first time, the ^{15}O(^{4}He, ^{4}He)^{15}O elastic scattering excitation function with the TTIK technique with ^{15}O RIB produced at EXOTIC. ^{19}Ne has attracted much interest due to the astrophysical importance of the ^{15}O(α,γ)^{19}Ne reaction, identified as a break-out from the Hot CNO cycle and a trigger for the following *rp*-capture populating masses higher than neon [14]. Neon isotopes exhibit strong α-clustering phenomena, even in the ground states. Therefore, the idea has been to investigate the possible cluster nature of the neutron-poor side of the neon isotopes, almost totally unknown. The final excitation functions reveal the presence of resonances, that have been analysed by using a full R-Matrix fit with Azure2 [15].

Another possibility offered by the EXOTIC facility is to perform measurements of astrophysical interest with RIBs impinging on solid or gas light targets in inverse kinematics. Among the different processes of stellar nucleosynthesis forming elements heavier than ^{9}Be, the *rp*-capture and αp processes, occurring in explosive astrophysical environments, are those than can be investigated by using the EXOTIC RIBs. By developing a radioactive ^{18}Ne beam through the ^{3}He(^{16}O,^{18}Ne)n reaction, the ^{18}Ne(α,p)^{21}Na process could be studied at astrophysical energies to provide a link between the Hot

CNO cycle and the *rp*-process [14]. Another measurement relevant to astrophysics can be performed at EXOTIC such as the ^{30}P(p,γ)^{31}S with a ^{30}P beam, essential for the production of heavy elements (from Si to Ca) in the explosion of ONe novae and in particular to explain the anomalously high ^{30}Si/^{28}Si rate measured in pre-solar grains of possible ONe novae origin [16]. Moreover, experiments based on the Trojan Horse Method (THM) [17] are considered. In this direction, a first experiment was performed by means of the THM for the ^{7}Be(n,α)^{4}He reaction to shed light on the cosmological lithium-problem, of interest in the Big Bang Nucleosynthesis [18]. The first results have been presented in this conference [19].

Recently, we tested also the possibility to perform measurements of fusion cross sections at near- and sub-barrier energies, by using our facility as a separator of evaporation residues from the incident beam, provided by the INFN-LNL Tandem XTU accelerator. The evaporation residues are transported and detected at the focal plane of the facility (for details, see [20]).

To summarize, the EXOTIC facility dedicated to the in-flight production of low-energy light RIBs and its experimental set-up is fully operational at INFN-LNL. Stimulating nuclear physics and nuclear astrophysics measurements can be carried out employing the produced RIBs, in the framework of international collaborations. Moreover, EXOTIC can be used as a velocity filter to perform fusion-evaporation experiments at near-and sub-barrier energies with stable beams and also with the neutron-rich RIBs delivered by the next generation ISOL-type facility SPES [21], in construction at INFN-LNL.

References

[1] V.Z. Maidikov et al., Nucl. Phys. **A 746**, 389c (2004);
[2] D. Pierroutsakou et al., Eur. Phys. J. Special Topics **150**, 47 (2007);
[3] F. Farinon et al., Nucl. Instr. and Meth. **B 266**, 4097 (2008);
[4] M. Mazzocco et al., Nucl. Instr. and Meth. **B 266**, 4665 (2008) and Nucl. Instr. and Meth. **B 317**, 223 (2013)
[5] D. Pierroutsakou et al., Nucl. Instr. and Meth. **A 834**, 46 (2016)
[6] C. Signorini et al., Eur. Phys. J. **A 44**, 63 (2010)
[7] M. Mazzocco et al., Phys. Rev. **C 82**, 054604 (2010), Phys. Rev. **C 92**, 024615 (2015); EPJ Web of Conferences **117**, 06006 (2016)
[8] N. Patronis et al., Phys. Rev. **C 85**, 024609 (2012); A. Pakou et al., Phys. Rev. **C 87**, 014619 (2013); Eur. Phys. J. **A 51**, 55 (2015); Eur. Phys. J. **A 51**, 90 (2015)
[9] E. Strano et al., Phys. Rev. **C 94**, 024622 (2016)
[10] O. Sgouros et al., Phys. Rev. **C 94**, 044623 (2016); Phys. Rev. **C 95**, 054609 (2017)
[11] L.F. Canto et al., Phys. Rep. **424**, 1 (2006); Phys. Rep. **596**, 1 (2015)
[12] M. Freer, Rep. Prog. Phys. **70**, 2149221 (2007)
[13] K. Artemov et al., Sov. J. Nucl. Phys. **52**, 408-411 (1990)
[14] M. Wiescher et al., Prog. Part. and Nucl. Phys. **59**, 51 (2007)
[15] D. Torresi et al., J. Phys.: Conf. Ser. **863**, 012026 (2017); Phys. Rev. **C 96**, 044317 (2017)
[16] J. José et al., Astrophys. J. **560**, 897 (2001)
[17] G. Baur, Phys. Lett. **B 178**, 135 (1986); R. E. Tribble et al., Rep. Prog. Phys. **77**, 106901 (2014)
[18] C. Broggini et al., J. Cosmol. Astropart. Phys. **6**, 30 (2012)
[19] L. Lamia et al., to be published
[20] E. Strano et al., Nucl. Instr. and Meth. **A 877**, 293 (2018)
[21] https://web.infn.it/spes/

EPJ Web of Conferences **165**, 01042 (2017) DOI: 10.1051/epjconf/201716501042
NPA8 2017

The ^{10}B(p,α)^{7}Be S(E)-factor from 5 keV to 1.5 MeV using the Trojan Horse Method

Sebastiana Maria Regina Puglia[1,2,⋆], *Claudio* Spitaleri[1,2], *Marco* La Cognata[1], *Livio* Lamia[1,2], *Carlo* Broggini[3], *Antonio* Caciolli[3,4], *Nelson* Carlin[5], *Silvio* Cherubini[1,2], *Alexandra* Cvetinovic[1], *Giuseppe* D'Agata[1,2], *Daniele* Dell'aquila[6,7], *Rosanna* Depalo[3,4], *Marisa* Gulino[1,8], *Giovanni Luca* Guardo[1], *Iolanda* Indelicato[1], *Ivano* Lombardo[6,7], *Roberto* Menegazzo[3], *Marcelo Gimenez* Munhoz[5], *Rosario Gianluca* Pizzone[1], *Giuseppe Gabriele* Rapisarda[1,2], *Valentino* Rigato[9], *Stefano* Romano[1,2], *Maria Letizia* Sergi[1], *Francisco* Souza[5], *Roberta* Sparta'[1,2], *Salvo* Tudisco[1,2], and *Aurora* Tumino[1,8]

[1] *INFN-Laboratori Nazionali del Sud, Catania, Italy*
[2] *Dipartimento di Fisica e Astronomia, University of Catania, Catania, Italy*
[3] *INFN-Padova, Padova, Italy*
[4] *Dipartimento di Fisica, University of Padova, Padova, Italy*
[5] *Departamento de Fisica Nuclear, Universitade de Sao Paulo, Brasil*
[6] *Dipartimento di Fisica, University of Napoli, Napoli, Ital*
[7] *INFN-Napoli, Napoli , Italy*
[8] *Facolta di Ingegneria e Architettura, Kore, University, Enna, Italy*
[9] *INFN-Laboratori Nazionali di Legnaro, Legnaro, Italy*

Abstract. The ^{10}B(p,α)^{7}Be reaction is the main responsible for the ^{10}B destruction in stellar interior [1]. In such environments this p-capture process occurs at a Gamow energy of 10 keV, and takes places mainly through a resonant state (Ex = 8.701 MeV) of the compound ^{11}C nucleus. Thus, a resonance right in the region of the Gamow peak is expected to significantly influence the behavior of the astrophysical S(E)-factor. The ^{10}B(p,α)^{7}Be reaction was studied via the Trojan Horse Method (THM) applied to the ^{2}H(^{10}B,α^{7}Be)n in order to extract the astrophysical S(E)-factor in a wide energy range, from 5 keV to 1.5 MeV.

1 Introduction

The study of the abundances of light elements Li, Be and B provides important indications about chemical evolution of the Galaxy, stellar structure and about the nucleosynthesis. In particular a comprehensive study of Li Be B abundances can confirm the presence of non-standard mixing processes in stellar envelopes [2]. These elements are destroyed by nuclear fusion reactions at different depths inside stars corresponding to temperatures of about 2.5×10^{6} K for Li, 3.5×10^{6} K for Be and 5×10^{6} K for B [1]. The most involved processes are (p,α) reactions, induced at Gamow energy E_G of a few keV, which require precise cross section measurements. In particular the THM allows to extract the cross section for a charged-particle-induced reaction by selecting the quasi-free contribution

⋆e-mail: puglia@lns.infn.it

to a suitable three body reaction. Moreover the THM allows one to measure the bare nucleus S(E)-factor for astrophysically relevant reactions without experiencing Coulomb penetrability effects. The details of the method are explained in [3–5]. In this work the THM [3] has been applied to the ^{2}H(^{10}B,α^{7}Be)n three-body reaction in order to measure the astrophysical S(E)-factor for the ^{10}B(p,α)^{7}Be reaction. The importance of this reaction is due to the presence of a resonance at about E_{cm}=10 keV corresponding to the Gamow peak for this reaction. The resonance corresponds to the ^{11}C*(8.701 MeV, J=5/2$^+$) excited state. Moreover the ^{10}B(p,α)^{7}Be reaction is of interest also for nuclear physics in fact it allows to investigate ^{11}C states presently poorly known [6][7] .

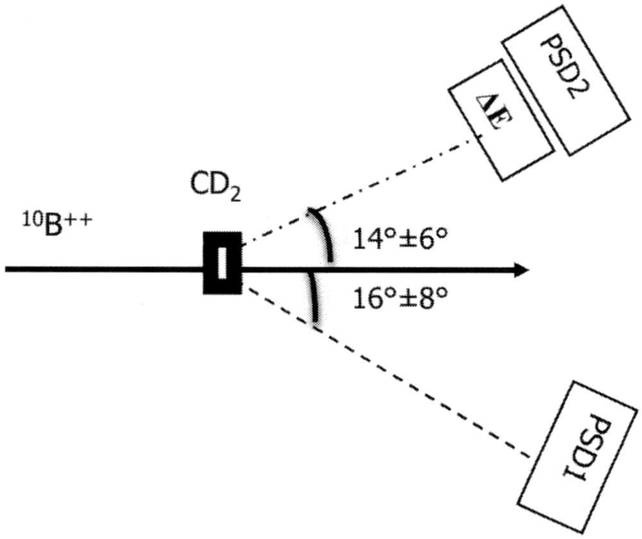

Figure 1. The experimental setup for the Pelletron Linac Institute experiment. The ΔE-E system devoted to ^{7}Be detection and a position sensitive detector (PSD) for alpha particle detection are shown.

2 Experimental setup

In order to study the ^{10}B(p,α)^{7}Be reaction two experiments were performed, one devoted to study the resonance at 10 keV in ^{10}B-p center of mass and the other one to investigate the ^{11}C high-energy excited states. The first experiment was performed at Pelletron Linac Laboratory in Sao Paolo (Brazil). The Tandem accelerator provided a 27 MeV ^{10}B beam and intensity up 1 nA, the beam energy was chosen in order to span the relative energy range from 1.5 MeV down to zero [8].The detection setup consisted of a position sensitive detector (PSD) and a telescope system having a proportional counter (PC) as ΔE and a standard PSD as E detector, the experimental setup for Sao Paolo measurement is shown in fig.1. The second experiment was performed at Laboratory Nazionali del Sud in Catania [9], the Tandem Van der Graff provided 24.5 MeV ^{10}B beam and intensity 1.5 nA, in this case the main goal was the study of the 10 keV resonance. The detection setup consisted of a ΔE-E system made up of a ionization chamber filled with butane gas and a PSD to measure the residual energy of the emitted particles. Two position sensitive detectors were placed at opposite sides with respect the beam direction, more detail are in [9]. The experimental setup was chosen to cover the whole quasi-free angular range [3] known from a Monte Carlo simulation. Energy and position signals for

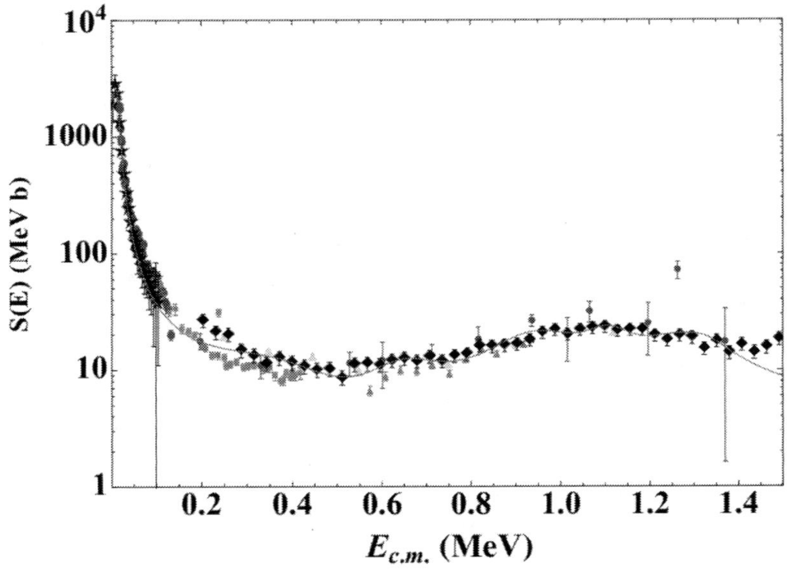

Figure 2. The resulting S(E)-factor in absolute units displayed as black diamonds obtained in [8] and the previous data sets [9] S(E)-factor shown as black stars.This S(E)-factor is compared with data from the ref[15] as blue circles, with the data from [14] corrected for the factor 1.83 as recommended in [15] and reported as red squares. Moreover thick-target data in [11] reported as purple circles, and recent works [12] and [13] S(E) -factors as green and grey triangles, respectively. R-matrix fit of the THM data is reported as red line

the detected particles were processed by standard electronics and sent to the acquisition system for the on-line monitoring and data storage.

3 Data analysis

After the calibration of the involved detectors, the first step in the analysis was the ^{2}H(^{10}B,α^7Be)n channel identification, which has been obtained by means of the comparison between the experimental and theoretical kinematical locus, as a standard THM procedure [10]. The next step in the analysis was to select the quasi-free contribution, in order to extract this contribution the data were studied as a function of the neutron momentum. The analysis procedure is fully described in [8][9]. The aim of this work was to study the astrophysical S(E)-factor from 5 keV to 1.5 MeV. For this purpose the two sets of experimental data were studied together. The ^{10}B(p,α)^{7}Be astrophyisical factor has been measured for the first time at the Gamow peak at the LNS experiment. In order to study the 10 keV resonance the energy resolution effects and selection of the events of interest for the THM investigation have been carefully evaluated together with the corresponding uncertainties [9]. The THM astrophysical factor in absolute units has been obtained by normalization to direct data [11–13] in a energy range from 200 to 1200 keV. Thanks to this wide normalization range we are able to reach a very low normalization uncertainty (4%). The effect of the reduced normalization error influences the 10 keV astrophysical factor, leading to a more accurate S_b (10 keV) = 3127 ± 583 MeV b, where the uncertainties include statistical, subthreshold level subtraction, normalization, and channel radius uncertainties [8]. In the Figure 2 is shown the comparison of the THM S(E)-factor displayed as black diamonds obtained in [8] with the all data reported in literature as explained in the figure caption. A very good agreement is found between the direct data and the THM ones after normalization. The availability of THM data up to 1.5 MeV represents a very good opportunity to revise the THM low energy S(E)-factor [8]. In fact by performing a weighted scaling of the THM S(E)-factor to the available direct data, the normalization error turned out to be equal to 4% which is significantly smaller than what is given in [9] namely 18%. Moreover we use R-matrix fit of the THM data (in fig 2 as red line), where the fitting parameters of the lowest laying resonances at 10 and 500 keV were fixed to those in [9], which is dominated by a 10 keV resonance. In order to improve this results a new measurement has been performed by using the THM in the same energy interval.

References

[1] A.M. Boesgaard et al. Astrophys. Journ. **621**, 991 (2005)
[2] L. Lamia et al. Astropys. Journ. **881**, 99 (2015)
[3] C. Spitaleri et al. Eur. Phys. J. A **52**, 4 (2016)
[4] G. Baur et al. Phys. Lett. B **178**, 135 (1986)
[5] R.E Tribble et al. Rep. Prog. Phys. **77**, 106901 (2014)
[6] M. Freer et al. Phys Rev C. **85**, 014304 (2012)
[7] H.Yamaguchi et al. Phys Rev C **87**, 034303 (2013)
[8] C. Spitaleri et al. Phys-ReV. C **95**, 035801 (2017)
[9] C. Spitaleri et al. Phys-ReV. C **90**, 035801 (2014)
[10] C. Spitaleri et al. Phys-ReV. C **69**, 055806 (2004)
[11] N. A. Roughton et al, At. Data Nucl. Tables **177**, 23 (1979)
[12] I. Lombardo et al. J. Phys. G: Nucl. Part. Phys. **43**, 045109 (2016).
[13] A. Caciolli et al. Eur. Phys. J. A **52**, 136 (2016)
[14] M.Youn et al. Nucl Phys A **533**, 321 (1991)
[15] C. Angulo et al. Nucl. Phys. A **656**, 3 (1999).

 DOI: 10.1051/epjconf/201716501043

X-ray burst studies with the JENSA gas jet target

Konrad Schmidt[1,2,★], *Kelly A.* Chipps[3], *Sunghoon* Ahn[1,2], *Jacob M.* Allen[4], *Sara* Ayoub[1,2,5], *Daniel W.* Bardayan[4], *Jeffrey C.* Blackmon[6], *Drew* Blankstein[4], *Justin* Browne[1,2,5], *Soomi* Cha[7], *Kyung Yuk* Chae[7], *Jolie* Cizewski[8], *Catherine M.* Deibel[6], *Eric* Deleeuw[1,2,5], *Orlando* Gomez[9], *Uwe* Greife[10], *Ulrike* Hager[1,5], *Matthew R.* Hall[4], *Katherine L.* Jones[11], *Antonios* Kontos[12], *Raymond L.* Kozub[13], *Eunji* Lee[7], *Alex* Lepailleur[8], *Laura E.* Linhardt[1], *Milan* Matos[14], *Zach* Meisel[2,15], *Fernando* Montes[1,2], *Patrick D.* O'Malley[4], *Wei Jia* Ong[1,2,5], *Steven D.* Pain[3], *Alison* Sachs[11], *Hendrik* Schatz[1,2,5], *Kyle T.* Schmitt[11], *Karl* Smith[11], *Michael S.* Smith[3], *Natã F.* Soares de Bem[13,16], *Paul J.* Thompson[11], *Rebecca* Toomey[8,17], and *David* Walter[8]

[1]*National Superconducting Cyclotron Laboratory, East Lansing, MI, USA*
[2]*JINA Center for the Evolution of the Elements, East Lansing, MI, USA*
[3]*Oak Ridge National Laboratory, Oak Ridge, TN, USA*
[4]*University of Notre Dame, Notre Dame, IN, USA*
[5]*Michigan State University, East Lansing, MI, USA*
[6]*Louisiana State University, Baton Rouge, LA, USA*
[7]*Sungkyunkwan University, Suwon, South Korea*
[8]*Rutgers University, Piscataway, NJ, USA*
[9]*Florida International University, Miami, FL, USA*
[10]*Colorado School of Mines, Golden, CO, USA*
[11]*University of Tennessee, Knoxville, TN, USA*
[12]*Massachusetts Institute of Technology, Cambridge, MA, USA*
[13]*Tennessee Technological University, Cookeville, TN, USA*
[14]*International Atomic Energy Agency, Vienna, Austria*
[15]*Ohio University, Athens, OH, USA*
[16]*Federal Center for Technological Education of Minas Gerais, Brazil*
[17]*University of Surrey, Guildford, England, UK*

Abstract. When a neutron star accretes hydrogen and helium from the outer layers of its companion star, thermonuclear burning enables the αp-process as a break out mechanism from the hot CNO cycle. Model calculations predict (α, p) reaction rates significantly affect both the light curves and elemental abundances in the burst ashes. The Jet Experiments in Nuclear Structure and Astrophysics (JENSA) gas jet target enables the direct measurement of previously inaccessible (α,p) reactions with radioactive beams provided by the rare isotope re-accelerator ReA3 at the National Superconducting Cyclotron Laboratory (NSCL), USA. JENSA is going to be the main target for the Recoil Separator for Capture Reactions (SECAR) at the Facility for Rare Isotope Beams (FRIB). Commissioning of JENSA and first experiments at Oak Ridge National Laboratory (ORNL) showed a highly localized, pure gas target with a density of $\sim 10^{19}$ atoms per square centimeter. Preliminary results are presented from the first direct cross section measurement of the $^{34}\mathrm{Ar}(\alpha, \mathrm{p})^{37}\mathrm{K}$ reaction at NSCL.

★e-mail: schmidtk@nscl.msu.edu

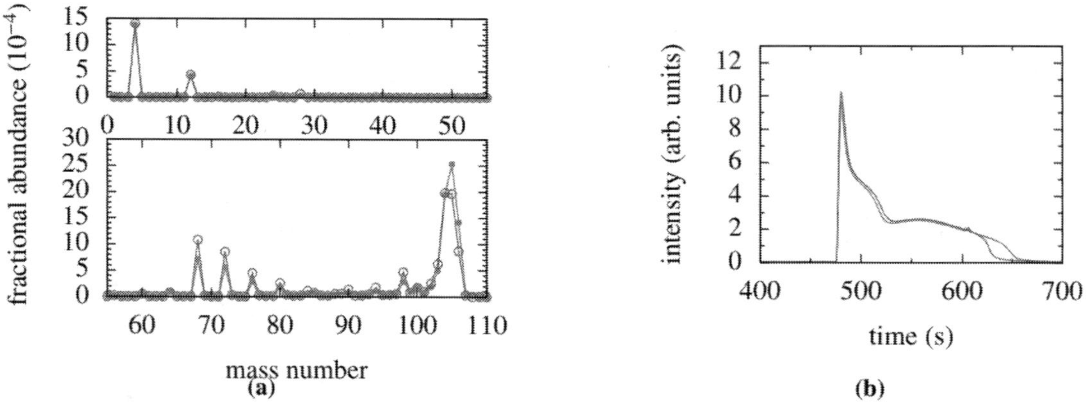

Figure 1: Final abundance plot **(a)** and light curves **(b)** from a single-zone X-ray burst calculation. The nominal rate of the ^{34}Ar(α, p)^{37}K reaction has been decreased (red) and increased (blue) by a factor of 100.

1 Introduction

A type I X-ray burst occurs when a neutron star accretes material from the outer layers of its binary companion. The transfered helium and hydrogen accumulates on the neutron star surface resulting eventually in a thermonuclear explosion, and triggering reaction chains from the triple-α reaction via the hot CNO cycle, breaking out in the αp-process, and leading into the rp-process [1, 2]. Sensitivity studies [3–5] show the αp-process reaction flow passing through the ^{34}Ar(α, p)^{37}K reaction. Only little is known experimentally about the reaction cross section. Theoretical calculations based on the Hauser-Feshbach statistical approach deviate from a recent indirect study [6] by a factor of about 100 at astrophysical energies.

In order to study the influence of the ^{34}Ar(α, p)^{37}K reaction, its nominal rate has been decreased and increased by a factor 100 in a single-zone X-ray burst calculation with an initial hydrogen abundance of 0.5. The reaction shows a broad effect on multiple species in the burst ashes (Figure 1a), as a higher rate increases the hydrogen to seed ratio of the subsequent rp-process. Furthermore, changing only that one reaction rate directly affects the burst light curve (Figure 1b). An increased reaction rate accelerates the reaction flow and the burst burns out faster.

In order to investigate discrepancies between theoretical calculations and indirect measurements, a direct study of the ^{34}Ar(α, p)^{37}K reaction in inverse kinematics has been performed at the National Superconducting Cyclotron Laboratory (NSCL). Since current rare isotope beam intensities are not sufficient to measure at astrophysically relevant energies, the reaction has been studied above the Gamow window, at E_{CM} = 5.8 and 6.1 MeV in the center of mass frame.

2 Experimental Setup

The rare isotope beam facility ReA3 [7] at NSCL on the campus of Michigan State University, USA, provided a radioactive ^{34}Ar beam with an average intensity of ~3000 particles per second (pps). A chemically pure, highly-localized helium target with high density and low energy straggling was generated by the Jet Experiments for Nuclear Structure and Astrophysics (JENSA) gas jet system. The recirculating gas system created a jet of ~10^{19} atoms/cm^2 [8] by compressing the gas to about 30 atmospheres and forcing it through a nozzle with a minimum inner diameter of 1 mm. This high pressure gas constituted the main target area where the (α, p) reaction took place. The jet streamed into a receiver system and the gas was pumped through multiple pumping stages leading back to the compressor. The vanishingly small fraction of the gas from the jet that streamed into the vacuum chamber was evacuated by a magnetically levitated turbo molecular pump. Further upstream along the beam line, four additional turbo molecular pumps reduced the pressure down to 10^{-7} mbar.

The jet was surrounded by multiple highly segmented silicon strip detectors, two Super Oak Ridge Rutgers University Barrels (SuperORRUBA) [9] upstream and downstream of the jet, as well as the Silicon Detector Array (SIDAR) [10] at forward angles. Upstream SuperORRUBA is composed of six single Micron style BB15 silicon detectors of 1-mm thickness. Downstream, four of the six BB15 detectors were stacked each with a Micron style BB10 silicon detector of 65-μm creating ΔE-E telescopes. SIDAR is arranged by five 65-μm and five 1-mm thick Micron style YY1 detectors as telescopes in a lampshade configuration.

A position sensitive ionization chamber (PSIC) [11] was installed downstream of the vacuum chamber, separated by a 3-μm thick Mylar ($H_8C_{10}O_4$) window, to measure beam current and beam composition, as well as reaction recoils. The PSIC was divided in four sections in order to measure the beam and recoil position in X (section 1) and Y (section 2) relative to the beam axis, as well as the energy loss in those first two sections, the intermediate energy loss ΔE (section 3), and the residual energy E_{res} (section 4). The PSIC was filled with isobutane (C_4H_{10}) at a pressure of about 20 mbar chosen to completely stop the beam inside the chamber.

Preamplifier signals from the silicon detectors were passed to the Application Specific Integrated Circuits (ASICs) data acquisition [12, 13] and combined with the Digital Data Acquisition System (DDAS) [14] output from the preamplified PSIC chamber signals, including 32 channels from each of the X- and Y-section, one channel from ΔE and one from E . Signals from both systems were then recorded with a synchronized time stamp.

Beam contaminants were exclusively caused by ^{34}Ar radioactive decay upstream of the JENSA target. The ^{34}Ar (~60%) rare-isotope beam was contaminated with the daughter ^{34}Cl (~30%) and the grand daughter ^{34}S (~10%). Recoils from the (α,p) reactions with ^{34}Ar and with beam contaminants will be disentangled by the different energy losses in each of the four sections in the PSIC.

3 Preliminary Results

Signals in the silicon detector telescope in coincidence with recoil signals from the PSIC are shown in Figure 2a. Particle identification in this ΔE-E spectrum clearly separates scattered α particles and protons from the (α,p) reaction. Note that the acceptance angle of the PSIC is optimized for proton recoils but recoils from α scattering are mostly cut off by the size of the PSIC entrance window. As a consequence, only α particles in coincidence with scattered recoils at small laboratory angles with respect to the beam axis are shown in Figure 2a.

The gated proton signals from Figure 2a follow very well the expected kinematic lines (Figure 2b). Since protons are emitted from decays into various excited states of ^{37}K and ^{37}Ar, as well as into the ground and the first excited state of ^{37}Cl, the measured proton signals in the silicon detector telescopes cover a wide energy spread (Figure 2b) instead of following a single kinematic curve. Data analysis including signals from the eight single silicon detectors (two downstream and six upstream of the jet position) is underway. The ^{34}Ar$(\alpha, p)^{37}$K reaction cross section will be determined from the well-known target thickness [8], detected protons and the effective beam current.

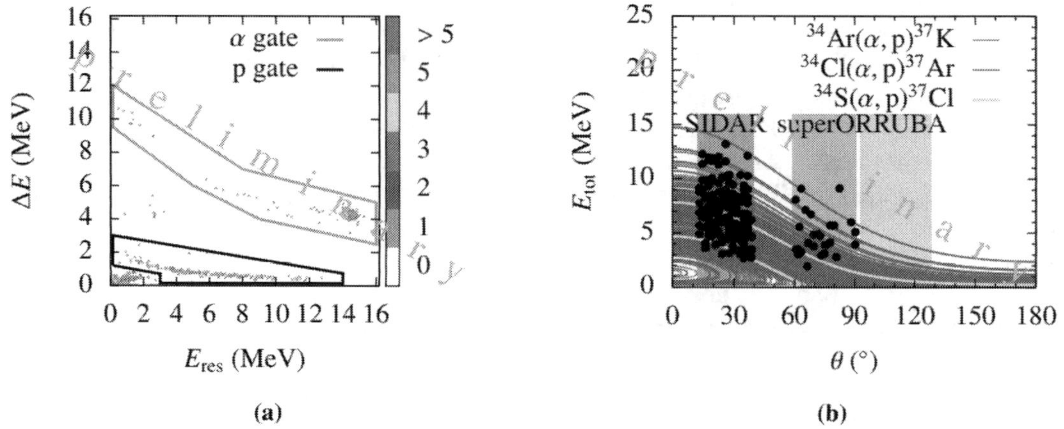

Figure 2: **(a)** Preliminary α and proton signals in coincidence with recoil signals in the PSIC. The PSIC covers the full recoil angle for (α, p) reactions with ^{34}Ar and its beam contaminations. Recoils from α scattering are only partly covered by the PSIC. **(b)** Protons from (a) and kinematic lines. In downstream SuperORRUBA (59° to 91°) only four detectors are arranged as telescopes and contribute to the ΔE-E gate, upstream superORRUBA exclusively comprises single layer detectors. Reactions on ^{34}S emit protons from transitions only to the ground and the first excited state whereas protons from reactions with ^{34}Ar and ^{34}Cl contribute to several excited states.

Acknowledgements

This research is supported by the National Science Foundation in part under Grant No. PHY-1430152 (JINA Center for the Evolution of the Elements), Grant No. PHY-1419765, Grant No. PHY-1404218, by the U. S. Department of Energy (DoE), Office of Nuclear Physics and National Nuclear Security Administration, and by the Laboratory Directed Research and Development Program of Oak Ridge National Laboratory, managed by UT-Battelle, LLC, for DoE.

References

[1] R. K. Wallace and S. E. Woosley, Astrophys. J. Suppl. Ser.**45**, 389 (1981)
[2] H. Schatz *et al.* Phys. Rep. **294**, 167 (1998)
[3] J. L. Fisker, H. Schatz, and F.-K. Thielemann, Astrophys. J. Suppl. Ser.**174**, 261 (2008)
[4] A. Parikh *et al.*, Astrophys. J. Suppl. Ser.**178**, 110 (2008)
[5] R. H. Cyburt *et al.*, Astrophys. J. **830**, 55 (2016)
[6] A. M. Long *et al.*, Phys. Rev. C **95**, 055803 (2017)
[7] D. Leitner *et al.*, Nucl. Instrum. Meth. B **317**, 235 (2013)
[8] K.A. Chipps *et al.*, Nucl. Instrum. Meth. A **763**, 553 (2014)
[9] D. W. Bardayan *et al.*, Nucl. Instrum. Meth. A **711**, 160 (2013)
[10] D. W. Bardayan *et al.*, Phys. Rev. C **63**, 065802 (2001)
[11] K. Y. Chae *et al.*, Nucl. Instrum. Meth. A **751**, 6 (2014)
[12] M. S. Wallace *et al.*, Nucl. Instrum. Meth. A **573**, 418 (2007)
[13] M. S. Wallace *et al.*, Nucl. Instrum. Meth. A **583**, 302 (2007)
[14] K. Starosta *et al.*, Nucl. Instrum. Meth. A **610**, 700 (2007)

EPJ Web of Conferences **165**, 01044 (2017) DOI: 10.1051/epjconf/201716501044
NPA8 2017

Study of nuclear physics input-parameters via high-resolution γ-ray spectroscopy

Philipp Scholz[1,⋆], *Felix* Heim[1], *Jan* Mayer[1], *Mark* Spieker[1], and *Andreas* Zilges[1]

[1] *Institut für Kernphysik, Universität zu Köln, Zülpicher Str. 77, 50937 Köln, Germany*

Abstract. For nucleosynthesis networks of isotopes heavier than those in the iron-peak region, reaction rates are often calculated within the Hauser-Feshbach statistical model. The accuracy of the predicted cross sections strongly depends on the uncertainties of the nuclear-physics input-parameters. These are nuclear-level densities, γ-ray strength functions, and particle+nucleus optical-model potentials. Here we report on the investigation of statistical properties of nuclei via radiative proton-capture reactions using high-resolution γ-ray spectroscopy.

1 Introduction

The Hauser-Feshbach (HF) statistical model is widely used to predict reactions rates for nucleosynthesis process like the *r* process [1] or the various *p* processes [2]. Uncertainties in the predictions of these reaction rates can vary by orders of magnitude depending on the nuclear-physics models which are entering the HF calculations. Therefore, one key element in understanding the nucleosynthesis in stellar environments is a robust basis of models for nuclear-physics input as nuclear-level densities (NLDs) and γ-ray strength functions (γSFs) [3].

In addition to their direct impact to reaction-network calculations, cross-section measurements of radiative charged-particle reactions can be a useful tool to investigate the statistical properties of nuclei. Using the high-efficiency γ-ray spectrometer HORUS at the Institute for Nuclear Physics at the University of Cologne [4], the γ-decay behavior of reaction products can be investigated with very high resolution. Here we report on two methods which can be used to study γSFs as well as NLDs; the measurement of partial (p,γ) cross sections and the study of two-step γ-ray cascades.

2 Radiative proton-capture on ^{92}Mo and two step γ-ray cascades

The isotope ^{92}Mo is the isotopically most abundant *p* nucleus [5] and, therefore, attracted a lot of attention to its production sites and mechanisms [2]. As the inverse reaction of the γ-induced proton emission on ^{93}Tc within the γ process, the ^{92}Mo(p,γ) reaction is one of the most studied (see Ref. [6] and references therein). However, previous experiments using different techniques solely focused on low proton energies close to or inside the astrophysically relevant energy region. At these low energies huge fluctuations in the measured cross-section values were reported which hampered the comparison

⋆e-mail: pscholz@ikp.uni-koeln.de

 DOI: 10.1051/epjconf/201716501044

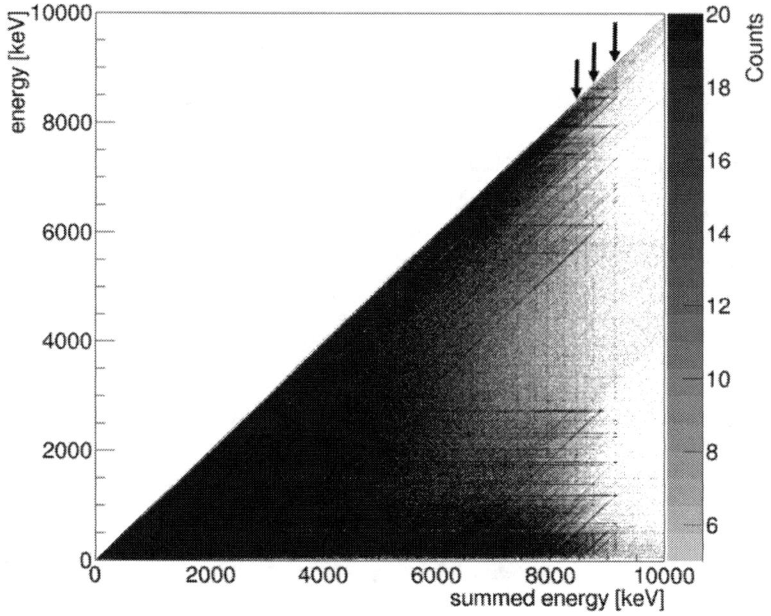

Figure 1. The extracted TSC matrix for ^{92}Mo for a proton energy of 5.1 MeV. Vertical cuts in the TSC matrix correspond to TSC spectra for the deexcitation to different levels in ^{93}Tc via two step γ-ray cascades. Two step cascade energies for the ground-state and the first two exited states are indicated by arrows.

to HF calculations and, thus, conclusion on nuclear-physics models. Therefore, cross-section measurements were performed at proton energies between 3.7 MeV and 5.3 MeV via the in-beam method at the HORUS γ-ray spectrometer [7]. The in-beam method with high-purity germanium detectors (HPGe) enables the determination of absolute total cross sections via the detections of all ground-state transitions in the reaction product. Additionally partial cross-sections for the radiative capture to excited levels can be determined (see, e.g, [8, 9] and references therein).

As it was shown for the ^{89}Y(p,γ)^{90}Zr reaction [10], partial cross-sections for the population of excited levels in the reaction product yield insight into the γ-strength of a nucleus. Different models for the γSF combined with a specific NLD model and a proton+nucleus optical model potential can describe total reaction cross-sections well enough although they might not be able to describe the statistical γ-decays during the deexcitation of the compound nucleus. In contrast to that, partial cross-sections yield information on the probability of γ-ray emission depending on the γ-ray energies.

For the case of ^{92}Mo(p,γ) it was shown that a combination of M1-strength from shell model calculations and E1-strength from the Skyrme-/Gogny-HFB+QRPA model used for the HF calculations describe the total cross section much better than the default settings of the TALYS code (Standard Lorentzian/Constant Temperature model) [7]. However, none of the models were able to successfully describe all partial cross sections. Please note, that the comparison of HF calculation results to partial- and total cross sections of radiative proton-capture reactions is not suitable for a direct determination of neither γSF nor NLD and, hence, other methods need to be used to confirm the results.

Another approach to verify γSF and NLD models simultanously is the method of two step cascades (TSC) [11]. Originally developed for radiative neutron-capture reactions, this technique was applied recently to radiative proton-capture reactions as well [12] and served as additional confirmation for results obtained via the Oslo Method [13]. The advantage in the use of protons instead of neutrons is the smaller influence of Porter-Thomas fluctuations on the measured spectra [14].

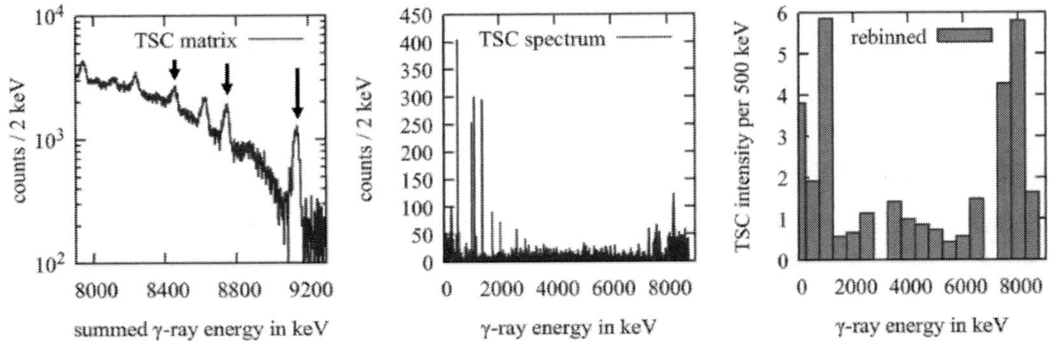

Figure 2. In the projection of the TSC matrix on to the axis of summed γ-ray energies (a), peaks for the population of different states via two-setp cascades in ^{93}Tc are visible. Via gating on of the summing peaks in the TSC matrix, the TSC spectrum (b) can be obtained showing the energies of the γ-rays as a part of the cascade. For the comparison to the results of modeling the TSC spectra based on different inputs for the γ-strength and level-density, TSC spectra are usually presented in wider energy bins, here 500 keV (c).

The idea of this technique is the investigation of statistical γ-decay behavior via the detection of coincident γ-rays. If a spectrum of individual γ-rays of a two step cascade from an inital state i to a final state f via an intermediate state m in the reaction product is measured, the intensity $I_{if}(E_1, E_2)$ in the so-called TSC spectrum must follow the equation

$$I_{if}(E_1, E_2) = \sum_{XL,XL',J_m^\pi} \frac{\Gamma_{im}^{XL}(E_1)}{\Gamma_i} \rho(E_m, J_m^\pi) \frac{\Gamma_{mf}^{XL'}(E_2)}{\Gamma_m} + \sum_{XL,XL',J_{m'}^\pi} \frac{\Gamma_{im'}^{XL}(E_1)}{\Gamma_i} \rho(E_{m'}, J_{m'}^\pi) \frac{\Gamma_{m'f}^{XL'}(E_2)}{\Gamma_{m'}} \quad (1)$$

where $\Gamma_{im}^{XL}(E)$ is the γ-decay width for a γ-transition of character and multipolarity XL and energy E and $\rho(E, J)$ is the level density for a specific spin J and parity π at the intermediate energy E_m [14]. Because the order of both γ-rays is indistinguishable, the term in Eq. 1 appears twice.

Since the digital data acquisition at the HORUS γ-ray spectrometer allows for the measurement of $\gamma\gamma$ coincidence data [4] the method can naturally be applied to data measured during previous experiments aiming for the determination of cross-section values for, *e.g.*, ^{92}Mo(p,γ).

At first, the so-called TSC matrix is extracted from the symmetric $\gamma\gamma$ coincidence matrix by sorting events with multiple hits in pairs into a 2D-histogram depending on the single and the total (summed) γ-ray energy (c.f. Fig. 1). As it can be seen in Fig. 1, the TSC matrix shows rather weak vertical lines amongst others at summed energies of 9.1 MeV, 8.7 MeV, and 8.4 MeV representing γ-ray cascades populating the ground-state, the first, and the second excited state in ^{93}Tc, respectively.

Figure 2 shows from the left to the right an example of how the final TSC spectra can be obtained from the 5.1 MeV measurement of the ^{92}Mo(p,γ) reaction. In Fig. 2 a) the projection of the TSC matrix onto the summed energies is plotted showing summing-peaks for the population of different levels in ^{93}Tc. Via gating on one of these summing-peaks, the spectrum of two step γ-ray cascades populating the specific level can be obtained (c.f. Fig. 2 b)). In the low energy and high-energy part of these TSC spectra discrete γ-ray transistion from excited states in ^{93}Tc to the ground state and the prompt γ-rays from the entry state to a intermediate state are visible, respectively. At medium energies, statistical γ-ray transitions appear whose intensities strongly depend on the γSF and NLD of ^{93}Tc. Once the TSC spectrum is corrected for coincidence-efficiencies, normalized onto the absolute number of proton captures, and rebinned, the symmetric nature of the TSC spectrum becomes apparent (see Fig. 2 c)) and the TSC intensity of the energy bins can be compared to Eq. 1. Usually this is done utilizing the DICEBOX algorithm [15] by varying the input for γSF and NLD. Via such comparisons structures like the low-energy upbend [12] or scissor modes [16] in the γSF have already been investigated.

3 Measurement of $^{63}\mathrm{Cu}(p,\gamma\gamma)^{64}\mathrm{Zn}$ and $^{65}\mathrm{Cu}(p,\gamma\gamma)^{66}\mathrm{Zn}$

Recently, the first radiative proton-capture experiment at the HORUS γ-ray spectrometer was specifically performed for applying the TSC method to the ^{63}Cu(p,$\gamma\gamma$)^{64}Zn and ^{65}Cu(p,$\gamma\gamma$)^{66}Zn reactions at 3.5 MeV and 2.0 MeV, respectively. The beam energies were chosen to be below the respective (p,n)-reaction thresholds to suppress any contribution from this channel. With Q-values of 7.7 MeV and 8.9 MeV, NLD and γSF models can be studied up to excitation energies of 11.2 MeV and 10.9 MeV, respectively.

In both cases, summing-peaks for the populations of several excited states in the reaction products were observed and can be used to extract TSC spectra. Until the the final TSC intensities can be compared to theory, a careful investigation of the background in the TSC spectra has to be carried out via GEANT4 simulations. However, the application of the TSC method to the data obtained with the HORUS γ-ray spectrometer seems promising and final results for the studied reactions should be available soon.

Acknowledgment

The authors thank A. Dewald and the accelerator staff at the Institute for Nuclear Physics in Cologne for providing excellent beams. This project was supported by the Deutsche Forschungsgemeinschaft under Contract No. ZI 510/8-1 and the ULDETIS project within the University of Cologne Excellence Initiative institutional strategy. J.M. and P.S. are supported by the Bonn-Cologne Graduate School of Physics and Astronomy.

References

[1] M. Arnould, S. Goriely, and K. Takahashi, Phys. Rep. **450** (2007) 97.
[2] M. Arnould and S. Goriely, Phys. Rep. **384** (2003) 1.
[3] T. Rauscher, Int. J. Mod. Phys. E **20** (2011) 1071.
[4] L. Netterdon *et al.*, Nucl. Instr. Meth. A **754** (2014)94.
[5] Michael Berglund and Michael E. Wieser, Pure Appl. Chem. **83** (2011) 397.
[6] G. Gyürky *et al.* Nucl. Phys. A **922** (2014) 112.
[7] J. Mayer *et al.*, Phys. Rev. C **93** (2016) 045809.
[8] A. Sauerwein *et al.*, Phys. Rev. C **86** (2012) 035802.
[9] S. Harissopulos. *et al.*, Phys. Rev C **93** (2016) 025804.
[10] L. Netterdon *et al.*, Phys. Lett. B **744** (2015) 358.
[11] F. Bečvář *et al.*, Phys. Rev. C **46** (1992) 1276.
[12] A. Voinov *et al.*, Phys. Rev. Lett. **93** (2004) 142504.
[13] A. Schiller, L. Bergholt, M. Guttormsen, E. Melby, J. Rekstad, and S. Siem, Nucl. Instrum. Methods Phys. Res. A **447**, 498 (2000).
[14] A. Voinov *et al.* EPJ Web of Conferences **93**, 01022 (2015).
[15] F. Bečvář, Nucl. Instr. Meth. A **417**, 434 (1998).
[16] F. Bečvář *et al.*, Nucl. Instr. Meth. B **261** (2007) 930.

EPJ Web of Conferences **165**, 01045 (2017) DOI: 10.1051/epjconf/201716501045
NPA8 2017

The ν process in the innermost supernova ejecta

Andre Sieverding[1,2,★], *Gabriel* Martínez Pinedo[1,2,], *Karlheinz* Langanke[2,1,], *J. Austin* Harris[3,4,], and *W. Raphael* Hix[4,5,]

[1] *Institute for Nuclear Physics (Theory Department), Darmstadt University of Technology, Schlossgartenstraße 2, 64289 Darmstadt, Germany*
[2] *Helmholtz Center for Heavy Ion Research (GSI), Planckstr. 1, 64259 Darmstadt, Germany*
[3] *Nuclear Science Division, Lawrence Berkeley National Laboratory, Berkeley, CA 94720, USA*
[4] *Physics Division, Oak Ridge National Laboratory, P.O. Box 2008, Oak Ridge, TN 37831-6354, USA*
[5] *Department of Physics and Astronomy, University of Tennessee, Knoxville, TN 37996-1200, USA*

Abstract. The neutrino-induced nucleosynthesis (ν process) in supernova explosions of massive stars of solar metallicity with initial main sequence masses between 13 and 30 $M_\odot$ has been studied with an analytic explosion model using a new extensive set of neutrino-nucleus cross-sections and spectral properties that agree with modern supernova simulations. The production factors for the nuclei ^{7}Li, ^{11}B, ^{19}F, ^{138}La and ^{180}Ta, are still significantly enhanced but do not reproduce the full solar abundances. We study the possible contribution of the innermost supernova eject to the production of the light elements ^{7}Li and ^{11}B with tracer particles based on a 2D supernova simulation of a 12 $M_\odot$ progenitor and conclude, that a contribution exists but is negligible for the total yield for this explosion model.

1 Introduction

Core Collapse Supernovae are the most luminous source of neutrinos in nature that we know today. All flavors of neutrinos are emitted from the hot and dense environment of a collapsing massive star. Even though neutrinos interact with matter only via the charged- and neutral current channels of the weak interaction, they appear in the Supernova environment in such tremendous numbers that they not only help to revive the explosion shock wave but also interact with the material in the outer layers of star, that has been processed by the hydrostatic burning stages during the life of the progenitor star. The neutrinos coming from the deep interior have high energies compared to the thermal energies encountered in the stellar mantle and can either be captured on nuclei in an inverse β decay or induce spallation reactions, i.e. lead to nuclear excitations that decay by particle emission. The effect of these interactions on the final composition of the ejecta is called the ν process and has been suggested to contribute to the production of ^{7}Li, ^{11}B and ^{19}F as well as ^{138}La and ^{180}Ta. This process has been included in most of the recent nucleosynthesis surveys (e.g. [1]).

★e-mail: asiever@theorie.ikp.physik.tu-darmstadt.de

Table 1. Production factors relative to solar abundances from reference [7], normalized to ^{16}O production averaged with a Salpeter IMF over the progenitor range 13-30 $M_\odot$. Shown are the results obtained without neutrinos, with the set of low neutrino energies that are in agreement with recent supernova simulations and with the traditional choice of high neutrino energies as described in the text. For ^{180}Ta is assumed that 35% of its amount at 200 s after the onset of the explosion remain in the long-lived isomeric state [8]. For a detailed discussion of this aspect see also [9].

Nucleus	^{7}Li	^{11}B	^{19}F	^{138}La	^{180}Ta
no ν	0.001	0.01	0.34	0.13	0.17
Low energies	0.05	0.26	0.42	0.56	0.47
High energies	0.36	0.99	0.56	1.19	0.84

2 The ν process in a simple 1D model

One of the first detailed studies of the ν process was presented in ref. [2] using an analytic approximation to describe the thermodynamic conditions of the supernova shock heating and expansion for a range of the compositional layers that are expected to be found in massive star as a result of the hydrostatic burning stages. Following this approach we have calculated the nucleosynthesis for a range of progenitor models with initial masses between 13 and 30 $M_\odot$ from ref. [1] with the same analytic model that is based on a radiation dominated shock and assuming a total thermal explosion energy of 10^{51}erg. The evolution of the nuclear abundances starting from the composition of the progenitor models is followed with a nuclear reaction network including 1988 species and an extensive set of neutrino-nucleus cross section partially based on experimental data for key reactions and including multi-particle emission channels. For more details see also ref. [3]. The neutrinos are assumed to follow Fermi-Dirac distributions with chemical potential $\mu_\nu = 0$.

Previous studies of the ν process (e.g. ref. [4]) have considered neutrinos with relatively high average energies of $\langle E_{\nu_e}\rangle = 12.6$ MeV), $\langle E_{\bar{\nu}_e}\rangle = 15.8$ MeV), and $\langle E_{\nu_{\mu,\tau}}\rangle = 18.8$ MeV. This choice of neutrino energies will in the following be denoted as high energies. Recent advances in supernova simulations that include more extensive and accurate neutrino opacities within the neutrino transport in multiple dimensions have lead to the conclusion that the neutrino energies could even be much lower than these values (e.g. [5]). Therefore we consider here a set of low energies with $\langle E_{\nu_e}\rangle =$ 8.8 MeV), $\langle E_{\bar{\nu}_e}\rangle = 12.6$ MeV), and $\langle E_{\nu_{\mu,\tau}}\rangle = 12.6$ MeV. Table 1 shows the resulting production factors normalized to ^{16}O production and averaged over the range of progenitors considered here with a Salpeter-like initial mass function (IMF) following $dn_*/dm_* \propto m_*^{-1.35}$. With the lower energies the effect of the ν process is reduced compared to the case of the more energetic spectra from previous studies. However, several of this nuclei can also have contributions from other processes, e.g. ref. [6] suggests that up to 70% of the solar abundance of ^{11}B can be provided by Cosmic Ray spallation in the interstellar medium. Hence, a production factor of around 0.3 from supernova ejecta is consistent with the solar abundance. The ν process cannot contribute significantly to the production of ^{7}Li when the lower neutrino energies are considered. The production of light elements in the outer regions can also be affected by neutrino flavor transformations, see e.g. [10, 11]. Supernovae of massive stars can contribute, but are probably not the primary source for the solar abundance of ^{19}F. The heavier nuclei ^{138}La and ^{180}Ta are still produced up to half of their solar abundance. A large fraction of the total yield of ^{11}B is produced in the C-rich layer via ^{12}C$(\nu, \nu' p/n)$. Another big contribution comes from the reaction chain ^{3}He$(\alpha,\gamma)^7$Be$(\alpha,\gamma)^{11}$C and ^{3}H$(\alpha,\gamma)^7$Li$(\alpha,\gamma)^{11}$B which is made possible by ^{4}He$(\nu, \nu' n/p)$. This process also produces ^{7}Li, mostly as ^{7}Be and it occurs mostly in the lower He shell but can also in the innermost regions where the composition results from an α rich freeze out from NSE in Si burning. Large parts of this material could eventually fall back and get accreted onto the neutron star. This effect is commonly accounted for in 1D calculation by assuming an artificial

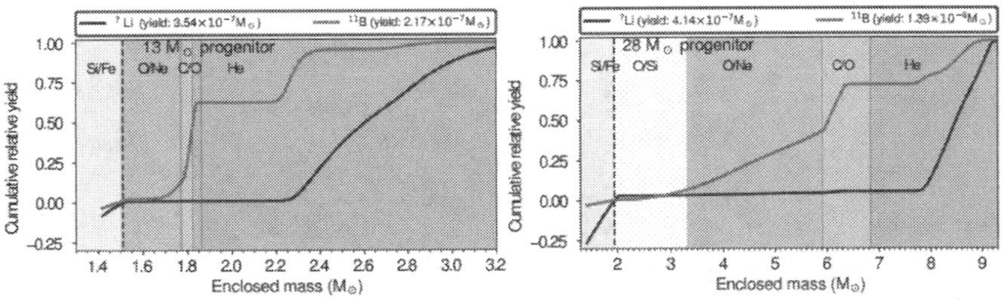

Figure 1. Cumulated relative yields for ^{7}Li and ^{11}B over the profile of the 13 and 28 M$_\odot$ models considered here. The dashed black horizontal line indicates the position of the mass cut. Everything below the mass cut is commonly not counted as part of the ejecta and would not contribute to the yield. This part is shown here as negative.

mass cut, below which material is not counted for the ejecta. Figure 1 shows the cumulated yield normalized to the total yield taking into account a mass cut at the edge of the Fe core. It shows that ^{7}Li is almost exclusively made in the He shell while ^{11}B has also important contributions from the C layer. The contribution from the α rich freeze out, which appears as negative when it is below the mass cut in Figure 1, is rather negligible in the 13 M$_\odot$ model but it increases for more massive progenitor stars. For the 28 M$_\odot$ progenitor an amount of more than 20% of the total yield of ^{7}Li could be contributed the innermost region that is hidden by the mass cut.

3 The ν process in 2D

The fate of the material from the innermost regions is sensitive to the details of the explosion dynamics. In two- and three dimensional simulations matter from this region below the mass cut can be ejected and convection could lead to a longer exposure of the material to the strong neutrino radiation close to the core. In order to explore the role of the ν process for the nucleosynthesis in the innermost ejecta, we have performed nuclear reaction network calculations for the thermodynamic histories of 4000 tracer particles from a self consistent 2D supernova simulation based on a 12 M$_\odot$ progenitor presented in ref. [12, 13]. The tracer particles are representative for the innermost 2 M$_\odot$ of the star, extending well into the lower part of the C-rich layer. Figure 2 shows the resulting mass fractions of ^{7}Li as a function of the initial position of the tracer particles and as projected onto the corresponding mass coordinate. The total yield is calculated by combining the results of the tracer particles for the inner region with the data from the 1D explosion model for the rest of the star. From these calculations we can see that material from the α-rich freeze out, that gets enriched in ^{7}Li and ^{11}B is ejected. However, for this progenitor the contribution to the total yield is negligible. Since we have seen in our 1D study that the production in the α rich freeze out can be more important for more massive progenitors, it is possible that a noticeable contribution from the innermost regions could result.

This work was partly supported by the Deutsche Forschungsgemeinschaft through contract SFB 1245, AH was supported by an Australian Research Council (ARC) Future Fellowship (FT120100363). WRH was supported by the U.S. Department of Energy, Office of Science, Offices of Nuclear Physics and the National Science Foundation Nuclear Theory Program (PHY-1516197). The Harris et al (2017) simulations were performed with resources of the National Energy Research Scientific Computing Center, supported by the U.S. DOE Office of Science under Contract No. DE-AC02-05CH11231.

DOI: 10.1051/epjconf/201716501045

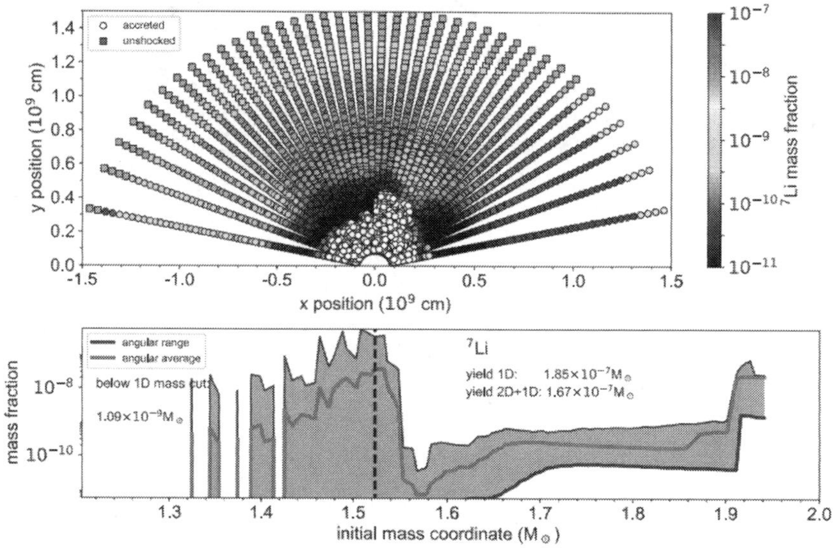

Figure 2. Mass fraction of ^{7}Li for the tracer particles from a self-consistent axisymmetric supernova simulation. The upper panel shows the mass fraction as color code for the initial position of the tracer particles. The lower panel shows the projection of these mass fractions onto the mass coordinate. Shown are is the range of mass fractions for different angles and the angular average.

References

[1] T. Rauscher, A. Heger, R.D. Hoffman, S.E. Woosley, Astrophys. J. **576**, 323 (2002)

[2] S.E. Woosley, D.H. Hartmann, R.D. Hoffman, W.C. Haxton, Astrophys. J. **356**, 272 (1990)

[3] A. Sieverding, L. Huther, G. Martínez-Pinedo, K. Langanke, A. Heger, *Neutrino nucleosynthesis in core-collapse Supernova explosions*, in *European Physical Journal Web of Conferences* (2016), Vol. 109 of *European Physical Journal Web of Conferences*, p. 06004

[4] A. Heger, E. Kolbe, W. Haxton, K. Langanke, G. Martínez-Pinedo, S.E. Woosley, Phys. Lett. B **606**, 258 (2005)

[5] A. Mirizzi, I. Tamborra, H.T. Janka, N. Saviano, K. Scholberg, R. Bollig, L. Hüdepohl, S. Chakraborty, Riv. del Nuovo Cim. **39**, 1 (2016)

[6] S.M. Austin, C. West, A. Heger, Phys. Rev. Lett. **112**, 111101 (2014)

[7] K. Lodders, Astrophys. J. **591**, 1220 (2003)

[8] P. Mohr, F. Käppeler, R. Gallino, Phys. Rev. C **75**, 012802 (2007)

[9] T. Hayakawa, T. Kajino, S. Chiba, G.J. Mathews, Phys. Rev. C **81**, 052801 (2010)

[10] T. Yoshida, T. Kajino, H. Yokomakura, K. Kimura, A. Takamura, D.H. Hartmann, Physical Review Letters **96**, 091101 (2006)

[11] G.J. Mathews, T. Kajino, W. Aoki, W. Fujiya, J.B. Pitts, Phys. Rev. D **85**, 105023 (2012)

[12] S.W. Bruenn, E.J. Lentz, W.R. Hix, A. Mezzacappa, J.A. Harris, O.E.B. Messer, E. Endeve, J.M. Blondin, M.A. Chertkow, E.J. Lingerfelt et al., ApJ **818**, 123 (2016)

[13] J.A. Harris, W.R. Hix, M.A. Chertkow, C.T. Lee, E.J. Lentz, O.E.B. Messer, ApJ **843**, 2 (2017)

EPJ Web of Conferences **165**, 01046 (2017) DOI: 10.1051/epjconf/201716501046
NPA8 2017

Search for Deeply Bound Kaonic Nuclear States with AMADEUS

Magdalena Skurzok[1,⋆], *Michael* Cargnelli[2], *Catalina* Curceanu[3], *Raffaele* Del Grande[3], *Laura* Fabbietti[4,5], *Carlo* Guaraldo[3], *Johann* Marton[2], *Pawel* Moskal[1], *Kristian* Piscicchia[6,3], *Alessandro* Scordo[3], *Michal* Silarski[1], *Diana Laura* Sirghi[3], *Ivana* Tucakovic[7], *Oton* Vazquez Doce[4,5], *Sławomir* Wycech[8], *Eberhard* Widmann[2], and *Johann* Zmeskal[2]

[1] *Institute of Physics, Jagiellonian University, 30-059 Krakow, Poland*
[2] *Stefan-Meyer-Institut für subatomare Physik, Vienna, Austria*
[3] *INFN, Laboratori Nazionali di Frascati, 00044 Frascati, Italy*
[4] *Excellence Cluster "Origin and Structure of the Universe", 85748 Garching, Germany*
[5] *Physik Department E12, Technische Universität München, 85748 Garching, Germany*
[6] *CENTRO FERMI - Museo Storico della Fisica e Centro Studi e Ricerche "Enrico Fermi", Roma, Italy*
[7] *Ruger Bosković Institute, Zagreb, Croatia*
[8] *National Centre for Nuclear Research, 00681 Warsaw, Poland*

Abstract. We briefly report on the search for Deeply Bound Kaonic Nuclear States with AMADEUS in the $\Sigma^0 p$ channel following K^- absorption on ^{12}C and outline future perspectives for this work.

1 Introduction

The existence or not of the Deeply Bound Kaonic Nuclear States (DBKNS) is, currently, one of the hottest topics in nuclear and hadronic strangeness physics, from both experimental and theoretical points of view. The existence of bound kaonic nuclear states, also called kaonic nuclear clusters, was first predicted in 1986 [1]. The mass of the $\Lambda(1405)$ reflects the strength of the $\bar{\mathrm{K}}\mathrm{N}$ interaction, thus influencing the possible formation of $\bar{\mathrm{K}}$ multi-nucleon bound states. Recent theoretical calculations, based on different approaches, deliver a wide range of binding energies and widths for the di-baryonic kaonic bound state ppK^- [2–7] while experimental results remain contradictory [8–14]. Moreover, the extraction of a ppK^- signal in K^- absorption experiments is strongly affected by the dynamics of the competing K^- multi-nucleon absorption processes. Therefore, in order to clarify this issue, experimental data are needed. This research is very important in understanding the fundamental laws of Nature and the Universe and can have important consequences in various sectors of physics, such as nuclear and particle physics, as well as astrophysics. The binding of the kaon in the nuclear medium may impact on models describing the structure of neutron stars through their Equation of State [15, 16] and collapsing binaries which are now known to be sources of gravitational waves [17]. Investigation of stable forms of strange matter like DBKNS in extreme conditions would be helpful for a better understanding of elementary kaon-nucleon interaction for low energies in non-perturbative quantum chromodynamics (QCD) and would contribute to solving crucial problems in hadron physics: hadron

⋆e-mail: magdalena.skurzok@uj.edu.pl

masses (related to the chiral symmetry breaking), hadron interactions in the nuclear medium and the structure of the dense nuclear matter.

The AMADEUS group has developed a method with a high probability for the discovery of DBKNS corresponding to K^-pp , K^-ppn and K^-ppnn kaonic nuclear clusters and their decays to Σ^0/Λp, Σ^0/Λd and Σ^0/Λt, respectively. The method is based on the exclusive measurement of the momentum, angular and invariant mass spectra for correlated Σ^0/Λp, Σ^0/Λd, Σ^0/Λt pairs [18]. Possible DBKNS may be produced with K^- stopped in helium or carbon and then decaying into the considered decay channels. The AMADEUS experiment is conceived to perform a systematic investigation of the low-energy $\bar{K}$N interaction, taking advantage of the kaons beams delivered by the DAΦNE collider [19]. The ongoing AMADEUS analyses refer to two data samples. One is represented by the data collected by the KLOE [20] collaboration during the 2004/2005 data taking, corresponding to ~ 1.74 fb^{-1}. The KLOE detector [21] is used as an active target, with the hadronic interaction of negative kaons with the materials of the apparatus being investigated; in particular K$^-$-^{9}Be absorptions in the DAΦNE thin Beryllium cylindrical layer and the DAΦNE aluminised beryllium pipe, K$^-$-^{12}C and K$^-$-H absorptions in the KLOE Drift Chamber (DC) inner wall (aluminised carbon fiber) [22], and K$^-$-^{4}He in the DC gas. Extremely rich experimental information is contained in this sample, with K$^-$ hadronic captures both at rest and in-flight [23]. In order to increase the statistics, and as an essential interpretation tool, a high purity carbon target (graphite) was produced in summer 2012 and installed inside the KLOE detector, between the beam pipe and the DC inner wall. The geometry of the target was optimized to maximize the kaon stopping power. The total collected integrated luminosity was ~90 pb^{-1}. Up till now sample of 37 pb^{-1} reconstructed data was analysed. The adopted procedure is to exclusively measure the correlated Σ^0/Λp, Σ^0/Λd and Σ^0/Λt pairs, searching for K^-pp , K^-ppn and K^-ppnn multi-nucleon, and related bound states, production. In the next section we present briefly results of data analysis devoted to the Σ^0p channel. A more detailed description can be found in Ref. [24].

2 Σ^0p data analysis

The analysis presented below is focused on studies of K^- absorption processes inside the Drift Chamber (DC) carbon wall leading to Σ^0p final states [24]. Selection of $\Lambda(1116)$ hyperons being the signature of the K^- hadronic interaction in ^{12}C was the first aim of the data analysis. Λs were identified by reconstruction of their decay vertex $\Lambda \rightarrow p + \pi^-$ (BR = 63.8%). The protons and negatively charged pions selection and identification was carried out using the dE/dx information in DC installed around the interaction point and the track momenta. The proton-pion invariant mass was determined under the p and π^- mass hypothesis. The constructed $p\pi^-$ invariant mass resulted in a value of 1115.753±0.002 MeV/c^2 and a statistical error of 0.5 MeV/c^2. Then, Λ-proton vertices were searched for and K^- absorption processes inside the DC wall were selected based on the radial position $\rho_{\Lambda p}$. The identification of Σ^0 candidates was carried out through their decay into $\Lambda\gamma$ pairs where γ quanta were identified in the Calorimeter [25]. Afterwards, a simultaneous global fit to the $\Sigma^0 p$ invariant mass, $\cos(\theta_{\Sigma^0 p})$ and the Σ^0 and the proton momenta was performed in order to extract contributions from the different absorption processes like 2 Nucleon Absorption (2NA), with final state interaction (FSI) or without (QF), 3NA and 4NA. For this purpose Monte Carlo simulations of the absorption processes in ^{12}C have been carried out. The final fit delivered the contribution of the different absorption processes. The fit results in the first measurement of the 2NA free from FSI which was found to be 12% of the total absorption cross-section. The contribution from the K^-pp bound state was included in the second fit of the experimental data. A Breit-Wigner shape was used to simulate the bound state contribution, scanning a grid of possible bindings (15-75 MeV/c^2) and widths (30-70 MeV/c^2). The best value of the total reduced χ^2 was achieved for the hypothesis of a binding energy of 45 MeV/c^2 and width 30 MeV/c^2 and are shown in Fig. 1. The K^-pp yield extracted from the fit is K^-pp/K^-_{stop}=$(0.044\pm0.009 stat^{+0.004}_{-0.005} syst)\cdot 10^{-2}$. However the 1σ significance of the result doesn't allow the observation of the bound state to be claimed.

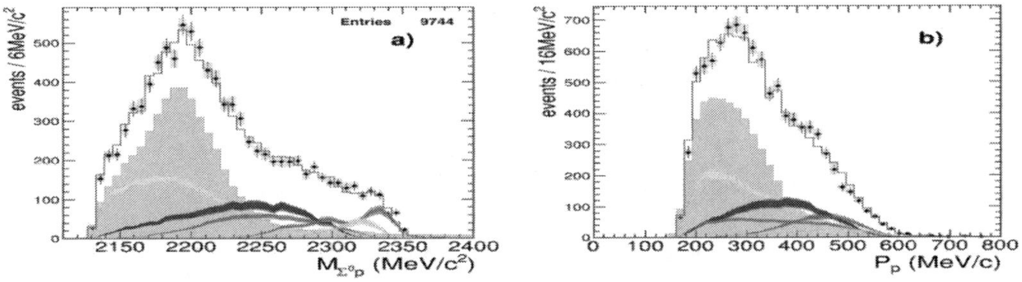

Figure 1. The simultaneous fit to the $\Sigma^0 p$ invariant mass (left) and proton momentum distributions (right) performed for kaons stopped in DC wall. The colors represent the following fit contributions: π^0 background (grey), 4NA with uncorrelated production of the $\Sigma^0 p$ final state (cyan), 3NA (blue), 2NA-FSI (magenta) and 2NA-QF (red). The black line shows the overall fit. Figure is adopted from [24].

3 Summary and conclusion

One of the AMADEUS's goals is to search for Deeply Bound Kaonic Nuclear States by studies of K^- absorption processes in various targets. In this work investigation of $\Sigma^0 p$ final state was presented. It results in the extraction of contributions from various few nucleon absorption processes. The contribution from a possible K^-pp bound state was determined, however the significance of the result is not sufficient to claim the observation of a bound system. Presently, the analysis of the channels like Λp, Λd and Λt is in progress. Moreover, a feasibility study [26, 27] is ongoing for the realization of a dedicated AMADEUS experimental setup, in order to deepen and extend the low energy anti-kaon nuclei interaction studies and obtain fundamental input for the understanding of the QCD with strangeness.

Acknowledgements

We acknowledge the KLOE Collaboration for their support and for having provided us with the data and the tools to perform the analysis presented in this paper. We also thank as well the DAΦNE staff for the excellent working conditions and permanent support. We acknowledge the CENTRO FERMI - Museo Storico della Fisica e Centro Studi e Ricerche 'Enrico Fermi', for the progect PAMQ. Part of this work was supported by the Austrian Science Fund (FWF): [P24756-N20]; Austrian Federal Ministry of Science and Research BMBWK 650962/0001 VI/2/2009; the Grantïn-Aid for Specially Promoted Research (20002003); Minstero degli Affari Esteri e della Cooperazione Internazionale, Direzione Generale per la Promozione del Sistema Paese (MAECI), Strange Matter project; Polish National Science Center, grant No. UMO-2016/21/D/ST2/01155.

References

[1] S. Wycech, Nucl. Phys. A450 (1986) 399c.
[2] T. Yamazaki *et al.*, Phys. Rev. C76 (2007) 045201.
[3] S. Wycech *et al.*, Phys. Rev. C79 (2009) 014001.
[4] N. Barnea *et al.*, Phys. Lett. B712 (2012) 132.

[5] E. Oset *et al.*, Nucl. Phys. A881 (2012) 127.
[6] A. Dote, T. Hyodo, W. Weise, Phys. Rev. C79 (2009) 014003.
[7] Y. Ikeda, T. Sato, Phys. Rev. C79 (2009) 035201.
[8] G. Agakishiev *et al.*, HADES Coll., Phys. Lett. B742 (2015) 242.
[9] M. Agnello *et al.* (FINUDA Coll.), Phys. Rev. Lett. 94 (2005) 212303.
[10] T. Yamazaki *et al.*, Phys. Rev. Lett. 104 (2010) 132502.
[11] Y. Ichikawa *et al.*, Prog. Theor. Exp. Phys. 021D01 (2015).
[12] A. O. Tokiyasu *et al.*, Phys.Lett. B728 (2014) 616.
[13] L. Fabbietti *et al.*, Nucl. Phys. A914 (2013) 60.
[14] T. Hashimoto *et al.*, Prog. Theor. Exp. Phys. 061D01 (2015).
[15] A. E. Nelson and D. B. Kaplan, Phys. Lett B192 (1987) 193.
[16] A. Scordo *et al.*, AIP Conf. Proc. 1735 (2016) 080015.
[17] B. P. Abbott *et al.*, Phys. Rev. Lett. 119 (2017) 161101.
[18] C. Curceanu *et al.* , Acta Phys. Polon. B46 (2015) 203.
[19] G. V. Vignola *et al.*, Conf. Proc. C 930517 (1993) 1993.
[20] F. Bossi *et al.*, KLOE coll.: Riv. Nuovo Cim. 31 (2008) 531.
[21] F. Ambrosino *et al.*, NIM A534 (2004) 403.
[22] M. Adinolfi *et al.*, [KLOE Collaboration], Nucl. Inst. Meth. A488 (2002) 51.
[23] K. Piscicchia *et al.*, PoS Bormio2013 (2013) 034.
[24] O. Vazques Doce *et al.*, Phys. Lett. B758 (2016) 134.
[25] M. Adinolfi *et al.* [KLOE Collaboration], Nucl. Inst. Meth. A482 (2002) 368.
[26] M. Bazzi *et al.*, Nuclear Instruments and Methods in Physics Research A 671 (2012) 125.
[27] M. Bazzi *et al.*, Journal of Instrumentation 8(05) January 2013.

Microscopic Study of the ^{6}Li(p, α)^{3}He Reaction at Low Energies

Alexander Solovyev[1,2,*] and *Sergey* Igashov[1]

[1]Dukhov Research Institute of Automatics (VNIIA), 22, Sushchevskaya str., Moscow 127055, Russia
[2]Pacific National University, 136, Tikhookeanskaya str., Khabarovsk 680035, Russia

Abstract. The ^{6}Li(p, α)^{3}He reaction important for nuclear astrophysics is studied in the framework of a microscopic approach based on a multichannel algebraic version of the resonating group model. Astrophysical *S*-factor for the reaction is calculated at low energies. The obtained result is compared with experimental data and other theoretical calculations.

1 Introduction

Reactions responsible for production and destruction of lithium isotopes are of great interest to nuclear astrophysics [1, 2]. Cross sections (astrophysical *S*-factors) for these reactions at low energies are required for understanding Big Bang nucleosynthesis and for solving, in particular, so-called "lithium depletion" problem in galactic stars including the Sun. Experimental measurements of these cross sections at astrophysically important energies as a rule meet principal difficulties due to the sizeable Coulomb barrier. That is why theoretical calculations based on microscopic approaches become a very important source of knowledge of these cross sections.

The aim of the present work is to calculate the ^{6}Li(p, α)^{3}He astrophysical *S*-factor at low energies, using a multichannel algebraic version of the resonating group model (AVRGM) [3]. This reaction plays an important role in nuclear astrophysics as the process of the ^{6}Li nuclei destruction during the primordial and stellar nucleosynthesis.

2 Brief description of the multichannel AVRGM

In the framework of the multichannel AVRGM, the total wave function of a system considered is written in the form of an expansion [4]

$$\Psi = \sum_{J^\pi M\, i\, l_i s_i \nu_i} C^{(i)}_{J^\pi M\, l_i s_i \nu_i} \Psi^{(i)}_{J^\pi M\, l_i s_i \nu_i} \tag{1}$$

in series of the AVRGM basis functions

$$\Psi^{(i)}_{J^\pi M\, l_i s_i \nu_i} = \mathrm{A}\left\{ \sum_{m_i+\sigma_i=M} C^{JM}_{l_i m_i\, s_i \sigma_i} \left[\varphi^{(1)_i}_{s_{1i}} \varphi^{(2)_i}_{s_{2i}}\right]_{s_i \sigma_i} f_{\nu_i l_i m_i}(\mathbf{q}_i) \right\}. \tag{2}$$

* Corresponding author: alexander.solovyev@mail.ru

Here A is the antisymmetrization operator; i is the number of the cluster channel characterized by the specified cluster structure (all the channels are assumed to be two-cluster ones); l_i, s_i, m_i, and σ_i are respectively the relative orbital momentum, the channel spin, and their projections in the ith cluster channel; J, M, and π are the total angular momentum, its projection, and the parity respectively (they are integrals of the motion and have identical values for all the cluster channels); $\varphi^{(1)_i}_{s_{1i}\sigma_{1i}}$ and $\varphi^{(2)_i}_{s_{2i}\sigma_{2i}}$ are respectively the internal wave functions for the first $(1)_i$ and second $(2)_i$ clusters with spins s_{1i}, s_{2i} and their projections σ_{1i}, σ_{2i} in the ith channel (these functions are assumed to be fixed and are chosen, as a rule, in the form of the translationally invariant oscillator shell model wave functions for the lowest states compatible with the Pauli exclusion principle); $C^{JM}_{lm\,s\sigma}$ is the Clebsch–Gordan coefficient; $f_{\nu lm}(\mathbf{q})$ is the oscillator function, ν is the number of oscillator quanta, $\mathbf{q}$ is the Jacobi vector.

Unknown quantities in the AVRGM approach are the expansion coefficients $C^{(i)}_{J^\pi M\, l_i s_i \nu_i}$ of the total wave function (1) over AVRGM basis functions (2). These coefficients for an expansion of a continuum wave function are found by solving a set of linear inhomogeneous algebraic equations, so-called the AVRGM equations set for continuum [4]. The solution of this set gives not only expansion coefficients but also elements of the scattering matrix $S^{J^\pi}(E_{\mathrm{c.m.}})$ (S-matrix), which are required to calculate cross section for a transfer reaction

$$\sigma_{i\to f}(E_{\mathrm{c.m.}}) = \frac{\pi}{k^2}\sum_{J^\pi}\frac{2J+1}{(2s_1+1)(2s_2+1)}\sum_{l_i l_f s_i s_f}\left|S^{J^\pi}_{f l_f s_f,\, i l_i s_i}(E_{\mathrm{c.m.}})\right|^2, \tag{3}$$

where indices i and f denote the initial and final channels of the reaction respectively; s_1 and s_2 are the spins of the colliding nuclei in the entrance channel, k and $E_{\mathrm{c.m.}}$ are the wave number and the energy of their relative motion in the center-of-mass system respectively.

3 Discussion of the results. Conclusions

The present study of the ^{6}Li(p, α)^{3}He reaction is first of all a benchmark for the implemented approach based on the multichannel AVRGM. It allows to demonstrate capability of the approach to describe charged-particle induced reactions involving p-shell nuclei. Two cluster configurations ^{6}Li + p and α + ^{3}He were adopted in the calculations. Consideration of the reaction at low energies makes possible to be restricted only to the s-waves in the entrance channel ^{6}Li + p.

The astrophysical S-factor for the ^{6}Li(p, α)^{3}He reaction calculated within the multichannel AVRGM, using the modified Hasegawa–Nagata potential [5] for description of nuclear interaction, is shown in Fig. 1 by solid line. Symbols in Fig. 1 are the experimental data (see [2] for references and [6–8]). The obtained result is seen to be in a reasonable enough agreement with the data. The calculation was performed without including any screening effects. That is why the energy behavior of the calculated astrophysical S-factor somewhat differs qualitatively from that of the experimental data below 100 keV, which are affected by the electron screening effects in this energy range [6] and consequently have values larger than those of the calculation at very low energies.

It should be noted that the ^{6}Li(p, α)^{3}He reaction was already considered on the basis of microscopic approaches, for example, in works [9, 10] (see Fig. 1 for curves). The former provides the RGM description of the considered reaction (dashed line) while the latter is based on a three-cluster model (dash-dotted line). In both the works, the Minnesota potential [11] as central part and the Reichstein–Tang spin-orbit interaction [12] as noncentral component of nuclear forces were utilized.

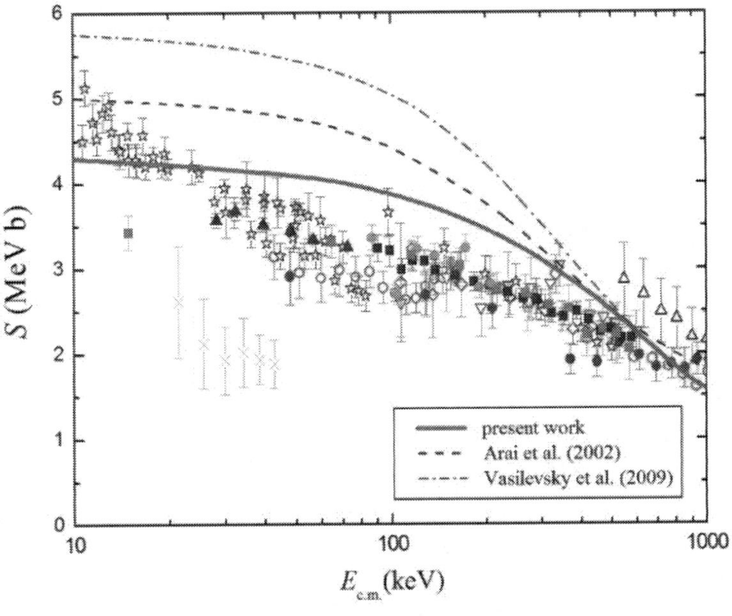

Fig. 1. Energy dependence of astrophysical S-factor for the ^{6}Li(p, α)^{3}He reaction.

The screening effects are also not included in the calculations [9, 10]. The results of these works qualitatively agree with each other and with ours but all three calculations differ quantitatively from each other at low energies. The calculation [10] gives the largest overestimation of the data compared to the others.

There are several ways to improve our study: using more complicated internal cluster wave functions, taking into account more cluster channels, inclusion of the screening effects, consideration of high energy range covering the resonance observed in the data near 1.6 MeV [2], etc. Most likely, these improvements will allow to refine the implemented AVRGM approach and to increase reliability of the obtained theoretical results.

This work was supported by the Russian Science Foundation (grant No. 16-12-10048).

References

1. E.G. Adelberger, A. García, R.G.H. Robertson et al., Rev. Mod. Phys. **83**, 195 (2011)
2. Y. Xu, K. Takahashi, S. Goriely et al., Nucl. Phys. A **918**, 61 (2013)
3. G.F. Filippov, I.P. Okhrimenko, Phys. Atom. Nucl. **32**, 480 (1980)
4. S.Yu. Igashov, Bull. Russ. Acad. Sci.: Phys. **65**, 817 (2001)
5. H. Kanada, T. Kaneko, S. Nagata, M. Nomoto, Prog. Theor. Phys. **61**, 1327 (1979)
6. S. Engstler, G. Raimann, C. Angulo et al., Z. Phys. A **342**, 471 (1992)
7. L. Lamia, C. Spitaleri, R.G. Pizzone et al., Astrophys. J. **768**, 1 (2013)
8. J.J. He, S.Z. Chen, C.E. Rolfs et al., Phys. Lett. B **725**, 287 (2013)
9. K. Arai, D. Baye, P. Descouvemont, Nucl. Phys. A **699**, 963 (2002)
10. V.S. Vasilevsky, F. Arickx, J. Broeckhove, T.P. Kovalenko, Nucl. Phys. A **824**, 37 (2009)
11. D.R. Thompson, M. LeMere, Y.C. Tang, Nucl. Phys. A **286**, 53 (1977)
12. I. Reichstein, Y.C. Tang, Nucl. Phys. A **158**, 529 (1970)

EPJ Web of Conferences **165**, 01048 (2017) DOI: 10.1051/epjconf/201716501048
NPA8 2017

Roles of nuclear weak rates on the evolution of degenerate cores in stars

Toshio Suzuki[1,2,⋆], *Naofumi* Tsunoda[3], *Yuhsuke* Tsunoda[3], *Noritaka* Shimizu[3], and *Takaharu* Otsuka[4]

[1] *Department of Physics, College of Humanities and Sciences, Nihon University, Sakurajosui 3-25-40, Setagaya-ku, Tokyo 156-8550, Japan*
[2] *National Astronomical Observatory of Japan, Osawa 2-21-1, Mitaka, Tokyo 181-8588, Japan*
[3] *Center for Nuclear Study, The University of Tokyo, Hongo, Tokyo 113-0033 and Wako, Saitama 351-0198, Japan*
[4] *RIKEN Nishina Center for Accelerated-Based Science, Hirosawa 2-1, Wako, Saitama 351-0198, Japan*

Abstract. Electron-capture and β-decay rates in stellar environments are evaluated with the use of new shell-model Hamiltonians for *sd*-shell and *pf*-shell nuclei as well as for nuclei belonging to the island of inversion. Important role of the nuclear weak rates on the final evolution of stellar degenerate cores is presented. The weak interaction rates for *sd*-shell nuclei are calculated to study nuclear Urca processes in O-Ne-Mg cores of stars with 8-10 $M_\odot$ (solar mass) and their effects on the final fate of the stars. Nucleosynthesis of iron-group elements in Type Ia supernova explosions are studied with the weak rates for *pf*-shell nuclei. The problem of the neutron-rich iron-group isotope over-production compared to the solar abundances is shown to be nearly solved with the use of the new rates and explosion model of slow defraglation with delayed detonation. Evaluation of the weak rates is extended to the island of inversion and the region of neutron-rich nuclei near ^{78}Ni, where two major shells contribute to their configurations.

1 Introduction

Roles of the rate of nuclear processes driven by electron-capture and β-decays on the final evolution of electron-degenerate cores of stars are investigated. Standard electron-capture and β-decay rates of nuclei presented in a recent literature [1] are those for (1) *sd*-shell nuclei by [2], (2) *pf*-shell of LMP [3], (3) $pf-g$-shell of LMSH [4], and (4) Fuller-Fowler-Newman (FFN) rates [5]. Besides them, an approximate two-parameter formula for the rates is obtained [6] to be used for nuclei, which have no available data tables.

Here, we study nuclear Urca processes in stars with O-Ne-Mg cores by employing a set of new weak rates computed by the USDB [7] for *sd*-shell nuclei. Nucleosynthesis of iron-group nuclei in Type Ia supernova explosions (SNe) (see [8] for a recent review on nucleosynthesis in Type Ia SNe and hydrodynamical explosion models) is discussed by using a new Hamiltonian GXPF1J [9] for *pf*-shell nuclei. We also extend our study to the weak rates for nuclei in the island of inversion, where

⋆e-mail: suzuki@phys.chs.nihon-u.ac.jp

sd-shell and *pf*-shell components are considerably admixed, as well as those for *pf* − *g*-shell nuclei near ^{78}Ni, which play important roles on nucleosynthesis in core-collapse SNe.

2 Weak rates in *sd*-shell and nuclear Urca processes in O-Ne-Mg cores

Stars with 8-10 $M_{\odot}$, where O-Ne-Mg cores form after carbon burning, may evolve in various ways such as O-Ne-Mg white dwarfs (WD), e-capture (EC) SNe or core-collapse (CC) SNe. The stellar fate is sensitive to their masses and nuclear e-capture and β-decay rates. Cooling of the O-Ne-Mg core by nuclear Urca processes determines whether the star ends up with ECSNe or CCSNe [10, 11].

Nuclear Urca processes for pairs of *sd*-shell nuclei, ^{23}Ne-^{23}Na and ^{25}Na-^{25}Mg, are found to be important for cooling the O-Ne-Mg core [10, 11]. Electron-capture and β-decay occur simultaneously at a certain density, called "Urca density", and stellar energy is lost by emissions of both ν and $\bar{\nu}$. At high densities electron chemical potential increases and e-capture on nuclei is ignited when it reaches the Q-value of the reaction. The Urca density determined by the Q-value of the reaction is found to be $\log_{10}\rho Y_e$ = 8.92 and 8.78 for ^{23}Ne-^{23}Na and ^{25}Na-^{25}Mg pairs, respectively.

Double e-capture reactions on ^{24}Mg or ^{20}Ne, which are succeeded by e-capture on ^{24}Na or ^{20}F, respectively, are important for heating the core in later stages of the stellar evolution [12, 13]. Electron-capture, β-decay and ν energy loss rates as well as energy production rates by γ emissions are tabulated for *sd*-shell nuclei using a fine grid for density and temperature in [13].

3 Electron-capture rates in *pf*-shell nuclei and nucleosynthesis in Type Ia SNe

New shell-model Hamiltonians in *pf*-shell, GXPF1 [14] and GXPF1J [9], can describe spin properties of *pf*-shell nuclei quite well. The Gamow-Teller (GT) strengths in Ni isotopes and M1 strengths in various *pf*-shell nuclei are also well described [15]. In particular, for ^{56}Ni, the GT strength distribution is nicely reproduced [15, 16]. In Type-Ia supernova explosions, a plenty amount of ^{56}Ni is produced. Production yields of neutron-rich nuclei and the lepton-to-baryon ratio (or proton fraction) Y_e depend sensitively on the rate of e-capture on ^{56}Ni. The ratio of the production yields of ^{58}Ni over ^{56}Ni can be reduced nearly by half for GXPF1J compared to KB3G [17].

The problem of over-production of ^{58}Ni, ^{54}Cr and ^{54}Fe compared to the solar abundance was discussed [18] with the use of the capture rates of FFN [5]. A possible solution of the problem with slower e-capture rates was discussed in [3, 19]. The problem could be solved by using slower e-capture rates of GXPF1J. Here, we use the W7 and WDD2 model [18] as explosion models of Type-Ia supernova explosions starting from C-O white dwarf with a mass 1.38 $M_{\odot}$. The W7 model proceeds by a fast deflagration while the WDD2 model proceeds by a slow deflagration with delayed detonation. Electron-capture rates of GXPF1J are used for *pf*-shell nuclei with $21 \leq Z \leq 32$ and those of KBF [3] otherwise. At high densities above a few 10^8 gcm^{-3}, slower e-capture processes on Ni and Fe isotopes lead to less production of neutron-rich nuclei (see [20] for the details). The results of final element abundances are shown in Ref. [20]. In the WDD2 (W7) model, over-production of neutron-rich Cr, Fe and Ni isotopes compared with the solar system abundances is suppressed within a factor of 2 (2-3). This is much smaller than the case of FFN weak rates, where the over-production factor becomes as large as up to 4-5.

4 Weak rates for nuclei in the island of inversion

Urca processes for nuclear pairs in the island of inversion [21] such as ^{31}Mg-^{31}Al and ^{33}Mg-^{33}Al pairs have been pointed out to be important for the cooling of neutron star crusts [22]. Large *sd* − *pf* shell admixtures are found in neutron-rich Ne, Na and Mg isotopes near N=20. Lowering of 2_1^+ states and enhancement of $E2$ transition strengths show evidence for the breaking of the magicity at N =20 [21]. Energy levels of 2_1^+ states and $B(E2 : 2_1^+ \rightarrow 0_{g.s.}^+)$ are sucessfully reproduced by SDPF-M Hamiltonian [23] with $sd-pf$-shell configurations. The important contributions from 2p-2h components are found in ^{30}Ne and ^{32}Mg.

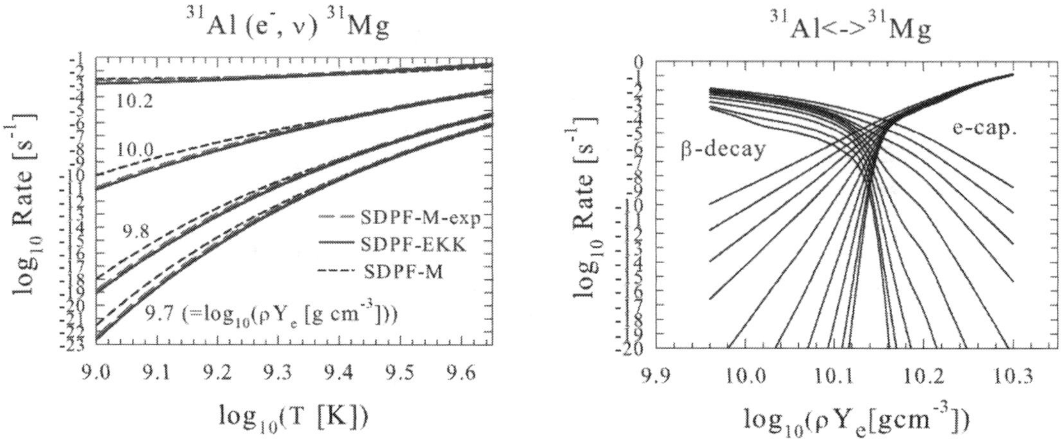

Figure 1. (Left) Electron-capture rates for ^{31}Al (e$^-$, ν) ^{31}Mg obtained with SDPF-M Hamiltonian (SDPF-M) and EKK method (SDPF-EKK). The rates obtained with SDPF-M including experimental data are denoted by SDPF-M-exp. (Right) Electron-capture and β-decay rates for ^{31}Al-^{31}Mg pair at various temperatures obtained by the EKK method.

We discuss the weak rates for ^{31}Mg-^{31}Al pair. The SDPF-M fails to reproduce the energy levels of ^{31}Mg, that is, 7/2$^-$ state becomes the ground state while the experimental g.s. is 1/2$^+$. The Urca density can not be clearly assigned for the weak rates for SDPF-M. This shortcoming can be improved for the effective interaction obtained by EKK (extended Kuo-Krenciglowa) method [24], which can properly treat Q-box calculations in two-major shells without divergence problems [25]. Energy levels in ^{31}Mg are well reproduced, and more important roles of p-h excitations are noticed compared with the SDPF-M case. The 4p-4h components are found to be as much as the 2p-2h components in ^{32}Mg. The weak rates are evaluated with the EKK method and they prove to be close to those obtained by taking into account the available experimental data as shown in Fig. 1. For the EKK approach, the Urca density can be assigned to be $\log_{10}\rho Y_e = 10.14$.

5 Summary

New weak rates for *sd*-shell and *pf*-shell nuclei are obtained with USDB and GXPF1J Hamiltonians, respectively. The rates for *sd*-shell are used to study evolutions of the O-Ne-Mg cores of stars with 8-10 $M_\odot$. Nuclear Urca processes for A = 23 and 25 nuclear pairs are found to be important for the cooling of the cores and determination of the fate of the stars whether they end up with ECSNe or CCSNe. The rates for *pf*-shell are used to study nucleosynthesis of iron-group elements in Type Ia SNe. Relatively smaller e-capture rates for GXPF1J compared with KB3G, KBF and FFN are found to considerably suppress the over-production of neutron-rich iron-group isotopes, especially in the WDD2 explosion model with detonation. The weak rates for GXPF1J are tabulated for further study of astrophysical processes [26].

Evaluation of new weak rates is extended to nuclei in the island of inversion, where substantial admixtures of both *sd*-shell and *pf*-shell components are noticed. Nuclear weak rates for ^{31}Mg-^{31}Al pair, important for Urca processes in neutron star crusts, are evaluated with a new interaction obtained by the EKK method. Study of nuclear structure and weak rates for $pf - g$-shell nuclei near ^{78}Ni, which play important roles on nucleosynthesis in CCSNe, is in progress. RPA calculations give e-capture rates close to the rates of the approximate formula in Ref. [1]. Shell-model calculations within $pf - g_{9/2}d_{5/2}$-shell configurations can explain the excitation energy of the first 2^+ state in ^{78}Ni [27], but the configuration space needs to be extended to include the full *gds*-shells to obtain reliable weak rates as dominant contributions come from spin-dipole forbidden transitions.

Nuclear weak rates, thus, play important role on the final evolution of degenerate cores in stars. Accurate evaluation of the weak rates is essential for the studies of various astrophysical processes sensitive to the rates.

The author would like to thank K. Nomoto, H. Toki, S. Jones, R. Hirschi, M. Honma, K. Mori, M. Famiano and T. Kajino for collaborations of the works in Sects. 2 and 3. This work has been supported in part by Grants-in-Aid for Scientific Research (C) 15K05090 of the MEXT of Japan.

References

[1] C. Sullivan et al., Astrophys. J. **816**, 44 (2016).
[2] T. Oda et al., At. and Nucl. Data Tables **56**, 231 (2002).
[3] K. Langanke and G. Martinez-Pinedo, At. And Nucl. Data Tables, **79**, 1 (2001).
[4] K. Langanke, E. Kolbe and D. J. Dean, Phys. Rev. C **63**, 032801 (2001).
[5] G. M. Fuller, W. Fowler and M. Newman, Astrophys. J. **252**, 715 (1982).
[6] K. Langanke, G. Martinez-Pinedo, 2003, Rev. Mod. Phys. **75**, 819 (2003).
[7] B. A. Brown and W. Richter, Phys. Rev. C **74**, 034315 (2006).
[8] I. R. Seitenzahl and D. M. Townsley, Handbook of Supernovae (Springer, 2017).
[9] M. Honma et al., J. Phys. Conf. Ser., **20**, 7 (2005).
[10] S. Jones et al., Astrophys. J. **772**, 150 (2013).
[11] H. Toki, T. Suzuki, K. Nomoto, S. Jones, R. Hirsci, Phys. Rev. C **88**, 015806 (2013).
[12] G. Martinex-Pinedo et al., Phys. Rev. C **89**, 045806 (2014).
[13] T. Suzuki, H. Toki and K. Nomoto, Astrophys. J. **817**, 163 (2016).
[14] M. Honma et al., Phys. Rev. C **69**, 034335 (2004).
[15] T. Suzuki et al., Phys. Rev. C **83**, 044619 (2011).
[16] M. Sasano et al., Phys. Rev. Lett. **107**, 202501 (2011).
[17] J. Poves et al., Nucl. Phys. A **694**, 157 (2001).
[18] K. Iwamoto et al., Astrophys. J. **125**, 439 (1999).
[19] F. Brachwitz et al., Astrophys. J. **536**, 934 (2001).
[20] K. Mori et al., Astrophys. J. **833**, 179 (2016).
[21] E. K. Warburton, J. A. Becker and B. A. Brown, Phys. Rev. C **41**, 1147 (1990).
[22] H. Schatz et al., Nature **505**, 65 (2014).
[23] Y. Utsuno et al., Phys. Rev. C **60**, 054315 (1999).
[24] N. Tsunoda et al., Phys. Rev. C **95**, 021304(R) (2017).
[25] N. Tsunoda, K. Takayanagi, Hjorth-Jensen and T. Otsuka, Phys. Rev. C **89**, 024313 (2014).
[26] M. Honma and T. Suzuki, http://www.phys.chs.nihon-u.ac.jp/suzuki/data4/link.html (2017).
[27] Y. Tsunoda et al., Phys. Rev. C **89**, 031301(R) (2014).

 DOI: 10.1051/epjconf/201716501049

^{3}He(α,γ)^{7}Be cross section in a wide energy range

Tamás Szücs[1,★], *György* Gyürky[1], *Zoltán* Halász[1], *Gábor Gy.* Kiss[1], and *Zsolt* Fülöp[1]

[1] *Institute for Nuclear Research (MTA Atomki), Debrecen, Hungary*

Abstract. The reaction rate of the ^{3}He(α,γ)^{7}Be reaction is important both in the Big Bang Nucleosynthesis (BBN) and in the Solar hydrogen burning. There have been a lot of experimental and theoretical efforts to determine this reaction rate with high precision. Some long standing issues have been solved by the more precise investigations, like the different S(0) values predicted by the activation and in-beam measurement. However, the recent, more detailed astrophysical model predictions require the reaction rate with even higher precision to unravel new issues like the Solar composition. One way to increase the precision is to provide a comprehensive dataset in a wide energy range, extending the experimental cross section database of this reaction. This paper presents a new cross section measurement between E_{cm} = 2.5 – 4.4 MeV, in an energy range which extends above the ^{7}Be proton separation threshold.

1 Introduction

The ^{3}He(α,γ)^{7}Be reaction is the starting point of the ppII and ppIII reaction branches in the solar hydrogen burning pp-chains, therefore its rate has sizeable impact on the solar ^{7}Be and ^{8}B neutrino production. Using the standard solar model [1], the flux of these neutrinos can be calculated. With the known solar parameters and reaction rates, we gain insight into the solar core through comparison of the measured and calculated neutrino spectrum. Recently, the solar neutrino detections reached a precision of a few percent [2, 3], which would allow these investigations. However, now the precision of the nuclear physics input has to catch up to have this unique tool for precise solar core diagnostics.

The reaction has its other major role in the Big Bang Nucleosynthesis (BBN). For a long time it has been a candidate for solving the primordial lithium problem [4]. However, in the last decade the LUNA collaboration measured the reaction cross section directly in the BBN energy range [5–8], excluding this possibility. Even though, the ^{3}He(α,γ)^{7}Be reaction cannot be the solution of the primordial Li problem, its BBN reaction rate has to be known with higher precision to test models using new physics e.g. [9].

There are a few methods to investigate experimentally the ^{3}He(α,γ)^{7}Be reaction. One of the possibilities is the so-called in-beam gamma spectroscopy. In this case the prompt gamma-rays produced in the reaction have to be detected. Anther possibility is the activation method. The reaction product ^{7}Be is radioactive with a half life of 53.22 days [10], and 10.44 % of the ^{7}Be decay goes to an excited state of ^{7}Li producing a 477.6 keV γ-ray. By counting this γ-ray the produced ^{7}Be activity can be derived and the reaction cross section can be determined.

★e-mail: szucs.tamas@atomki.mta.hu

In the 1998 Solar fusion cross section compilation [11] several cross section measurements were listed for the $^3He(\alpha,\gamma)^7Be$ reaction, however, there was a discrepancy between the cross section values obtained with the two methods. Later the LUNA collaboration performed an experiment, where the reaction cross section was measured simultaneously by both methods [6]. The new data showed perfect agreement between the two methods, indicating some hidden systematic uncertainty in the previous measurements.

Recently a third technique became available to determine the $^3He(\alpha,\gamma)^7Be$ reaction cross section. Namely, the direct detection of the ^7Be recoils. This measurement was done by the ERNA collaboration [12], where additionally a few prompt gamma and activation data points were recorded at overlapping energies, and consistent results were found from the three methods. This experiment extended the experimental database up to $E_{cm} = 3.1$ MeV. Only one former dataset contains data above $E_{cm} = 1.5$ MeV [13]. These data exhibit a different slope of the the S-factor towards higher energies as pointed out in the 2011 Solar fusion cross section compilation [14]. To resolve the discrepancy two new datasets have been measured at different laboratories [15, 16], and both were in agreement with the ERNA data.

As a summary of the current situation, most of the recent reaction cross section measurements have focused on the low energies below $E_{cm} = 1.5$ MeV, including the most recent one [17]. However, there is no experimental data above $E_{cm} = 3.1$ MeV. It was suggested recently, that the R-matrix models have to be tested on higher energy datasets [18]. In addition, there are conflicting datasets for the $^6Li(p,\gamma)^7Be$ reaction [19, 20] having impact on the level scheme of ^7Be, which has to be understood for a precise extrapolation toward the solar energies.

To address the need for experimental data outlined above, a new measurement is being performed across a wide energy range, extending the experimental database of the $^3He(\alpha,\gamma)^7Be$ reaction cross section with a dataset in the energy range of $E_{cm} = 2.5 - 4.4$ MeV.

2 Experimental details

In this work the $^3He(\alpha,\gamma)^7Be$ reaction cross section is measured with the activation technique. The lower energy points overlap with the energy range of the former studies, and spans slightly above the proton separation threshold of ^7Be, thus it can be compared also with the $^6Li(p,\gamma)^7Be$ reaction cross sections.

A thin window gas cell target was used for the investigations. The cell was slightly modified, but it is similar to the one used in Atomki for a previous measurement campaign [16]. The ^3He gas is confined between two aluminium foils with 10 μm nominal thickness. The first one served as an entrance foil for the beam, the second one was used as an exit and catcher foil for the produced ^7Be. After the exit foil the beam was stopped in a water cooled beam stop. With this double window configuration less beam power is absorbed in the gas volume, therefore less pressure change was observed during the irradiations.

The number of target atoms is determined from the initial gas pressure and temperature knowing the length of the cell. Without beam the gas pressure was constant over several days. During the irradiations the gas pressure was continuously monitored, and a slight increase was observed. This additional pressure did not decrease after the beam was stopped, therefore it is assumed to be outgassing of internal surfaces of the cell. The additional pressure was considered in the effective energy calculation.

The thickness of the entrance foil was measured by α-energy loss. A triple isotope alpha source was placed below the foil in an α-spectrometer, and the energy of the α-particles penetrating through the foil was measured by an ion-implanted silicon detector. From the measured energy loss, the thickness was determined using SRIM [21].

The beam energies varied between 7.3 – 11.3 MeV to measure at the targeted center-of-mass energies. The energy loss of the particles in the entrance foil and the target gas itself was calculated from SRIM [21].

After the beam defining aperture the whole irradiation chamber and gas cell was isolated to allow the beam current measurement by charge integration. Typical irradiation length was 20 – 22 hours with an α^{++} current of 1 μA.

After the irradiation the catcher foil was transported to a HPGe detector with 100 % relative efficiency in a commercial 10 cm thick 4π lead shielding. For the countings 1 cm sample to detector end-cap distance was used. The activity of a given catcher was measured multiple times to check the activity decrease. The half-life of the reaction product was consistent with the literature half-life of ^{7}Be. The detector efficiency calibration was done with calibrated multi-line gamma sources at far geometries, where the true coincidence summing is negligible. Then the close distance efficiency was determined by a stronger ^{7}Be source. A scaling factor was derived between the two geometries, and used later in the analysis.

3 Preliminary results and outlook

The obtained cross sections range between between 6 – 12 μbarn. In Fig. 1. a few preliminary data points are shown together with the data from [12]. In the overlapping energy region the agreement between the new and former data is clearly seen.

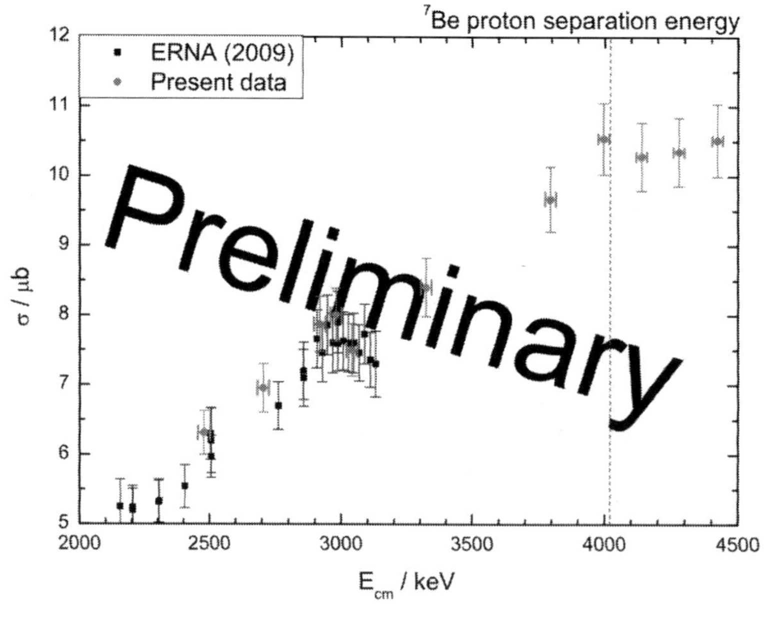

Figure 1. Preliminary cross section data are indicated by red dots, black points are the data from [12]. Horizontal error bars represent the energy uncertainty. Vertically the total uncertainties are displayed including statistical and systematic ones.

The average trend of the data is following the lower energy trend up to the ^{7}Be proton separation threshold. Above the threshold, the excitation function becomes more flat. The opened ^{3}He(α,p)^{6}Li reaction channel is not strong close to the threshold, and does not take away sizeable yield.

The data taking and analysis is ongoing. The whole energy range will be populated with additional points, especially above the threshold energy, where a new ^{6}Li(p,γ)^{7}Be resonance was suggested [20]. The energy region of the resonance measured by the ERNA collaboration [12] will also be investigated and populated with more data points.

References

[1] A. Serenelli, C. Peña Garay, W.C. Haxton, Phys. Rev. D **87**, 043001 (2013)

[2] B. Aharmim, S.N. Ahmed, J.F. Amsbaugh, J.M. Anaya, A.E. Anthony, J. Banar, N. Barros, E.W. Beier, A. Bellerive, B. Beltran et al. (SNO Collaboration), Phys. Rev. C **87**, 015502 (2013)

[3] G. Bellini, J. Benziger, D. Bick, G. Bonfini, D. Bravo, M.B. Avanzini, B. Caccianiga, L. Cadonati, F. Calaprice, P. Cavalcante et al. (Borexino Collaboration), Phys. Rev. D **89**, 112007 (2014)

[4] B.D. Fields, Annu. Rev. Nucl. Part. Sci. **61**, 47 (2011)

[5] D. Bemmerer, F. Confortola, H. Costantini, A. Formicola, G. Gyürky, R. Bonetti, C. Broggini, P. Corvisiero, Z. Elekes, Z. Fülöp et al. (LUNA Collaboration), Phys. Rev. Lett. **97**, 122502 (2006)

[6] F. Confortola, D. Bemmerer, H. Costantini, A. Formicola, G. Gyürky, P. Bezzon, R. Bonetti, C. Broggini, P. Corvisiero, Z. Elekes et al. (LUNA Collaboration), Phys. Rev. C **75**, 065803 (2007)

[7] G. Gyürky, F. Confortola, H. Costantini, A. Formicola, D. Bemmerer, R. Bonetti, C. Broggini, P. Corvisiero, Z. Elekes, Z. Fülöp et al. (LUNA Collaboration), Phys. Rev. C **75**, 035805 (2007)

[8] H. Costantini, D. Bemmerer, F. Confortola, A. Formicola, G. Gyürky, P. Bezzon, R. Bonetti, C. Broggini, P. Corvisiero, Z. Elekes et al. (LUNA Collaboration), Nucl. Phys. A **814**, 144 (2008)

[9] M. Pospelov, J. Pradler, Annu. Rev. Nucl. Part. Sci. **60**, 539 (2010)

[10] D. Tilley, C. Cheves, J. Godwin, G. Hale, H. Hofmann, J. Kelley, C. Sheu, H. Weller, Nucl. Phys. A **708**, 3 (2002)

[11] E.G. Adelberger, S.M. Austin, J.N. Bahcall, A.B. Balantekin, G. Bogaert, L.S. Brown, L. Buchmann, F.E. Cecil, A.E. Champagne, L. de Braeckeleer et al., Rev. Mod. Phys. **70**, 1265 (1998)

[12] A. Di Leva, L. Gialanella, R. Kunz, D. Rogalla, D. Schürmann, F. Strieder, M. De Cesare, N. De Cesare, A. D'Onofrio, Z. Fülöp et al., Phys. Rev. Lett. **102**, 232502 (2009), erratum: Phys. Rev. Lett. 103, 1599032 (2009)

[13] P.D. Parker, R.W. Kavanagh, Phys. Rev. **131**, 2578 (1963)

[14] E.G. Adelberger, A. García, R.G.H. Robertson, K.A. Snover, A.B. Balantekin, K. Heeger, M.J. Ramsey-Musolf, D. Bemmerer, A. Junghans, C.A. Bertulani et al., Rev. Mod. Phys. **83**, 195 (2011)

[15] M. Carmona-Gallardo, B.S.N. Singh, M.J.G. Borge, J.A. Briz, M. Cubero, B.R. Fulton, H. Fynbo, N. Gordillo, M. Hass, G. Haquin et al., Phys. Rev. C **86**, 032801 (2012)

[16] C. Bordeanu, G. Gyürky, Z. Halász, T. Szücs, G. Kiss, Z. Elekes, J. Farkas, Z. Fülöp, E. Somorjai, Nucl. Phys. A **908**, 1 (2013)

[17] A. Kontos, E. Uberseder, R. deBoer, J. Görres, C. Akers, A. Best, M. Couder, M. Wiescher, Phys. Rev. C **87**, 065804 (2013)

[18] R.J. deBoer, J. Görres, K. Smith, E. Uberseder, M. Wiescher, A. Kontos, G. Imbriani, A.D. Leva, F. Strieder, Phys. Rev. C **90**, 035804 (2014)

[19] R.M. Prior, M.C. Spraker, A.M. Amthor, K.J. Keeter, S.O. Nelson, A. Sabourov, K. Sabourov, A. Tonchev, M. Ahmed, J.H. Kelley et al., Phys. Rev. C **70**, 055801 (2004)

[20] J. He, S. Chen, C. Rolfs, S. Xu, J. Hu, X. Ma, M. Wiescher, R. deBoer, T. Kajino, M. Kusakabe et al., Phys. Lett. B **725**, 287 (2013)

[21] J.F. Ziegler, M. Ziegler, J. Biersack, Nucl. Instrum. Methods Phys. Res., Sect. B **268**, 1818 (2010), http://www.srim.org

α-induced reaction cross section measurements on ^{197}Au

Tamás Szücs[1,★], *György* Gyürky[1], *Zoltán* Halász[1], *Gábor Gy.* Kiss[1], and *Zsolt* Fülöp[1]

[1] *Institute for Nuclear Research (MTA Atomki), Debrecen, Hungary*

Abstract.
The γ-process is responsible for creating the majority of the isotopes of heavier elements on the proton rich side of the valley of stability. The γ-process simulations fail to reproduce the measured solar system abundance of these isotopes. The problem can lie in the not well known astrophysical scenarios where the process takes place, or in the not sufficiently known nuclear physics input. To improve the latter part, α-induced reaction cross section measurements on ^{197}Au were carried out at Atomki. With this dataset new experimental information will become available, which can be later used as validation of the theoretical cross section calculations used in the γ-process simulations.

1 Introduction

The majority of the nuclei beyond iron are produced by neutron capture reactions in the s- and r-processes [1]. There are a few dozens of isotopes on the proton rich side of the valley of stability which cannot be produced by these processes. Those are the so-called p-nuclei [2] mainly produced via the γ-process [3], which is a series of photo-emission of neutrons and charged particles converting the pre-existing s- and r-process seed nuclei. In the γ-process network the charged particle photo-emissions occur mostly on unstable nuclei, thus the network calculations cannot use direct experimental reaction rate values. These rates are usually derived from Hauser-Feshbach statistical model [4] calculations. The input parameters of these calculations have to be validated. To maximize the experimental constraint on the stellar rate of the photodisintegration reactions, those should be derived from the inverse radiative capture reaction cross sections [5]. The experimental data of α-capture reactions in the relevant energy region are however still scarce [6].

In this work we present a new experimental campaign targeting α-induced reaction cross section measurements on ^{197}Au close to the astrophysically relevant energy region.

2 Reactions to be investigated

Beside the radiative capture on ^{197}Au, the (α,n) and (α,2n) reactions take also place in the investigated energy range of $E_\alpha = 14 - 20$ MeV. All the three reaction channels lead to radioactive nuclei with half-lives ranging from a few hours to a few days (see table 1.), thus can be investigated by the activation technique.

The ^{197}Au(α,2n)^{199}Tl and ^{197}Au(α,n)^{200}Tl reaction cross sections were determined by counting the γ-rays released during the decay of the reaction products (see table 1.). These cross sections were measured by many groups mostly at higher energies e.g. [10]. The ^{197}Au(α,γ)^{201}Tl reaction was

★e-mail: szucs.tamas@atomki.mta.hu

Table 1. α-induced reactions on ^{197}Au and decay parameters of the reaction products [7–9].

Reaction	(α,2n)		(α,n)		(α,γ)	
Threshold	17.14 MeV		9.94 MeV		1.56 MeV	
Product	^{199}Tl		^{200}Tl		^{201}Tl	
Half-life	7.42 ± 0.08 h		26.1 ± 0.1 h		73.01 ± 0.04 h	
	E / keV	I / %	E / keV	I / %	E / keV	I / %
X-ray	70.82	45.5 ± 2.5	70.82	40.4 ± 1.7	70.82	44.6 ± 0.6
γ-rays	158.4	5.0 ± 0.5	367.9	87.0	167.4	10.00 ± 0.06
	208.2	12.3 ± 1.3	579.3	13.7 ± 1.2		
	247.3	9.3 ± 1.0	828.3	10.8 ± 1.0		
	455.5	12.4 ± 1.4	1205.6	30 ± 3		

measured so far down to E_α = 17.9 MeV [10] by counting the only one intense γ-ray of the reaction product (see table 1.), well above the astrophysically relevant energy region of 7 – 11 MeV [11].

3 Experimental details

The α-induced reactions were investigated in the energy range of E_α = 14 – 20 MeV, with two types of targets. For the energies above the (α,2n) reaction threshold 0.1 – 0.3 μm thick gold layers evaporated on thin aluminium foils were used. The foil with 0.6 cm radius was placed at 4 cm distance from the evaporation boat, thus the evaporated layer assumed to be uniform within 1% which was considered in the thickness uncertainty. The thickness of the gold layers were measured by weighting the aluminium foil before and after the evaporation. For the lower energies self supporting gold foils of 0.6 – 0.8 μm thickness were used to have increased number of target atoms. The thickness of these foils was determined by measuring the energy loss of α-particles from a triple isotope α-source. From the measured energy loss, the thickness was determined using the SRIM [12] tables. Additionally, the thickness and uniformity of all targets will be determined via PIXE and RBS methods.

The irradiations were done at the cyclotron of Atomki. After the beam defining aperture and a secondary electron suppressor the whole irradiation chamber was electrically isolated from beamline to allow the beam current measurement by charge integration. Typical irradiation length was 24 – 28 hours with an α^{++} current of 1.0 – 2.5 μA.

After the irradiations the target was transported to a LEPS (low energy photon spectrometer) detector. This kind of HPGe detector is equipped with a thin germanium crystal having better resolution for low energy γ-rays and X-rays besides lower efficiency for disturbing high energy background. The detector is placed in a 4π shielding consisting of copper, cadmium and lead layers to suppress the environmental background [13]. For the countings 3 cm sample to detector end-cap distance was used. The detector efficiency calibration was done with calibrated multi-line gamma sources at 10 cm distance, where the true coincidence summing is negligible. Then the close distance efficiency was determined by counting a sample irradiated with 19.5 MeV α-beam in both geometry. A scaling factor was derived between the two geometries which includes the correction for true-coincidence summing-effect, and used later in the analysis.

The γ-ray from the decay of ^{201}Tl was only visible above 17.5 MeV in our case, because of the very intense 367.9 keV γ-line of the (α,n) reaction product. Below this energy the ^{197}Au(α,γ)^{201}Tl cross section was determined by X-ray counting. Although all the reaction products releases the same X-rays, the contribution from ^{199}Tl and ^{200}Tl can be well estimated from their known activity measured by their γ-rays, and than that can be subtracted. To get a measurable signal after the subtraction more than 4 half-life of ^{201}Tl should elapse, but thanks to the higher X-ray branching (see table 1.) the signal is still visible (see figure 1).

DOI: 10.1051/epjconf/201716501050

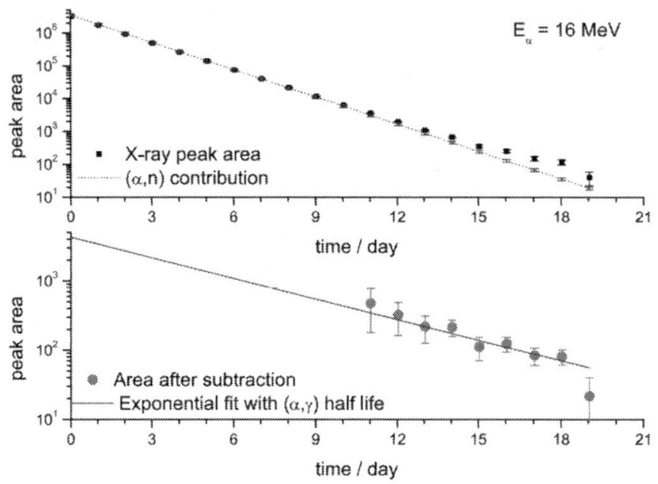

Figure 1. Example X-ray decay curves. In the upper panel the black dots indicate the measured X-ray peak area in one day. The blue dashed line with error bars is the calculated contribution from the ^{200}Tl with known activity.
The lower panel shows the X-ray peak area after subtraction. Detectable signal can be seen after 10 days, and the signal is still visible until the 19th day. These nine points were fitted by an exponential with the known half life of ^{201}Tl, thus the activated target atoms at the end of the irradiation was determined.

4 Outlook

As can be seen in figure 1. the measurement is feasible at beam energy of 16MeV, which is already closer to the astrophysically relevant energy region than any previous investigations. Already a few hundred nb cross section has been successfully measured, one order of magnitude lower than the previous $4 \pm 1\,\mu$b [10]. Knowing the laboratory background of the LEPS detector and the measured count rates, an additional factor of 10 – 30 lower (α,γ) cross section is foreseen to be measurable. Regarding the $(\alpha$,n) cross section, it is anticipated to be measured at lower energies with higher precision as the most recent investigation by Basunia *et al* [10]. The future comparison of the results with statistical model calculation will give an information about the goodness of the used alpha-nucleus optical model potential.

References

[1] E.M. Burbidge, G.R. Burbidge, W.A. Fowler, F. Hoyle, Rev. Mod. Phys. **29**, 547 (1957)
[2] T. Rauscher, N. Dauphas, I. Dillmann, C. Fröhlich, Z. Fülöp, G. Gyürky, Rep. Prog. Phys. **76**, 066201 (2013)
[3] S.E. Woosley, W.M. Howard, Astrophys. J. Suppl. S. **36**, 285 (1978)
[4] W. Hauser, H. Feshbach, Phys. Rev. **87**, 366 (1952)
[5] G.G. Kiss, T. Rauscher, G. Gyürky, A. Simon, Z. Fülöp, E. Somorjai, Phys. Rev. Lett. **101**, 191101 (2008)
[6] T. Szücs, I. Dillmann, R. Plag, Z. Fülöp, Nucl. Data Sheets **120**, 191 (2014), http://www.kadonis.org/pprocess
[7] B. Singh, Nuclear Data Sheets **108**, 79 (2007)
[8] F.G. Kondev, S. Lalkovski, Nucl. Data Sheets **108**, 1471 (2007)
[9] F.G. Kondev, Nucl. Data Sheets **108**, 365 (2007)
[10] M.S. Basunia, H.A. Shugart, A.R. Smith, E.B. Norman, Phys. Rev. C **75**, 015802 (2007)
[11] T. Rauscher, Phys. Rev. C **81**, 045807 (2010)
[12] J.F. Ziegler, M. Ziegler, J. Biersack, Nucl. Instrum. Methods Phys. Res., Sect. B **268**, 1818 (2010), http://www.srim.org
[13] T. Szücs, G.G. Kiss, Z. Fülöp, AIP Conf. Proc. **1595**, 173 (2014)

Commissioning of the BRIKEN beta-delayed neutron detector for the study of exotic neutron-rich nuclei

A. Tolosa-Delgado[1,⋆], *J. Agramunt*[1], *D. S. Ahn*[2], *A. Algora*[1,12], *H. Baba*[2], *S. Bae*[3], *N. T. Brewer*[4], *R. Caballero Folch*[5], *F. Calvino*[6], *P. J. Coleman-Smith*[7], *G. Cortes*[6], *T. Davinson*[9], *I. Dillmann*[5], *C. Domingo-Pardo*[1], *A. Estrade*[10], *N. Fukuda*[2], *S. Go*[11,2], *C. J. Griffin*[9], *R. Grzywacz*[11], *J. Ha*[3], *O. Hall*[9], *L. Harkness-Brennan*[8], *T. Isobe*[2], *D. Kahl*[2], *G. G. Kiss*[2,12], *M. Kogimtzis*[7], *S. Kubono*[2], *M. Labiche*[7], *I. Lazarus*[7], *J. Lee*[13], *J. Liu*[13], *G. Lorusso*[2,14], *K. Matsui*[2,15], *K. Miernik*[4,16], *F. Montes*[17], *B. Moon*[18], *A.I. Morales*[1], *N. Nepal*[10], *S. Nishimura*[2], *R. D. Page*[8], *V. H. Phong*[2,22], *Z. Podolyak*[19], *V. F. E. Pucknell*[7], *B. C. Rasco*[4], *P. Regan*[19,14], *A. Riego*[6], *B. Rubio*[1], *K.P. Rykaczewski*[4], *Y. Saito*[5], *H. Sakurai*[2,15], *Y. Shimizu*[2], *J. Simpson*[7], *P.-A. Söderström*[2], *D. W. Stracener*[4], *T. Sumikama*[2], *H. Suzuki*[2], *J. L. Tain*[1], *M. Takechi*[21], *H. Takeda*[2], *A. Tarifeño-Saldivia*[6], *S. L. Thomas*[20], and *P. Woods*[9]

[1] *Instituto de Fisica Corpuscular (CSIC-Universitat de Valencia), E-46980 Paterna, Spain*
[2] *RIKEN Nishina Center, Wako, Saitama 351-0198, Japan*
[3] *Department of Physics & Astronomy, Seoul National University, Seoul, 08826, Republic of Korea*
[4] *Physics Division, Oak Ridge National Laboratory, Oak Ridge, TN 37831, USA*
[5] *TRIUMF, 4004 Wesbrook Mall, Vancouver, British Columbia, V6T 2A3, Canada*
[6] *Universitat Politecnica de Catalunya (UPC), Barcelona, Spain*
[7] *STFC Daresbury Laboratory, Daresbury, Warrington WA4 4AD, UK*
[8] *University of Liverpool, Liverpool, L69 7ZE, United Kingdom*
[9] *University of Edinburgh, Edinburgh, EH9 3JZ, United Kingdom*
[10] *Central Michigan University, Mount Pleasant, MI 48859, USA*
[11] *Department of Physics & Astronomy, University of Tennessee , Knoxville, TN 37996-1200 USA*
[12] *Institute of Nuclear Research of the Hungarian Academy of Sciences, Debrecen Pf. 51, H-4001, Hungary*
[13] *Department of Physics, The University of Hong Kong, Pokfulam Road, Hong Kong, China*
[14] *National Physical Laboratory, Teddington, UK TW11 0LW*
[15] *Department of Physics, University of Tokyo, Hongo, Bunkyo-ku, Tokyo 113-0033, Japan*
[16] *Faculty of Physics, University of Warsaw, Warsaw PL-02-093, Poland*
[17] *National Superconducting Cyclotron Laboratory, Michigan State University, East Lansing, MI 48824, USA*
[18] *Department of Physics, Korea University, Seoul 136-701, Republic of Korea*
[19] *Department of Physics, University of Surrey, Guildford GU2 7XH, United Kingdom*
[20] *STFC Rutherford Appleton Laboratory, Harwell Campus, Didcot, Oxfordshire, OX11 0QX, UK*
[21] *Department of Physics, Niigata University, Niigata 950-2102, Japan*
[22] *Department of Nuclear Physics, Faculty of Physics, VNU University of Science, Hanoi, Vietnam*

Abstract. The commissioning of a new setup for β-delayed neutron measurements was carried out successfully in November-2016, at the RIKEN Nishina Center in Japan. The β-decay half-lives and P_n branching ratios of several isotopes in the ^{78}Ni region were measured. Details of the experimental setup and the first results are given.

⋆e-mail: Alvaro.Tolosa@ific.uv.es

DOI: 10.1051/epjconf/201716501051

1 Motivation

Beta-delayed neutron emission (βn) is a form of radioactive decay in which an electron, an anti-neutrino and one or more neutrons are emitted. This process arises if the energy window of the decay Q_β is greater than the neutron separation energy S_n of the daughter. The probability in each decay of emitting neutrons is called the P_n value.

This form of decay plays a key role in the synthesis of chemical elements in the Universe via the rapid neutron capture process, or r-process [1]. The r-process proceeds far from the valley of nuclear stability, and leads to very neutron-rich nuclei that then decay to the line of stability. Most of these nuclei are βn emitters. The initial abundance distribution is shaped by the decay half-life $T_{1/2}$, but the neutron emission affects the final abundances in two ways: on the one hand it shifts the decay path to lower masses and on the other hand it provides a source of neutrons for late captures that will shift it in the opposite way[2]. Thus, a good knowledge of P_n and $T_{1/2}$ values of the nuclei between the line of stability and the r-process path is needed for a correct understanding of the observed abundances.

Besides the astrophysical interest, the measurement of new P_n values and half-lives is also useful for refining nuclear structure models, constraining parameters of the models and testing β-strength functions [3]. Our current understanding of the process of β-delayed multiple neutron emission, as well as the competition between different open channels (gamma, one-neutron, two-neutron, etc), is incomplete because of the scarcity of experimental data [4].

With these ideas in mind, the BRIKEN (Beta-delayed-neutron-measurements at RIKEN) [5] collaboration was established, with the aim of greatly improving the quantity and quality of data the decays of very neutron-rich nuclei close to the r-process path.

2 Experimental setup

A detection system for direct measurements of βn-decay has been installed at the Radioactive Isotope Beam Factory (RIBF) in RIKEN, which currently provides the highest intensities of the most neutron-rich nuclei produced by in-flight fission.

The detection system is composed of the Advanced Implantation Detector Array (AIDA) [6], and a neutron counter. It is placed at the F11 focal plane of the BigRIPS+ZeroDegree spectrometer [7], which serves to identify the isotopes that are implanted in AIDA.

The mission of AIDA is to register ion implantations and the subsequently emitted β particles. It consists of a stack of six Si DSSDs. There are two different electronic branches, for high- and low-energy events respectively. Offline, after consistency checks, the high energy events are identified as implants, and low energy events as β particles. The highly segmented silicon strip detectors grant a high detection efficiency and minimize accidental implant-beta correlations. A digital eventless data acquisition system (DAQ) reads out the individual strips (1536 in total) for further processing.

2.1 The neutron counter

The neutron counter surrounding AIDA consists of a large polyethylene (PE) matrix with an array of 140 ^{3}He tubes, owned by several collaborating laboratories, including ORNL, UPC-Barcelona, GSI-Darmstadt and RIKEN. The neutron energies are moderated in collisions in the PE and the neutrons are absorbed by the ^{3}He in an exothermic reaction. The energy deposited in the tube produces

 DOI: 10.1051/epjconf/201716501051

a signal that is amplified and acquired by a digitizer working in self-triggering mode. The data are recorded by the eventless DAQ system *Gasific* developed at IFIC-Valencia [8].

The PE moderator geometry was optimized [9] in order to reach a high detection efficiency, that is as constant as possible for neutron energies from 1 keV to few MeV. The latter is required to minimize systematic errors in the derived P_n for decays with different neutron energy distributions.

Given the very large number of ^{3}He tubes and their different characteristics (length, pressure, size) a parametrized Monte Carlo optimization algorithm was developed in order to determine their best position inside the matrix, and achieve a high and flat neutron efficiency. The configuration finally chosen includes also two Clover-type HPGe detectors. This adds high-resolution γ-ray spectroscopy capability with the selectivity provided by neutron tagging, giving additional nuclear structure information.

The GEANT4 [10] simulated efficiency was crosschecked using a ^{252}Cf neutron source. The experimental efficiency for this source is 61 ± 2%, which is in good agreement with the simulated efficiency of 62%. This corresponds to an average efficiency of about 68% for neutrons below 1.5 MeV.

3 Measurements

The BRIKEN setup received the first radioactive beam on November 2016. A high intensity (40 pnA) primary beam of ^{238}U, with energy 345 MeV/u, hit a 5 mm Be target producing a secondary beam of exotic neutron-rich nuclei. The BigRIPS spectrometer setting was centred on ^{76}Ni. Figure 1 shows an identification plot of the accumulated implants during the experiment.

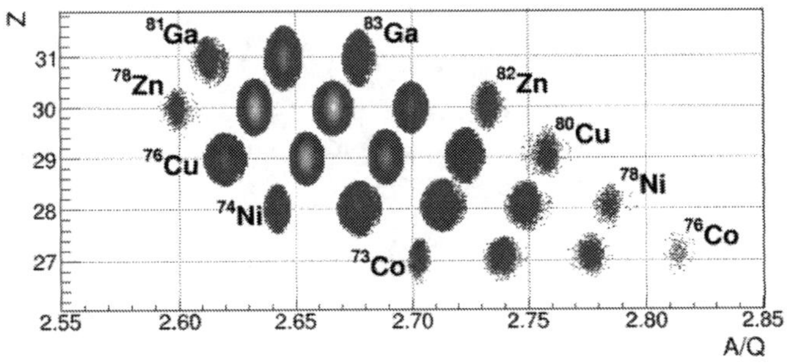

Figure 1. Particle identification plot of the implanted ions.The x-axis of the plot represents the particle mass over its charge, and the y-axis represents the atomic number Z. Given the high energy during the identification, the atoms are completely stripped of their electrons and their charge coincides with their atomic number.

Using these data, specific software was developed using the ROOT package [11] for merging offline the data from the three independently running but synchronized DAQs: BigRIPS, AIDA and BRIKEN. This software sorts the input data by time stamp, and at the same time includes in the output stream time-correlated data for preselected parameters; also, at this point the identification (atomic number Z and mass over charge ratio A/Q) of the implant is assigned by time coincidence with BigRIPS events. The output can be used directly in the analysis procedure where complex correlations, vetoes and tags can be applied. Figure 2 shows some of the correlated spectra that can be built using this software.

The P_n value, the number of parent decays, and eventually the half-lives, are obtained from a simultaneous fit of the implant-decay curves using the solution of the Bateman equations [12]. The neutron efficiency is a fixed input parameter. The usual assumption that the β efficiency is independent of the species of nucleus in the decay chain, and thus can be ignored, was tested and it was found that it requires a demanding minimization of thresholds for β signals in AIDA.

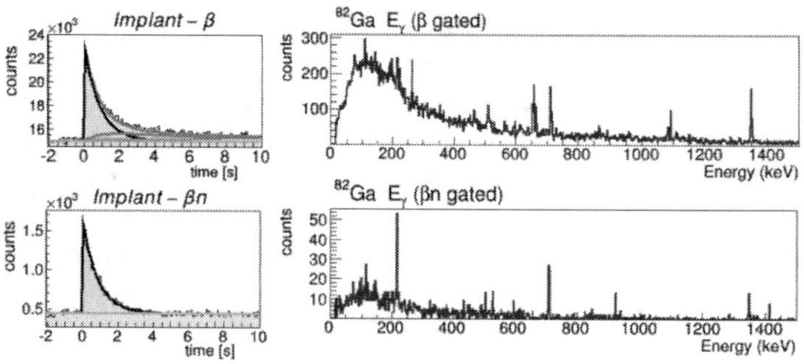

Figure 2. For each identified isotope (here ^{82}Ga), the implantation-decay time histogram can be built from the merger output without (top left) or with β-neutron coincidence time window (bottom left). Gamma-ray energy spectra in coincidence with β particles (top- and bottom-right) can be built integrating over a range of implantation-decay times.

4 Results and conclusions

A total of 15 P_n values could be determined for Ni, Cu, Zn and Ga isotopes. Good agreement was found with values in literature [13] for ^{76}Cu, ^{77}Cu and ^{82}Ga, where there are several existing measurements, thus confirming the reliability of the data. For the first time P_n values were determined for ^{76,77}Ni and ^{80}Cu. In the decay of ^{78}Ni a total of 29 β-neutron correlations were detected for the first time. More precise results were obtained for the remaining cases, some significantly different from previously published values. Overall these results demonstrate the good performance of the setup.

In May-June 2017, the first experimental BRIKEN campaign took place successfully. It covered part of the currently approved proposals. Although the data are still in the preliminary stages of the analysis we can report that several tens of new P_{1n} and P_{2n} values will be determined for the first time. Additional experiments will follow in the near future.

Acknowledgement

This work has been supported by the Spanish MINECO under grants FPA2011-24553, FPA2011-28770-C03-03, FPA2014-52823-C2-1/2, SEV-2014-0398; by STFC (UK); by FP7/EURATOM Contract No. 605203; by JSPS KAKENHI grant No. 17H06090; by NSERC grants SAPIN-2014-00028 and RGPAS 462257-2014 at TRIUMF; by U.S. DOE grant DE-AC05-00OR22725; by Polish NSC UMO-2015/18/E/ST2/002. Work partially done within IAEA-CRP for Beta Delayed Neutron Data.

References

[1] E. M. Burbidge, G. R. Burbidge, W. A. Fowler & F. Hoyle. Rev. Mod. Phys. **29**, 547 (1957)
[2] A. Arcones & G. Martinez-Pinedo. Phys. Rev. C **83**, 045809 (2011).
[3] P. Moller, B. Pfeiffer, and K.-L. Kratz, Phys. Rev. C **67**, 055802 (2003)
[4] R. M. Mumpower et al. Phys. Rev C **94**, 064317 (2016)
[5] https://www.wiki.ed.ac.uk/display/BRIKEN/Home
[6] T. Davinson et al., http://www2.ph.ed.ac.uk/~td/AIDA/
[7] T. Kubo et al., Prog. Theor. Exp. Phys. **2012**, 03C003 (2012)
[8] J. Agramunt et al., Nucl. Instrum. Methods Phys. Res. A **807**, 69 (2016)
[9] A. Tarifeno-Saldivia et al., J. Instrum. **12**, 04006 (2017)
[10] S. Agostinelli, et al. Nucl. Instrum. Methods Phys. Res., Sect. A, **506**, 250 (2003)
[11] R. Brun et al., Nucl. Instrum. Methods Phys. Res. A, **389**, 81 (1997)
[12] K. Skrable et al., Health Physics, **27**, 155 (1974)
[13] https://www.nndc.bnl.gov/ensdf/

EPJ Web of Conferences **165**, 01052 (2017)
NPA8 2017
DOI: 10.1051/epjconf/201716501052

Determining the $^{13}C(\alpha, n)^{16}O$ absolute cross section through the concurrent application of ANC and THM and astrophysical consequences for the s-process in AGB-LMSs

Oscar Trippella[1,2,*] and *Marco* La Cognata[3,**]

[1] *Department of Physics and Geology, University of Perugia, Via A. Pascoli, 06123 Perugia, Italy*
[2] *I.N.F.N., Section of Perugia, Via A. Pascoli, 06123 Perugia, Italy*
[3] *I.N.F.N., Laboratori Nazionali del Sud, Via S. Sofia 62, 95123 Catania, Italy*

Abstract. The $^{13}C(\alpha, n)^{16}O$ reaction is considered to be the most important neutron source for the s-process main component in low-mass asymptotic giant branch stars. No direct experimental data exist at very low energies and measurements performed through direct techniques show inconsistent results, mostly in their absolute values. In this context, we reversed the usual normalization procedure combining two indirect approaches, the asymptotic normalization coefficient and the Trojan Horse Method, to unambiguously determine the absolute value of the $^{13}C(\alpha, n)^{16}O$ astrophysical $S(E)$-factor in the most relevant energy-region for astrophysics. Adopting the new reaction rate for the n-source in the NEWTON s-process nucleosynthesis code, astrophysical calculations show only limited variations, less than 1%, for those nuclei whose production is considered to be totally due to slow neutron captures.

1 Introduction

The s-process main component, responsible for the synthesis of neutron-rich nuclei with $86 \leq A \leq 209$, is produced during the asymptotic giant branch (AGB) phase of low-mass stars (LMSs) through a series of subsequent n-captures and β-decays [1]. In this astrophysical environment, the main neutron source is considered to be the $^{13}C(\alpha, n)^{16}O$ reaction providing n-densities of the order of $10^6 - 10^8$ cm^{-3} at typical energies of 8 keV in radiative conditions. However, the cross section of the $^{13}C(\alpha, n)^{16}O$ reaction still remains uncertain in the Gamow window (140 – 230 keV) because of the contribution of a broad resonance, corresponding to the $1/2^+$ excited state of ^{17}O, close to the α-threshold. Several experiments adopting both direct and indirect techniques have been performed in the last decades to determine the $^{13}C(\alpha, n)^{16}O$ cross section at the energies of astrophysical interest. In particular, the ASFIN (AStroFIsica Nucleare) group of LNS (Catania, Italy) performed an experiment at the Florida State University (USA), adopting the indirect Trojan Horse Method (or THM, see [2] for more details). The $^{13}C(^6Li, n^{16}O)d$ reaction was measured in quasi-free kinematical conditions using a 7.82 MeV ^{6}Li beam. In this way, d is spectator to the $^{13}C(\alpha, n)^{16}O$ process and it can be extracted at astrophysical energies free of Coulomb suppression and the electron screening effect [3, 4].

*e-mail: oscar.trippella@pg.infn.it
**e-mail: lacognata@lns.infn.it

DOI: 10.1051/epjconf/201716501052

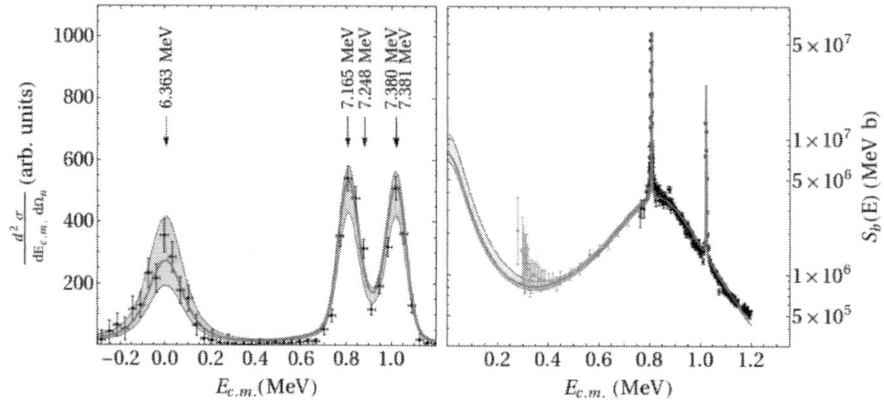

Figure 1. *Left panel* – New HOES (Half-Of-Energy-Shell) R-matrix fit (red shaded area) superimiposed on THM data (black points). *Right panel* – R-matrix $S(E)$-factor (red shaded region) superimposed on direct data from [7] and [8] (orange and black points, respectively). The red bands in each panel account for both statistical and normalization uncertainties.

Moreover, direct measurements in the literature cover only energies down to 300 keV and they are also affected by large systematic errors due to the spread in absolute values even at high energies [5], while indirect data crucially depend by the normalization procedure to direct ones in an energy region covered by both types of experiments. In this context, we have reversed the usual normalization procedure combining two indirect approaches, the asymptotic normalization coefficient (ANC) and the THM [6], to unambiguously determine the absolute value of the $^{13}C(\alpha, n)^{16}O$ astrophysical factor.

2 New normalization procedure from indirect to direct data

In order to define an absolute and unique value for the $^{13}C(\alpha, n)^{16}O$ data, we combined the recent and precise ANC calculation (3.6±0.7 fm^{-1} by [9]) and the full width for the threshold resonance (136 keV in [10]) from literature into a HOES R-matrix fit of the THM data [4] (see red curves and full symbols, respectively, in the left panel of Fig. 1). The first resonance on the left represents the $1/2^+$ excited states of ^{17}O at 6.363 MeV corresponding to 4.7 keV over the α-threshold, because this state was recently recognized by [10] to be centred at positive energies, while for long time it was considered as a sub-threshold resonance. In this way, we have to calculate the ANC of this resonance in the case of unbound states $C^2 = (-1)^{l_B} e^{\pi\eta_{aA}} e^{2i\delta^p_{l_B j_B J_B}(k_{aA(R)})} \frac{\mu_{aA}\Gamma_{aA}}{k_{aA(R)}}$, [11] where η is the Sommerfeld parameter, δ the potential (non-resonant) scattering phase shift at the real resonance relative momentum k, and Γ represents the partial width of the excited state of ^{17}O mentioned above. In this way, we obtained the γ values of the two peaks (each of them refer to two separate, but with similar central energy, resonances as shown by the four arrows in the left panel of Fig. 1) at higher energies in order to discriminate among the different direct datasets [5] and to unambiguously define (right panel of Fig. 1) the absolute value of the $S(E)$-factor. The R-matrix astrophysical factor (see the red band in right panel of Fig. 1), calculated adopting the same resonance parameters used in the corresponding left panel for the normalization procedure, shows that the measurements by [7] and [8] for the low-energy region (black and orange points, respectively) represent the only direct data set compatible with the THM $S(E)$-factor and the ANC of the $1/2^+$ state measured by [9].

 DOI: 10.1051/epjconf/201716501052

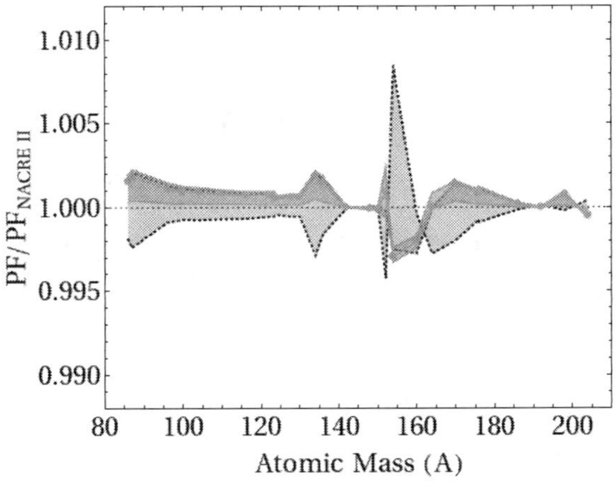

Figure 2. Production factors for the *s*-only nuclei of the main component calculated adopting the new THM rate for the $^{13}C(\alpha, n)^{16}O$ reaction (orange shaded region) or the one suggested by [5] (gray band). The horizontal dashed line, referring to predictions obtained using the recommended values of [5], is used as a reference.

3 Astrophysical consequences

Starting from the THM $S(E)$-factor described in the previous section, we calculated a very accurate reaction rate for the $^{13}C(\alpha, n)^{16}O$ neutron source and we introduced it into a code for *s*-process nucleosynthesis in AGB-LMSs [1, 12]. Because of the different normalization procedure, the reaction rate presented in this work is different, up to a factor of 1.5, with the one calculated in the precedent analysis of THM data [4]. On the contrary, it is in agreement, within errors, with the recommended value of the NACRE II compilation [5] in the most interesting temperature-region for astrophysics ($0.05 \leq T_9 \leq 0.30$), but the level of uncertainties is strongly reduced in our work. Fig. 2 shows the ratio between production factors (PF) of the main component (from Sr to Bi) *s*-only nuclei calculated by means of the NEWTON post-process code [1, 12] adopting the rate for the $^{13}C(\alpha, n)^{16}O$ reaction presented by [6] (orange band) and the one suggested in NACRE II (black shaded area [5]). In calculations of Fig. 2 we adopted a ^{13}C reservoir of about 5.0×10^{-3} $M_{\odot}$, as suggested by [13], assuming the case of a LMS with 1.5 $M_{\odot}$ and [Fe/H] = −0.15 experiencing nine thermal instabilities. In this scenario, we do not expect significant variations, less than 1%, for those isotopes which are produced exclusively by slow neutron captures.

References

[1] M. Busso, R. Gallino, G.J. Wasserburg, Ann. Rev. of Astron. and Astrophys. **37**, 239 (1999)
[2] C. Spitaleri, A.M. Mukhamedzhanov, et al., Physics of Atomic Nuclei **74**, 1725 (2011)
[3] M. La Cognata, C. Spitaleri, O. Trippella, et al., Physical Review Letters **109**, 232701 (2012)
[4] M. La Cognata, C. Spitaleri, O. Trippella, et al., Astrophys. J. **777**, 143 (2013)
[5] Y. Xu, K. Takahashi, S. Goriely, et al., Nuc. Phys. A **918**, 61 (2013), 1310.7099
[6] O. Trippella, M. La Cognata, Astrophys. J. **837**, 41 (2017), 1702.01589
[7] J.K. Bair, F.X. Haas, Phys. Rev. C **7**, 1356 (1973)
[8] H.W. Drotleff, A. Denker, H. Knee, et al., Astrophys. J. **414**, 735 (1993)
[9] M.L. Avila, G.V. Rogachev, E. Koshchiy, et al., Phys. Rev. C **91**, 048801 (2015), 1503.08743
[10] T. Faestermann, P. Mohr, R. Hertenberger, et al., Phys. Rev. C **92**, 052802 (2015), 1511.02642
[11] A.M. Mukhamedzhanov, Phys. Rev. C **86**, 044615 (2012), 1209.2158
[12] O. Trippella, M. Busso, E. Maiorca, et al., Astrophys. J. **787**, 41 (2014), 1403.6256
[13] O. Trippella, M. Busso, S. Palmerini, et al., Astrophys. J. **818**, 125 (2016), 1512.06777

EPJ Web of Conferences **165**, 01053 (2017) DOI: 10.1051/epjconf/201716501053
NPA8 2017

The solar Lithium problem: is the explanation due solely to mixing or also to the e^--capture decay rate of ^{7}Be?

Diego Vescovi[1,2,⋆], *Maurizio* Busso[1,2], *Sara* Palmerini[1,2], and *Oscar* Trippella[1,2]

[1] *Department of Physics and Geology, University of Perugia, Via A. Pascoli, 06123 Perugia, Italy*
[2] *I.N.F.N., Section of Perugia, Via A. Pascoli, 06123 Perugia, Italy*

Abstract. The nucleosynthesis of ^{7}Li is one of the most crucial problems in nuclear astrophysics, as its observations in several sites are hard to be explained. Concerning the Sun, the most common interpretations of the low Li abundance invoke either burning in early stages or non-convective mixing below the envelope. Here we apply a diffusive mechanism of mixing, together with a recent estimate of the rate for e^--captures on ^{7}Be, to establish whether the solar Li destruction should be attributed to purely pre-Main Sequence (MS) nuclear processes or if the coupling of mixing and nucleosynthesis on the MS can account for it. Our preliminary results indicate that, whether Li survives the pre-MS phase, the changes of the ^{7}Be e^--capture rate do not affect its production/destruction. The low Li abundance should then depend only on diffusion processes from the bottom of the convective envelope to the lowerlying tachocline zone. We suggest that, if diffusive processes occurred over the age of the Sun, they required diffusive mass transfers of a few 10^{-13} $M_\odot$/yr to explain the Li drop. This is a high estimate: future works will tell us if it is realistic or not. In this second case, pre-MS burning would remain the only alternative.

1 Introduction

The ^{7}Li abundance in the solar photosphere is a factor of about 160 lower than in meteorites [1]. This discrepancy ("the solar Li problem") may mean that nuclear processes in pre-MS phases affected strongly the Li abundance, which fact was found to be possible in models where convection is extended as compared to what the Schwarzschild criterion predicts by some forms of overshooting [2]. Models without them fail in destroying Li in pre-MS by more than a factor of around 7 and need that, during the MS itself, Li is exposed to conditions where p-captures effectively destroy it. On the MS, the temperature of sub-convective layers remain always rather low, but Li is a fragile nucleus and a weak burning integrated over 4.56 Gyr of life is enough for its consumption. Below the convective envelope, indeed, Li abundances as low as 10^{-15} or less are found [3]. Processes of diffusive mixing below the formal inner border of solar convection are therefore sometimes invoked as an alternative to pre-MS burning [4]. Thanks to helioseismology, we know that the internal solar structure is characterized by two different profiles of rotational velocity: a differential rotation in the convective envelope and a rigid-body one in the innermost regions. The transition region between these two regimes is called *tachocline*: it is the layer where a differential rotation is established and the solar dynamo with

⋆e-mail: diegontt92@gmail.com

 DOI: 10.1051/epjconf/201716501053

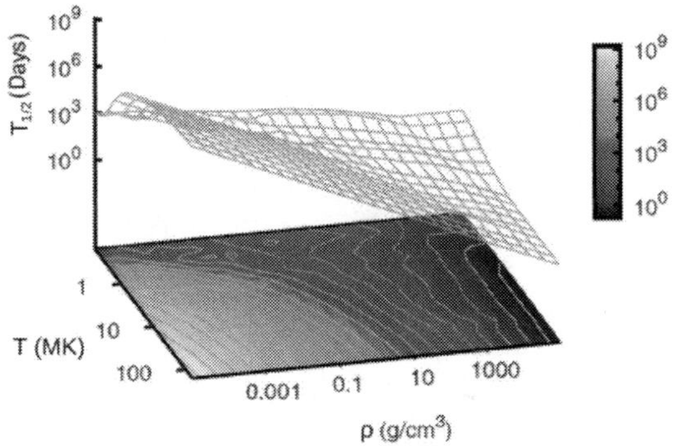

Figure 1. The half-life for ^{7}Be decay through e$^-$-captures, as a function of the plasma temperature T and the plasma density ρ, as derived in [5] and used in the present work in place of the previous one by [6].

all its ensuing buoyancy phenomena have their roots. The tachocline is limited above by the convective border, at $r \simeq 0.70\ R_\odot$, and has a thickness of about $0.04\ R_\odot$ [2]. The limited knowledge we have on the hydrodynamics of matter diffusion through the tachocline is added to our poor understanding of some nuclear reactions for the synthesis of ^{7}Li, in particular of the electron captures on ^{7}Be that directly feed it [5]. For these reasons we have investigated the possible variations due to the use of a new rate. In particular, in this work, we will use the rate for Be decay proposed by [5]. This was previously computed and applied for typical conditions of evolved stars, but we are allowed to use it also for the solar case. In fact, comparing the values of the electron density at the Be nucleus, representing the crucial parameter to calculate the e$^-$-capture rate, produced by this new method and the one from a traditional Debye-Hückel approximation, they differ by less than 1.5% at solar condition (see Table 1 of [5] and references therein).

2 Deep mixing and e$^-$-captures on ^{7}Be

Let us therefore assume a limited Li burning in the pre-MS phases. Using the Schwarzschild criterion, we actually found an initial MS concentration only a factor of 2 below the meteoritic value. Despite the limited masses involved in the tachocline, mixing through its layers would then be required to reduce the Li abundance. The temperature at the base of solar convection is indeed too low for any Li burning and the surface abundance must have been diluted with regions where instead Li consumption occurred, in order to explain its large drop. In this note, we have computed the mixing and nucleosynthesis processes that materials circulating through the inner border of solar convection must undergo. In these calculations, we adopted the rate for Be decay proposed by [5] and the corresponding half-life is shown in Figure 1 as a function of temperature and density. The treatment of the combined occurrence of mixing and burning is the same described in [7]. Namely, a slow circulation is added by imposing transport of envelope material at a mass-flow rate $\dot{M}$ down to some depth corresponding to the base of tachocline zone. Our post process code for deep-mixing calculations was then applied to a solar evolutionary sequence calculated with the FRANEC code [3]. We found that in the Sun, which is characterized by low mixing rate and low temperature, any nuclear effect from ^{3}He+^{4}He and the subsequent ^{7}Be decay is marginal. Therefore Li abundances are affected only by mixing (see panel a of Figure 2). We also found that the same rate for ^{7}Be production is extremely low at the temperatures of the tachocline ($\simeq 2.5\times10^6$ K), so that Be is not produced and hence does not decay to Li in the circulating material, which stays in the tachocline layers for less than 30% of the mixing time.

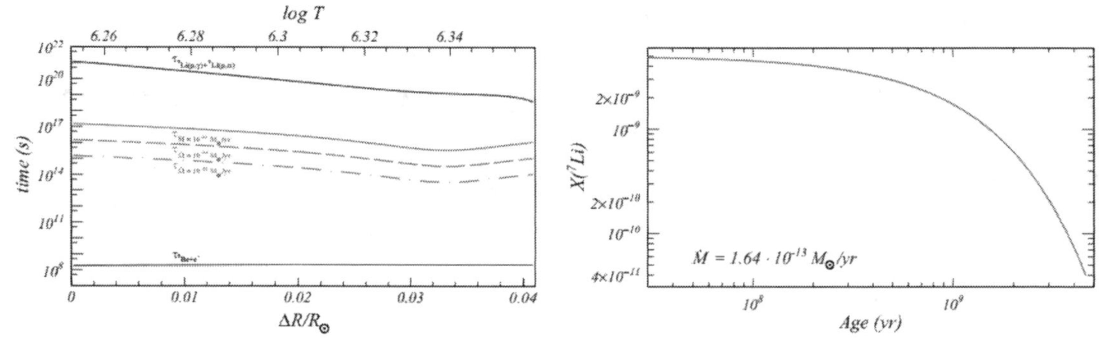

(a) (b)

Figure 2. (a) Comparison between the timescale for Li destruction by *p*-captures, for Li production from e^--captures on ^{7}Be, and the overturn time of mixing, assuming different circulation rates, as indicated. The plot refers to the solar tachocline zone. On the lower abscissa, ΔR represents the distance from the top of the tachocline; the upper abscissa shows the corresponding temperature. Here the competition, unlike the AGB stars cases, is almost negligible, so we cannot obtain enhanced/reduced Li abundances due to nuclear production/destruction. (b) The destruction of Li in a deep-mixing process, during which material is brought from the solar envelope to the *tachocline* and vice versa. The process can be described as a slow circulation with an average mass transfer around 10^{-13} $M_\odot$/yr.

3 A diffusive mixing solution?

The situation is therefore quite different with respect to normal radiative burning on the MS, so that Li is not affected by nuclear processing along the path. The main issue is that the decrease in the sub-convective Li, obtained by solar models without any circulation, derives by the exposition of the local material, for extremely long times, to the same stable conditions. In the matter arriving from the convective envelope, instead, the small extension of the tachocline allows the nuclei to stay in the warmer zones for a very limited time, insufficient to permit any significant burning. These results would indicate that a solution to the solar Li problem (see panel b of Figure 2) can rely only on a pure dilution, where the higher Li concentration of the envelope is decreased by being mixed with the low abundance already present in the inner layers. Such a mixing would require a relatively high circulation rate (a few 10^{-13} $M_\odot$/yr) to reduce the Li abundance to the observed value.

We acknowledge L. Piersanti for providing us the solar model starting from we performed our calculations [3].

References

[1] M. Asplund, N. Grevesse, A.J. Sauval et al., ARAA **47**, 481 (2009), 0909.0948
[2] F. Thévenin, A.V. Oreshina, V.A. Baturin et al., A&A **598**, A64 (2017), 1612.01331
[3] L. Piersanti, O. Straniero, S. Cristallo, A&A **462**, 1051 (2007), astro-ph/0611229
[4] J.D. Do Nascimento, J.S. da Costa, J.R. de Medeiros, A&A **519**, A101 (2010), 1006.3861
[5] S. Simonucci, S. Taioli, S. Palmerini, M. Busso, ApJ **764**, 118 (2013), 1212.5737
[6] E.G. Adelberger et al., Reviews of Modern Physics **83**, 195 (2011), 1004.2318
[7] S. Palmerini, S. Cristallo, M. Busso et al., ApJ **741**, 26 (2011), 1107.2844

EPJ Web of Conferences **165**, 01054 (2017) DOI: 10.1051/epjconf/201716501054
NPA8 2017

Constraining the ^{19}Ne(p,γ)^{20}Na Reaction Rate Using a Direct Measurement at DRAGON

R. Wilkinson[1,⋆], *G. Lotay*[1,2], *A. Lennarz*[3], *C. Ruiz*[3], *G. Christian*[4,5,6], *C. Akers*[3,⋆⋆], *W. N. Catford*[1], *A. A. Chen*[7], *D. Connolly*[3], *B. Davids*[3], *D. A. Hutcheon*[3], *D. Jedrejcic*[8], *A. M. Laird*[9], *L. Martin*[3], *E. McNeice*[7], *J. Riley*[9], and *M. Williams*[3,9]

[1] *Department of Physics, University of Surrey, Guildford, GU2 7XH, United Kingdom*
[2] *National Physical Laboratory, Teddington, Middlesex, TW11 0LW, United Kingdom*
[3] *TRIUMF, Vancouver, British Columbia, V6T 2A3, Canada*
[4] *Cyclotron Institute, Texas A&M University, College Station, Texas 77843-3366, USA*
[5] *Department of Physics and Astronomy, Texas A&M University, College Station, Texas 77843-3366, USA*
[6] *Nuclear Solutions Institute, Texas A&M University, College Station, Texas 77843-3366, USA*
[7] *Department of Physics and Astronomy, McMaster University, Hamilton, Ontario L8S 4M1, Canada*
[8] *Colorado School of Mines, Golden, Colorado 80401, USA*
[9] *Department of Physics, The University of York, York YO10 5DD, United Kingdom*

Abstract. A direct measurement of the ^{19}Ne(p,γ)^{20}Na reaction has been performed in inverse kinematics at the DRAGON recoil separator, at an energy ~ 10 keV higher than previous measurements. The key resonance in the ^{19}Ne $+p$ system relevant for ONe novae and Type-I X-ray burst temperatures have been successfully measured for the first time. Preliminary estimates of the resonance energy and strength are reported as $E_{c.m.} \approx 458$ keV and $\omega\gamma \approx 18$ meV. These results are consistent with previous direct measurements, but disagree with the most recent study of the ^{19}Ne(p,γ)^{20}Na reaction rate. These preliminary results will be finalised after a forthcoming negative log-likelihood analysis.

Classical novae and X-ray bursts are among the most common types of explosive stellar phenomena observed in our Galaxy. These cataclysmic binary systems are stellar environments with high temperatures and densities, thought to be active regions of ongoing nucleosynthesis which energetically impact the surrounding interstellar medium (ISM).

Classical novae are binary systems where hydrogen-rich material is accreted from a main sequence or red giant branch star onto the surface of a companion white dwarf. As the accreted material mixes with the heavier elements already present on the surface of the white dwarf, a thermonuclear runaway is triggered, which leads to explosive mass ejections from the white dwarf's surface, with the process recurring over a time-scale of ~ 10^4 - 10^5 years [1]. The composition of the white dwarf star has a significant impact on the composition of the nova ejecta; novae with more massive oxygen-neon (ONe) white dwarfs are characterised by higher peak temperatures (~ 0.4 GK) than less massive carbon-oxygen (CO) white dwarfs (~ 0.2 GK), and are thus thought to synthesise elements up to the Si-Ca mass region [2]. Therefore, determining the nature of the underlying white dwarf is crucial to meaningfully compare theoretical calculations with astronomical data.

⋆e-mail: r.wilkinson@surrey.ac.uk
⋆⋆Present address: Rare Isotope Science Project, Institute for Basic Science, Daejeon, 305-811, Republic of Korea

Highly ionized fluorine lines have been observed in the optical spectra of Nova Mon 2012 [3], the detection of which is thought to be one of only four isotopic signatures from novae that could indicate a high-mass underlying ONe white dwarf [4]. In explosive stellar phenomena, ^{19}F is produced through the ^{17}O$(p,\gamma)^{18}$F$(p,\gamma)^{19}$Ne$(\beta^+)^{19}$F reaction sequence. However, this process can be bypassed at the high peak temperatures in ONe novae via the ^{19}Ne$(p,\gamma)^{20}$Na$(\beta^+)^{19}$Ne reaction sequence. A nova nucleosynthesis sensitivity study by Iliadis et al. [5] has shown variations of a factor of 100 in the ^{19}Ne$(p,\gamma)^{20}$Na reaction rate can change the final abundance of ^{19}F by up to a factor of 7. Therefore, the ^{19}Ne$(p,\gamma)^{20}$Na reaction rate must be constrained to understand astronomical data from ONe novae.

X-ray bursts are cataclysmic binary systems similar to novae, but with a neutron star in place of the white dwarf. This results in much higher peak temperatures ($\sim$ 1.5 GK) and a much reduced burst recurrence time. Between outbursts, Type-I X-ray bursts produce energy via the β-limited hot CNO cycles. However, due to the higher peak temperatures achieved in X-ray bursts, during an outburst it becomes energetically feasible to "breakout" into the $rp-$process; a series of sequential, rapid proton capture reactions which has been suggested to synthesise elements up to the Sn-Te mass range[6]. The reaction sequence thought to dominate breakout into the $rp-$process is ^{15}O$(\alpha,\gamma)^{19}$Ne$(p,\gamma)^{20}$Na, thus the ^{19}Ne$(p,\gamma)^{20}$Na reaction is thought to be an important factor in understanding the behaviour of Type-I X-ray bursts. Moreover, it is thought that the strength of this reaction sequence helps to determine both the conditions for ignition to occur, as well as the burst recurrence time[7].

In the temperature range relevant for ONe novae and Type-I X-ray bursts, the ^{19}Ne$(p,\gamma)^{20}$Na reaction is thought to be dominated by the contribution of a single, low-lying resonance located ~450 keV above the proton-emission threshold energy in ^{20}Na; equivalent to an excited state energy of $E_x = \sim$ 2650 keV. However, for over twenty years, the exact strength and energy of this resonance state have been a subject of intense debate [8-21].

Previous direct measurements of the ^{19}Ne$(p,\gamma)^{20}$Na reaction have utilised radioactive beams of ^{19}Ne [12-16], but none of these studies have identified the resonance of interest or measured its strength. Consequently, these studies have only been able to report upper limits of 18 meV[14], 21 meV[15], and 15 meV[16] for the resonance strength of the state of interest. Additionally, Page et al. [14], concluded that the dominant remaining uncertainty in the rate was the unknown spin-parity assignment of the state of interest, which had been given conflicting 1^+[8] and 3^+[9] assignments. However, the recent non-observation of the resonant state in a β-delayed proton study of ^{20}Mg, which would otherwise be highly populated by an allowed Gamow-Teller transition if the state had $J^\pi = 1^+$, strongly indicates that the state has spin-parity 3^+[20]. Moreover, the same study re-examined data from a previous high precision (^{3}He,t) study[10] in conjunction with the most recent, precise ^{20}Na proton threshold energy[17] to infer a more precise energy for the resonance of interest of $E_{c.m.} = 457(3)$ keV. Interestingly, a theoretical study by Fortune et al. derived a lower limit on the resonance strength, assuming a spin-parity assignment of 3^+, of 16-18 meV[18]; extremely close to the upper limits set by direct measurements.

The most recent study of the ^{19}Ne$(p,\gamma)^{20}$Na reaction rate was performed by Belarge et al.[21]. A radioactive ion beam (RIB) of ^{19}Ne was used to populate states above the proton-emission threshold in ^{20}Na via a (d,n) reaction. The key resonance in the ^{19}Ne$(p,\gamma)^{20}$Na reaction was found to decay via strong proton decay branches at 440 keV and 200 keV, to the ground and first excited states. An angular distribution analysis of the decay to the first excited state found the resonant state of interest to have a spin-parity of 3^+, in agreement with Ref. [20]. Proton partial widths for both decays were determined from a couple-channel Born approximation. These were used, with a γ-ray partial width determined from the life-time of a proposed mirror state in ^{20}F at 2966.11(3) keV[22], to determine resonance strengths of proton captures to the ground and first excited state of $69^{\pm 15_{stat.}}_{\pm 17_{sys.}}$ meV and $21^{\pm 5_{stat.}}_{\pm 6_{sys.}}$ meV. These values of the resonance strength, contradict every previous direct measurement

[14-16] and the reported resonance energy is inconsistent with previous indirect studies [12,21]. Consequently, the increase in resonance strength and decrease in resonance energy reported in Ref [21] leads to a ^{19}Ne$(p,\gamma)^{20}$Na reaction rate that is a factor ~20 higher than previous evaluations [5,10].

The subject of the present work is a new direct experimental study of the ^{19}Ne$(p,\gamma)^{20}$Na reaction performed ~10 keV higher in energy than previous direct measurements. An intense RIB of ^{19}Ne was used to measure the energy and strength of the key resonance in the ^{19}Ne+p system at ONe novae and X-ray burst temperatures for the first time. The experiment was performed at the DRAGON recoil separator[23], at the ISAC-I radioactive beam facility at TRIUMF. An intense RIB of ^{19}Ne^{1+} was produced by bombarding a SiC target with a primary beam of 500 MeV protons. Mass ~19 a.m.u. products were extracted, ionized in a FEIBAD ion source[24] and filtered by a high-resolution mass spectrometer. This beam was then initially accelerated up to 150 A·keV using a radio-frequency quadrupole (RFQ) accelerator and stripped to a 5^+ charge state using a thin carbon foil. The ^{19}Ne^{5+} beam was then accelerated to an energy of 9.24 MeV and impinged upon a windowless gas target of H_2 at a pressure of 5 Torr for 89 hours. The incoming beam energy and gas target pressure were chosen to cover a center-of-mass energy range of E_{cm} = 447 – 465 keV across the gas target volume. Surrounding the windowless gas target is a highly efficient array of bismuth germanate (BGO) detectors[23], used to detect the radiative capture γ-rays while ^{20}Na^{6+} recoils were transmitted to the focal plane of DRAGON. These recoils were identified using both a local time-of-flight (TOF) system, consisting of two micro-channel plates (MCPs) detecting secondary electron emission from thin Diamond-like Carbon (DLC) foils, and an ionization chamber filled with isobutane gas. These particles could be easily distinguished from scattered and/or charge-changed ("leaky") beam through use of a dual TOF technique[25].

The gas target stopping power was determined using measurements with gas in/out of the target system at the nominal beam energy used during the experiment, and the charge state distribution for the recoils of interest was measured using a stable beam of ^{23}Na. Determining the number of incoming beam particles was achieved by measuring β decays in two 6mm thick plastic scintillators located part-way through DRAGON, where the 6^+ component of the unreacted beam was stopped. The average ^{19}Ne beam intensity over 89 hours was found to be ~ 7 x 10^6 pps, amounting to a total integrated beam flux of ~2.4 x 10^{12}. It should be noted that the isobaric beam contaminant ^{19}F was also present at a level of ~ 2 x 10^7 pps, but that reactions involving ^{19}F could be easily distinguished from those involving ^{19}Ne using the ionization chamber.

Figure 1 shows the γ-gated MCP TOF vs separator TOF spectrum obtained for the incident beam energy E_{beam} = 486 A·keV. A distinct cluster of 15 events, free from background, indicate the presence of an A = 20 radiative capture resonance. By additionally requiring the observation of a ^{20}Na event in the ionization chamber, a preliminary total of 9 "golden" ^{19}Ne $+p$ recoil-γ coincidences is obtained, as shown in Figure 2. Moreover, it is clear from Fig. 2 that any events arising from the contaminant ^{19}F$(p,\gamma)^{20}$Ne reaction are easily separable from the ^{20}Na recoils of interest.

A key strength of using the DRAGON recoil separator to study astrophysical resonant reaction rates is that it provides an independent measurement of the resonance strength - from the yield of detected recoils at the focal plane detectors - and the resonance energy - from the spatial distribution along the beam axis of the detected radiative capture γ-rays in the BGO array surrounding the gas target. This is of particular importance in the current study, where both the resonance strength and the resonance energy are not well defined. However, a determination of the resonance energy in the current study was made more complicated due to the low number of detected recoil-γ coincidences. Consequently, the position along the beam axis where the mean number of radiative capture γ-rays were detected is not very accurately constrained.

However, it was still possible to extract a preliminary value of the resonance energy by using the arithmetic mean position along the beam axis where the radiative capture γ-rays were detected.

DOI: 10.1051/epjconf/201716501054

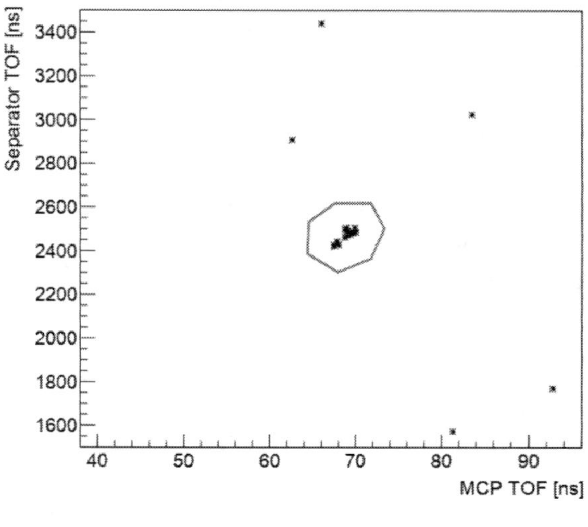

Figure 1. Dual TOF spectrum showing MCP TOF vs TOF through the DRAGON recoil separator. The red polygon indicates the expected location of A = 20 recoils.

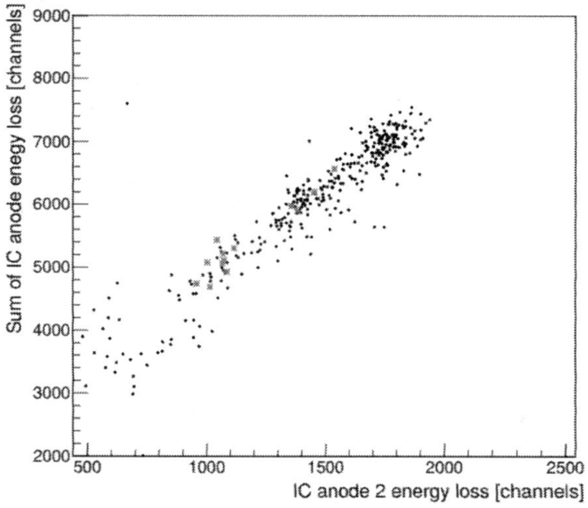

Figure 2. Total energy loss in the ionization chamber vs energy loss in the second anode of the ionization chamber. Black dots indicate singles events observed from all production runs. Red stars indicate events within the red polygon in Figure 1.These events appear to take up two main regions; one, around the center of the Figure, corresponding to contaminant $^{19}F(p,\gamma)^{20}Ne$ reactions; and the other, to the south-west of the center of the Figure, corresponding to the $^{19}Ne(p,\gamma)^{20}Na$ reactions of interest.

Converting this mean position to a resonance energy, using the methodology set out in Ref. [26], yields a preliminary resonance energy of ~458 keV. Using the thick target resonance strength formula [25], a preliminary estimate of the resonance strength of ~18 meV could also be determined.

In summary, we have performed a direct measurement of the $^{19}Ne(p,\gamma)^{20}Na$ reaction in inverse kinematics at the DRAGON recoil separator, at an energy ~ 10 keV higher than previous direct measurements. The key resonance in the $^{19}Ne + p$ system at ONe novae and Type-I X-ray burst temperatures has been successfully measured for the first time. Preliminary estimates of the resonance energy and strength are $E_{c.m.} \approx 458$ keV and $\omega\gamma \approx 18$ meV, respectively. These results are in stark contrast with the most recent study of the $^{19}Ne(p,\gamma)^{20}Na$ reaction rate [21], but are more consistent with the upper limits from historical direct measurements. Of particular note, the preliminary resonance energy is in good agreement with that inferred in the study by Wallace et al. [20]. Going forward, a negative log-likelihood analysis[27] will be performed to constrain the resonance strength and energy from this study more precisely. Furthermore, it would be of interest to check the impact of the final updated resonance energy and strength on the nucleosynthesis of both novae and X-ray bursts through the use of hydrodynamic simulations.

References

[1] J. José, M. Hernanz, C. Iliadis, Nucl. Phys. A **777**, 550 (2006)
[2] S. Starrfield, C. Iliadis, W.R. Hix, F.X. Timmes and W.M. Sparks, ApJ **692**, 1532 (2009)
[3] S. N. Shore *et al.*, A&A **553**, A123 (2013)
[4] J. José and M. Hernanz, ApJ **494**, 680 (1998)
[5] C. Iliadis *et al.*, Astrophys. J. Suppl. Ser. **142**, 105 (2002)
[6] H. Schatz *et al.*, Phys. Rev. Lett. **86**, 3471 (2001)
[7] B. Davids *et al.*, ApJ **735**, 40 (2011)
[8] L.O. Lamm *et al.*, Nucl. Phys. A **510**, 503 (1990)
[9] N.M. Clarke *et al.*, J. Phys. G: Nucl. Part. Phys. **19**, 1411 (1993)
[10] M.S. Smith *et al.*, Nucl. Phys. A **536**, 333 (1992)
[11] A. Piechaczek *et al.*, Nucl. Phys. A **584**, 509 (1995)
[12] R. Coszach *et al.*, Phys. Rev. C **50**, 1695 (1994)
[13] C. Michotte *et al.*, Phys. Lett. B **381**, 402 (1996)
[14] R. D. Page *et al.*, Phys. Rev. Lett. **73**, 3066 (1994)
[15] G. Vaancraeynest *et al.*, Phys. Rev. C **57**, 2711 (1998)
[16] M. Couder *et al.*, Phys. Rev. C **69**, 022801(R) (2004)
[17] C. Wrede *et al.*, Phys. Rev. C **81**, 055503 (2010)
[18] H.T. Fortune, R. Sherr and B.A. Brown, Phys. Rev. C **61**, 057303 (2000)
[19] J.P. Wallace and P.J. Woods *et al.*, Phys. Rev. C **86**, 068801 (2012)
[20] J.P. Wallace *et al.*, Phys. Lett. B **712**, 59 (2012)
[21] J. Belarge *et al.*, Phys. Rev. Lett. **117**, 182701 (2016)
[22] D.R. Tilley, C. Cheves, J. Kelly, S. Raman and H. Weller, Nucl. Phys. A **636**, 249 (1998)
[23] D.A. Hutcheon *et al.*, Nucl. Instr. Meth. In Phys. Rev. A **498**, 190 (2003)
[24] P.G. Bricault *et al.*, Rev. Sci. Instrum. **83**, 02A914 (2012)
[25] J. Fallis *et al.*, Phys. Rev. C **88**, 045801 (2013)
[26] D.A. Hutcheon, C. Ruiz, J. Fallis *et al.*, Nucl. Instr. Meth. In Phy. Rev. A **689**, 70 (2012)
[27] L. Erikson *et al.*, Phys. Rev. C **81**, 045808 (2010)

EPJ Web of Conferences **165**, 01055 (2017) DOI: 10.1051/epjconf/201716501055
NPA8 2017

Constraining the rp-process by measuring ^{23}Al(d,n)^{24}Si with GRETINA and LENDA at NSCL

Clemens Wolf[1], *Christoph* Langer[1], *Fernando* Montes[2,6], *Jorge* Pereira[2,6], *Sunghoon* Ahn[2,6], *Sara* Ayoub[2,7], *Daniel* Bazin[2], *Peter* Bender[2], *Alex* Brown[2], *Justin* Browne[2,7], *Heather* Crawford[3], *Eric* Deleeuw[2,7], *Brandon* Elman[2,7], *Stefan* Fiebiger[1], *Alexandra* Gade[2,7], *Panagiotis* Gastis[4], *Samuel* Lipschutz[2,7], *Brenden* Longfellow[2,7], *Filomena* Nunes[2,7], *Wei-Jia* Ong[2,7], *Terri* Poxon-Pearson[2,7], *Georgios* Perdikakis[2,4], *René* Reifarth[1], *Hendrik* Schatz[2,6,7], *Konrad* Schmidt[2,6], *Jaclyn* Schmitt[2,6], *Chris* Sullivan[2], *Rachel* Titus[2,6], *Dirk* Weisshaar[2], *Philip* Woods[5], *Juan Carlos* Zamora[2] and *Remco* G.T. Zegers[2,7]

[1] *Goethe University, Frankfurt, Germany*
[2] *National Superconducting Cyclotron Laboratory, East Lansing, MI, USA*
[3] *Lawrence Berkeley National Laboratory, Berkeley, CA, USA*
[4] *Central Michigan University, Mounts Pleasant, MI, USA*
[5] *University of Edinburgh, Edinburgh, Great Britain*
[6] *JINA Center for the Evolution of the Elements, East Lansing, MI, USA*
[7] *Michigan State University, East Lansing, MI, USA*

Abstract. The ^{23}Al(p,γ)^{24}Si stellar reaction rate has a significant impact on the light-curve emitted in X-ray bursts. Theoretical calculations show that the reaction rate is mainly determined by the properties of direct capture as well as low-lying 2^+ states and a possible 4^+ state in ^{24}Si. Currently, there is little experimental information on the properties of these states.

In this proceeding we will present a new experimental study to investigate this reaction, using the surrogate reaction ^{23}Al(d,n) at 47 AMeV at the National Superconducting Cyclotron Laboratory (NSCL). We will discuss our new experimental setup which allows us to use full kinematics employing the Gamma-Ray Energy Tracking In-beam Nuclear Array (GRETINA) to detect the γ-rays following the de-excitation of excited states of the reaction products and the Low Energy Neutron Detector Array (LENDA) to detect the recoiling neutrons. The S800 was used for identification of the ^{24}Si recoils. As a proof of principle to show the feasibility of this concept the Q-value spectrum of ^{22}Mg(d,n)^{23}Al is reconstructed.

1 Introduction

X-ray bursts are produced when a neutron star accretes H- and He-rich material from a companion star. During infall of the material onto the neutron star gravitional energy is released which produces a constant X-ray flux. Observations of accreting neutron stars show that on top of the constant flux occasionally bright peaks occur with typical duration of 10-100s and a recurrence time of hours to days. While the neutron star accretes, the material gets compressed and heats up. At a certain point

the temperature of this thin layer gets high enough for the layer to become thermonuclear unstable. What follows is a thermonuclear runaway, involving (α,p) ,(α,γ) and (p,γ)-reactions and beta-decays, which produces the characteristic light curve. It has been shown that this light curve can be used to extract stellar parameters [1]. In order to extract information of the underlying system, one needs to correctly simulate the X-ray bursts behavior using nuclear physics information such as masses and reaction rates. Recent extensive burst simulations show that the uncertainty in the ^{23}Al(p,γ)^{24}Si-rate has a large impact on the overall light curve [2].
The ^{23}Al(p,γ)^{24}Si reaction has a relatively low Q-value of only 3.29 (20) MeV, and it is therefore mainly determined by resonant capture into low-lying resonances in ^{24}Si. Earlier attempts to reconstruct the structure of ^{24}Si are not conclusive. A low-lying 2^+ resonance has been identified at 3.441 (10) MeV [4] and 3.410 (16) MeV in [4] and [5], respectively. According to the mirror nucleus ^{24}Ne and USD shell model calculations [5], another low-lying resonance between 150-200 keV might exist, which has not been observed yet and could potentially have a significant effect on the reaction rate. We therefore performed an experiment in inverse and complete kinematics to study the surrogate reaction ^{23}Al(d,n)^{24}Si to populate the ^{24}Si states of interest in the astrophysically relevant ^{23}Al(p,g)^{24}Si reaction. First preliminary results will be presented in the next sections, proving the feasibility of this novel technique.

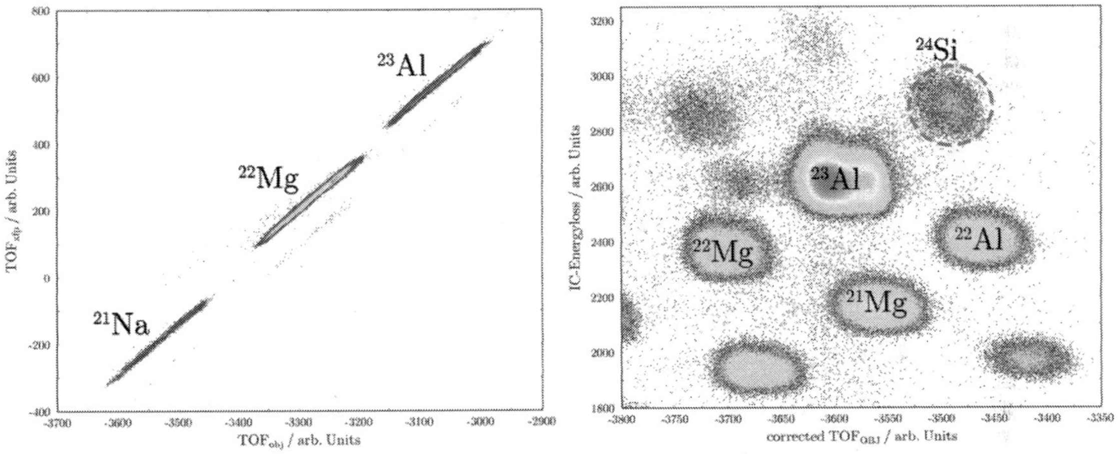

Figure 1. Incoming particle identification (ToF between the extended focal plan scintillator of the A1900 (xfp) plotted against ToF for fragments between the object scintillator of the S800) of runs without CD_2 target inserted (left). Reactions products identified in the S800 focal plane after the ^{23}Al beam hit the CD_2 target (right).

2 Experimental Details

Spectroscopic information of states in ^{24}Si near the proton separation energy is needed for ^{23}Al(p,γ) rate calculations. Thus, the main challenge in the experiment is to measure these quantities for all states up to 4 MeV in ^{24}Si, using a transfer method to populate the sates of interest (see [3] for more information).
Our new experimental concept combines the state-of-the-art γ-ray detector GRETINA [6, 7] with the neutron detector LENDA [8, 9]. This setup gives us the opportunity to perform a measurement in full kinematics and obtain very precise level energies from the gamma data and the properties of the

states from the angular distribution of the recoiling neutrons. The experiment was carried out at the National Superconducting Cyclotron Laboratory (NSCL). A primary beam of ^{22}Mg at 180 MeV/A impinged on a ^{9}Be target producing ^{23}Al via multi-particle exchange reactions. The ^{23}Al beam was then selected using the A1900 fragment separator and transported to the S800 vault. The experiment setup at the target location included GRETINA at its pivot point and LENDA at backward angles. The beam was focused on a 110.5 mg/cm^2 CD_2 target in the center of GRETINA to produce ^{24}Si by the desired ^{23}Al(d,n)-reaction. Due to the fact that half of the hemisphere was covered by the support structure of GRETINA a new double row holding structure for LENDA was used to cover an angle from 115-175 degree with respect to the beam axis. The first and second row with 12 LENDA bars each were placed at 1 and 1.1 m distance to the target, respectively. This was the best compromise between a good time of flight resolution and a high angular coverage of the 24 30x4.5x2.5cm bars for high detection efficiency.
Due to reactions on carbon atoms in the CD_2 target, runs with a 78 mg/cm^2 thick C target were used to substract background. During 7.5 days of beam time we were able to collect 96 h of data with beam on the CD_2 target and 35 h of data with the carbon target.

3 Beam production and particle identification

Figure 1 shows the incoming particle identification plot for the runs without the CD_2 target inserted. The beam composition was dominated by ^{22}Mg, resulting in a purity of ^{23}Al of only 13 %. Using the time of flight dependency between this two scintillators in the runs with target allows us to gate only on incoming ^{23}Al isotopes. The right side of Figure 1 shows the outgoing particle identification diagram gated on incoming ^{23}Al when using the CD_2 target. During the experiment we detected 11400 ^{24}Si isotopes when using the CD_2 target and 3400 ^{24}Si isotopes when using the C target.

4 Test case: ^{22}Mg(d,n)^{23}Al

In order to understand and test this new experimental technique, a reaction with higher statistics was investigated: ^{22}Mg(d,n)^{23}Al. The structure of ^{23}Al offers other advantages: there is only one excited state at 1616 keV that decays via γ-ray emission, all other states have a 100% proton-decay branching, since the proton separation threshold is very low (140 keV) and the first excited state is observed at 550 keV. The γ data of GRETINA show that the population of the state at 1616 keV via the (d,n)-reaction is negligible. All other excited states decay to ^{22}Mg before arriving at the spectrograph. Thus, only the ground state of ^{23}Al is populated through the (d,n) reaction when gating on outgoing ^{23}Al fragments in the S800 focal plane.
Figure 2 shows the angle of the detected neutrons of the ^{22}Mg(d,n)^{23}Al as a function of the kinetic energy which was determined using time-of-flight. For comparison the expected kinematics line range are also shown. The kinematic line range is a consequence of the thick target. This results in a large spread of different reaction energies in the target, which can not be reconstructed uniquely. Nevertheless, the good agreement with the experimental data shows that the employed technique is successful and we are able to investigate the transfer reaction with this method.
The Q-value spectrum can be reconstructed by:

$$(Q - Ex_{out}) = E^{N}_{kin}(1 + \frac{m_N}{m_{out}}) - E^{in}_{kin}(1 - \frac{m_{in}}{m_{out}}) - 2\sqrt{\frac{m_N m_{in} E^{in}_{kin} E^{N}_{kin}}{m^2_{out}}} \cos\theta_{lab} \quad (1)$$

where m_N, m_{in} and m_{out} are the mass of the neutron, incoming and outgoing fragment. E^{in}_{kin}, E^{N}_{kin} are the kinetic energies of the incoming fragment and the neutron and Ex_{out} is the excitation energy of the outgoing fragment. By using the energy E^{in}_{kin}, which we expect in the middle of the target, the reconstructed Q-value spectrum is rather wide (Figure 2 red line), because of the large energy loss through the target. Using event-by-event reconstruction of the isotope trajectories in the S800, a more precise determination of the energy loss in the target was obtained. This, in turn, permits the extraction of the reaction vertex in the target and results together with an approximation of the incoming beam energy by a time-of-flight measurement of each fragment between two scintillators before the target in a much more precise E^{in}_{kin} value event-by-event. With this approach, we managed to obtain a resolution of 1.1 MeV (Figure 2 blue line). A GEANT4 simulation shows that this resolution is dominated by the time resolution of the time-of-flight measurement of the neutrons which is about 951 ± 12 ps (FWHM). Analysis of the ^{24}Si data is currently under-way.

This work was supported by US National Science Foundation under Grant Numbers PHY-08-22648, and PHY-1430152 (JINA Center for the Evolution of the Elements). The research leading to these results has received funding from the European Research Council under the European Unions's Seventh Framework Programme (FP/2007-2013) / ERC Grant Agreement n. 615126

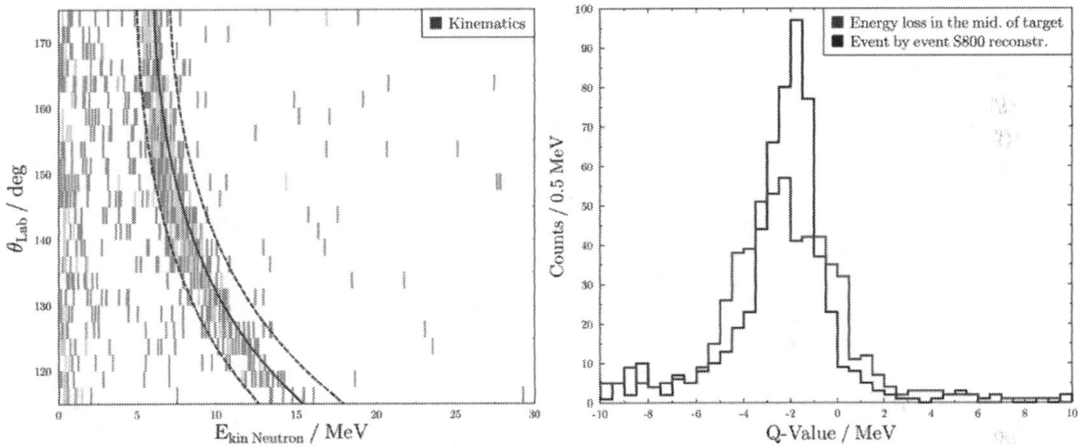

Figure 2. Left: Kinetic energy as a function of detection angle for $^{22}Mg(d,n)^{23}Al$-reaction cuts. Expected kinematics line of this reaction is shown in red. Right: Reconstructed Q-value spectra of this reaction before and after the additional correction.

References

[1] Zamfir et al. *The Astrophysical Journal*, 749:69 (2012)
[2] Cyburt et al. *The Astrophysical Journal*, 830:55 (2016)
[3] Langer et al. *Physical Review Letters*, 113(3):032502 (2014)
[4] Schatz et al. *Physical Review Letters*, 79(20):3845 (1997)
[5] Yoneda et. al. *Physical Review C*, 74(2):021303 (2006)
[6] Paschalis et al. *Nuclear Instruments and Methods in Physics Section A*, 709:44-55 (2013)
[7] Weisshaar et al. *Nuclear Instruments and Methods in Physics Section A*, 847:187-198 (2017)
[8] Perdikakis et al. *Nuclear Instruments and Methods in Physics Section A*, 686:117-124 (2012)
[9] Lipschutz et al. *Nuclear Instruments and Methods in Physics Section A*, 815:1-6 (2016)

Nuclear astrophysics projects at the low-energy RI beam separator CRIB

H. Yamaguchi[1,★], *D. Kahl*[1,2], *S. Hayakawa*[1], *L. Yang*[1,3], *H. Shimizu*[1], *Y. Sakaguchi*[1], *K. Abe*[1], *T. Nakao*[1,4], *T. Suhara*[5], *N. Iwasa*[6], *A. Kim*[7,8], *D.H. Kim*[8], *S.M. Cha*[7], *M.S. Kwag*[7], *J.H. Lee*[7], *E.J. Lee*[7], *K.Y. Chae*[7], *Y. Wakabayashi*[9], *N. Imai*[1], *N. Kitamura*[1], *P. Lee*[10], *J.Y. Moon*[11,12], *K.B. Lee*[11], *C. Akers*[11], *H.S. Jung*[12], *N.N. Duy*[13,14], *L.H. Khiem*[13], *C.S. Lee*[10], *T. Hashimoto*[11], *S. Kubono*[9], *T. Kawabata*[15], *T. Teranishi*[16], *Y.K. Kwon*[11], and *D.N. Binh*[17,1]

[1] *Center for Nuclear Study (CNS), University of Tokyo, RIKEN campus, 2-1 Hirosawa, Wako, Saitama 351-0198, Japan*
[2] *School of Physics and Astronomy, the University of Edinburgh, Peter Guthrie Tait Road, Edinburgh EH9 3BF, UK*
[3] *China Institute of Atomic Energy, P.O. Box 275(10), Beijing 102413, China*
[4] *Advanced Science Research Center, Japan Atomic Energy Agency, Tokai, Ibaraki 319-1195, Japan*
[5] *Matsue College of Technology, Matsue, Shimane 690-8518, Japan*
[6] *Department of Physics, Tohoku University, Aoba, Sendai, Miyagi 980-8578, Japan*
[7] *Department of Physics, Sungkyunkwan University, 2066 Seobu-ro, Jangan-gu, Suwon, Korea*
[8] *Department of Physics, Ewha Womans University, Seoul 120-750, Korea*
[9] *The Institute of Physical and Chemical Research (RIKEN), 2-1 Hirosawa, Wako, Saitama 351-0198, Japan*
[10] *Department of Physics, Chung-Ang University, Seoul 156-756, Korea*
[11] *Institute for Basic Science, 70, Yuseong-daero 1689-gil, Yuseong-gu, Daejeon 305-811, Korea*
[12] *Wako Nuclear Science Center (WNSC), KEK, 2-1 Hirosawa, Wako, Saitama 351-0198, Japan*
[13] *Institute of Physics, Vietnam Academy of Science and Technology, 10 Dao Tan, Ba Dinh, Hanoi, Vietnam*
[14] *Dong Nai University, Le Quy Don Street, Tan Hiep Ward, Bien Hoa City, Dong Nai, Vietnam*
[15] *Department of Physics, Kyoto University, Kita-Shirakawa, Kyoto 606-8502, Japan*
[16] *Department of Physics, Kyushu University, 6-10-1 Hakozaki, Fukuoka 812-8581, Japan*
[17] *30 MeV Cyclotron Center, Tran Hung Dao Hospital, Hoan Kiem District, Hanoi, Vietnam*

Abstract. Studies on nuclear reactions relevant for astrophysics have been performed using the radioactive-isotope (RI) beams at the low-energy RI beam separator CRIB, operated by Center for Nuclear Study (CNS), the University of Tokyo. A type of measurement to study astrophysical reactions at CRIB is by the elastic resonant scattering with the thick-target method in inverse kinematics. We introduce the α resonant scattering with ^{7}Be beam, related to the astrophysical ^{7}Be(α,γ) reactions, which is relevant in the hot p-p chain and νp-process in supernovae. Other α resonant scattering measurements with ^{30}S, ^{10}Be, ^{15}O, and ^{18}Ne beams have been performed at CRIB, using the thick-target method.

★e-mail: yamag@cns.s.u-tokyo.ac.jp

1 Introduction

CRIB [1, 2] is a radioactive-isotope (RI) beam separator operated by Center for Nuclear Study (CNS), the University of Tokyo, installed at the RIBF facility of RIKEN Nishina Center. CRIB can produce low-energy (< 10 MeV/u) RI beams by the in-flight technique, using primary heavy-ion beams accelerated at the AVF cyclotron of RIKEN (K=70). Most of the RI beams are produced via direct reactions such as (p, n), (d, p) and $(^3\text{He}, n)$, taking place at an 8-cm-long gas target with a maximum pressure of 760 Torr. A cryogenic target system, in which the target gas can be cooled down to about 90 K, is currently available, and an intense ^{7}Be beam of 2×10^8 pps was produced using the system [3]. A list of typical parameters of RI beams produced at CRIB is found in [4]. New RI beams recently developed at CRIB are ^{16}N (1×10^6 pps), ^{10}Be (2×10^4 pps), ^{15}O (1×10^6 pps), and ^{26}Al (1×10^5 pps). The ^{26}Al beam was developed to obtain an isomeric ^{26m}Al beam, related to the production rate of the galactic γ rays from ^{26}Al.

A recent major topic of our interest is the measurement of alpha-induced reactions. Several (α, p) reactions, such as $^{14}\text{O}(\alpha, p)$ [5], $^{11}\text{C}(\alpha, p)$ [6], $^{21}\text{Na}(\alpha, p)$, and $^{18}\text{Ne}(\alpha, p)$ have been studied by direct measurement at CRIB. Measurements on the elastic resonant scatterings with a helium target and beams of ^{7}Li [7] and ^{7}Be [8], ^{30}S, ^{10}Be, ^{15}O, and ^{18}Ne have also been performed. These measurements provide information on the resonances to evaluate astrophysical (α, γ) reaction rates, and also on the nuclear cluster structure of the compound nuclei.

2 ^{7}Be+α elastic resonant scattering

Here we introduce the ^{7}Be+α scattering measurement as a typical resonant scattering experiment at CRIB. The measurement allows us to evaluate the rate of the $^7\text{Be}(\alpha, \gamma)$ reaction, which is considered to play an important role in the hot *p*-*p* chain and related reaction sequences [9]. In the νp-process in core-collapse supernovae [10], the $^7\text{Be}(\alpha, \gamma)$ reaction may contribute as much as the triple-α process to the synthesis of elements heavier than boron at the relevant temperature of $T_9 = 1.5$–3, according to a theoretical calculation [11]. The Gamow energy window at the highest temperature $T_9 = 3$ corresponds to the excitation energy E_{ex}=8.2–9.6 MeV in ^{11}C.

We performed the measurement of the ^{7}Be+α resonant elastic and inelastic scatterings with the thick-target method in inverse kinematics at CRIB [8]. A low-energy ^{7}Be beam at 14.7 MeV was produced with a typical intensity of 1–2×10^5 per second. The main measurement was performed for 4 days, injecting 2.9×10^{10} ^{7}Be particles into the helium target. We obtained an excitation function of the elastic scattering with several peaks, corresponding to the resonance structure in ^{11}C. The obtained excitation is shown in the left panel of Figure 1. An R-matrix analysis was performed to deduce the parameters of the resonances, as the calculated curve also shown in the figure. The resonances observed in the present work might contribute to the astrophysical $^7\text{Be}(\alpha, \gamma)^{11}\text{C}$ reaction rate at high temperature, $T_9 > 1.5$. The resonant reaction rates were calculated for three resonances using analytical formula described in [12], and plotted in Fig. 1. In conclusion, the resonances at 8.90 MeV and 9.20 MeV have a possibility to give significant contributions to the reaction rate for $T_9 = 1.5$–3, although they are evaluated as one order of magnitude smaller contribution than the dominant 8.420-MeV resonance.

3 ^{10}Be+α elastic resonant scattering

Another study was performed on the ^{10}Be+α system, mainly on the interest of an exotic cluster structure, while $^{10}\text{Be}(\alpha, n)$ or $^{10}\text{Be}(\alpha, \gamma)$ reaction may play a role in the big-bang nucleosynthesis [13] or other high-temperature environments.

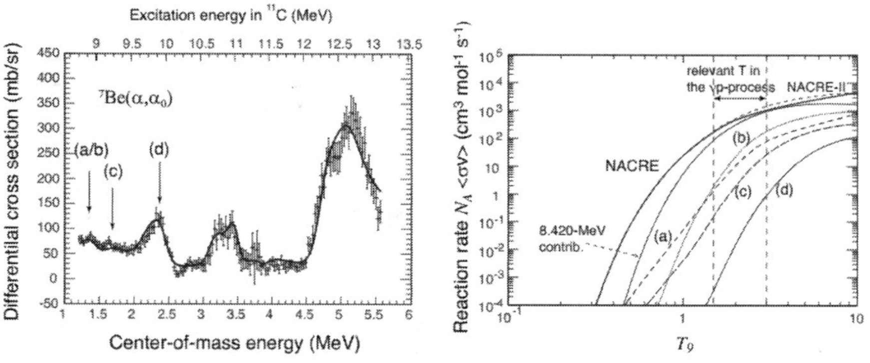

Figure 1. Excitation function of ^{7}Be+α elastic scattering with an R-matrix fit curve (left panel) and resonant reaction rate of ^{7}Be(α, γ) for the 8.90, 9.20, and 9.97-MeV resonances, calculated by the analytical formula. The evaluation by NACRE and NACRE-II are shown for comparison. The contribution by the 8.420-MeV resonance, included in NACRE, is also shown.

In 1956, Morinaga [14] came up with the novel idea of a particular cluster state: the linear-chain cluster state (LCCS). Now the LCCS is commonly considered as extreme and exotic, due to its presumed propensity to exhibit bending configurations. A theoretical prediction of LCCS in ^{14}C was made by Suhara and En'yo [15, 16] with an antisymmetrized molecular dynamics (AMD) calculation, yielding a prolate band ($J^\pi = 0^+, 2^+, 4^+$) that has a configuration of an LCCS at a few MeV or more above the ^{10}Be+α threshold.

We applied the ^{10}Be+α resonant scattering method in inverse kinematics to identify the predicted LCCS band in ^{14}C [17]. The ^{10}Be beam had a typical intensity of 2×10^4 particles per second, and the beam purity was better than 95%. The ^{10}Be beam at 25.8 MeV impinged on the gas target, which was a chamber filled with helium gas at 700 Torr and covered with a 20-μm-thick Mylar film as the beam entrance window. The measured ^{10}Be beam energy at the entrance of the helium gas target, after the Mylar film, was 24.9±0.3 MeV. α particles recoiling to the forward angles were detected by ΔE-E detector telescopes. We obtained an excitation function of the ^{10}Be+α resonant elastic scattering for 13.8–19.1 MeV. We performed an R-matrix calculation to deduce the resonance parameters, and we

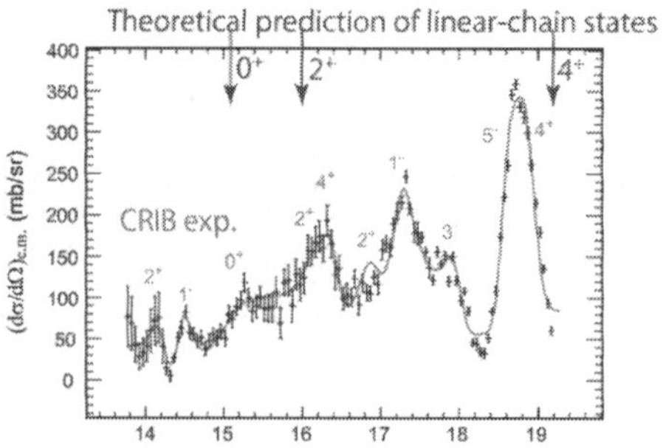

Figure 2. Excitation function of the ^{10}Be+α resonant scattering for θ_{lab}=0–8°.

identified three resonances well-corresponding to the predicted LCCS band; J^π are identical, and their energies and spacings are consistent with the theoretical prediction. The J^π for the 2^+ and 4^+ levels were assigned with a complex analysis of unresolved doublets, but they were supported by the angular distribution as well. We claim this as the strongest indication of the LCCS ever found. It can be also shown that both sets of level energies can be plotted almost on a line, $E_J = E_0 + \hbar^2/2\mathfrak{I}(J(J+1))$, where $\mathfrak{I}$ is the moment of inertia of the nucleus. The linearity allows us to interpret the levels as a rotational band, and the low $\hbar^2/2\mathfrak{I} = 0.19$ MeV implies the nucleus could be strongly deformed, consistent with the interpretation of an LCCS.

The experiments were performed at RI Beam Factory operated by RIKEN Nishina Center and CNS, the University of Tokyo. We are grateful to the RIKEN and CNS accelerator staff for their help. This work was partly supported by JSPS KAKENHI (No. 25800125, 15K17662, 16K05369, and 16H03980) from the Ministry of Education, Culture, Sports, Science and Technology (MEXT) of Japan, and the National Research Foundation Grant funded by Korea Government (Grants Nos. NRF-2009-0093817, NRF-2016R1D1A1A09917463, NRF-2016R1A5A1013277, and NRF-2016K1A3A7A09005579).

References

[1] S. Kubono, Y. Yanagisawa, T. Teranishi, S. Kato, T. Kishida, S. Michimasa, Y. Ohshiro, S. Shimoura, K. Ue, S. Watanabe et al., Eur. Phys. J. **A13**, 217 (2002)
[2] Y. Yanagisawa, S. Kubono, T. Teranishi, K. Ue, S. Michimasa, M. Notani, J.J. He, Y. Ohshiro, S. Shimoura, S. Watanabe et al., Nucl. Instrum. Meth. Phys. Res., Sect. A **539**, 74 (2005)
[3] H. Yamaguchi, Y. Wakabayashi, G. Amadio, S. Hayakawa, H. Fujikawa, S. Kubono, J. He, A. Kim, D. Binh, Nucl. Instrum. Meth. Phys. Res., Sect. A **589**, 150 (2008)
[4] D. Kahl, T. Hashimoto, N.N. Duy, S. Kubono, H. Yamaguchi, D.N. Binh, A.A. Chen, S. Cherubini, S. Hayakawa, J.J. He et al., AIP Conference Proceedings **1594**, 163 (2014)
[5] A. Kim, N.H. Lee, M.H. Han, J.S. Yoo, K.I. Hahn, H. Yamaguchi, D.N. Binh, T. Hashimoto, S. Hayakawa, D. Kahl et al., Phys. Rev. C **92**, 035801 (2015)
[6] S. Hayakawa, S. Kubono, D. Kahl, H. Yamaguchi, D.N. Binh, T. Hashimoto, Y. Wakabayashi, J.J. He, N. Iwasa, S. Kato et al., Phys. Rev. C **93**, 065802 (2016)
[7] H. Yamaguchi, T. Hashimoto, S. Hayakawa, D.N. Binh, D. Kahl, S. Kubono, Y. Wakabayashi, T. Kawabata, T. Teranishi, Phys. Rev. C **83**, 034306 (2011)
[8] H. Yamaguchi, D. Kahl, Y. Wakabayashi, S. Kubono, T. Hashimoto, S. Hayakawa, T. Kawabata, N. Iwasa, T. Teranishi, Y. Kwon et al., Phys. Rev. C **87**, 034303 (2013)
[9] M. Wiescher, J. Görres, S. Graff, L. Buchmann, F.K. Thielemann, Astrophys. J. **343**, 352 (1989)
[10] C. Fröhlich, P. Hauser, M. Liebendörfer, G. Martínez-Pinedo, F.K. Thielemann, E. Bravo, N. Zinner, W. Hix, K. Langanke, A. Mezzacappa et al., Astrophys. J. **637**, 415 (2006)
[11] S. Wanajo, H.T. Janka, S. Kubono, Astrophys. J. **729**, 46 (2011)
[12] C. Angulo *et al.*, Nucl. Phys. A **656**, 3 (1999)
[13] A. Coc, S. Goriely, Y. Xu, M. Saimpert, E. Vangioni, Astrophys. J. **744**, 158 (2012)
[14] H. Morinaga, Phys. Rev. **101**, 254 (1956)
[15] T. Suhara, Y. Kanada-En'yo, Phys. Rev. C **82**, 044301 (2010)
[16] T. Suhara, Y. Kanada-En'yo, Phys. Rev. C **84**, 024328 (2011)
[17] H. Yamaguchi, D. Kahl, S. Hayakawa, Y. Sakaguchi, K. Abe, T. Nakao, T. Suhara, N. Iwasa, A. Kim, D. Kim et al., Physics Letters B **766**, 11 (2017)

DOI: 10.1051/epjconf/201716501057

Recent Ultra High Energy neutrino bounds and multimessenger observations with the Pierre Auger Observatory

Enrique Zas[1,⋆] for The Pierre Auger Collaboration[2,⋆⋆]

[1] *Depto. Física de Partículas & Instituto Galego de Física de Altas Enerxías, Universidade de Santiago de Compostela, 15782 Santiago de Compostela, Spain*
[2] *Observatorio Pierre Auger, Av. San Martín Norte 304, 5613 Malargüe, Argentina*

Abstract. The overall picture of the highest energy particles produced in the Universe is changing because of measurements made with the Pierre Auger Observatory. Composition studies of cosmic rays point towards an unexpected mixed composition of intermediate mass nuclei, more isotropic than anticipated, which is reshaping the future of the field and underlining the priority to understand composition at the highest energies. The Observatory is competitive in the search for neutrinos of all flavors above about 100 PeV by looking for very inclined showers produced deep in the atmosphere by neutrinos interacting either in the atmosphere or in the Earth's crust. It covers a large field of view between $-85°$ and $60°$ declination in equatorial coordinates. Neutrinos are expected because of the existence of ultra high energy cosmic rays. They provide valuable complementary information, their fluxes being sensitive to the primary cosmic ray masses and their directions reflecting the source positions. We report the results of the neutrino search providing competitive bounds to neutrino production and strong constraints to a number of production models including cosmogenic neutrinos due to ultra high energy protons. We also report on two recent contributions of the Observatory to multimessenger studies by searching for correlations of neutrinos both with cosmic rays and with gravitational waves. The correlations of the directions of the highest energy astrophysical neutrinos discovered with IceCube with the highest energy cosmic rays detected with the Auger Observatory and the Telescope Array revealed an excess that is not statistically significant and is being monitored. The targeted search for neutrinos correlated with the discovery of the gravitational wave events GW150914 and GW151226 with advanced LIGO has led to the first bounds on the energy emitted by black hole mergers in Ultra-High Energy Neutrinos.

1 Introduction

Cosmic Rays (CR) are energetic protons and nuclei that reach the Earth giving a wealth of information about the Universe around us. Galactic CR are believed to be produced by Fermi acceleration in Supernovae shocks. Ultra-High Energy CR (UHECR) reach astonishing large energies above 10 EeV (1 EeV=10^{18} eV) producing interactions with the atmospheric nuclei with center of mass energies well

⋆e-mail: enrique.zas@usc.es
⋆⋆e-mail: auger_spokespersons@fnal.gov

 DOI: 10.1051/epjconf/201716501057

exceeding those reached in accelerator experiments. They can travel intergalactic distances and give information on the sources where they are produced and on the interactions they suffer as they travel to Earth. For instance, the most energetic particles are believed to be little deviated by the galactic and intergalactic magnetic fields and their direction could help to unveil the highly uncertain strength and pattern of these magnetic fields. Extracting information from UHECR is however an extremely challenging task because we do not yet know the sources of UHECR, we have limited information about the primary mass composition (and electric charge) [1] and, in addition, the models we have to describe their interactions are not sufficiently accurate [2].

High energy cosmic rays induce Extensive Air Showers (EAS) in the atmosphere as ultra-relativistic particles are produced in successive colisions with air molecules. Above EeV energies they reach *shower maximum* containing over 10^{10} particles that spread over several km of radius after traversing a depth of order 800 g cm^{-2}. The Pierre Auger Observatory consists of a 3000 km^2 array of water-Cherenkov detectors (SD) in a hexagonal grid of 1.5 km spacing which is used to detect the particle front of EAS and a Fluorescence Detector (FD) that monitors the emissions form excited Nitrogen atoms as the shower develops over this surface. Located in Mendoza, Argentina, at an average latitude of 35.2° S and at a mean atmospheric depth (altitude) of 875 g cm^{-2} (1400 m) [10] it has been running since its construction started in 2004. The FD gives a calorimetric measurement of the energy deposition in the atmosphere tracking the shower development which combined with the SD area gives the largest and most precise detector ever built to measure EeV cosmic rays. [1]

The spectrum measurements made at the Observatory have revealed a supression of the flux at energies above 40 EeV [4] which cannot be unambiguosly attributed to photoproduction with the Cosmic Microwave Background (CMB), the Greisen-Zatsepin-Kuz'min cut-off, because composition measurements indicate a transition to increasing mass primaries as the energy increases above about 2 EeV [1] which is also suggestive of an intrinsic maximum acceleration energy in the sources which is proportional to the nucleus charge, Z. Such findings are taking long to permeate the community, strongly predjudiced in favor of a pure proton composition as supported by TA Collaboration claims [5]. A joint analysis has revealed that TA and Auger data are $X_{\max}$-consistent: while Auger data are not compatible with pure protons at 10 EeV in any of the available hadronic models, the TA data, having larger uncertainties, are compatible with both protons and Auger composition results [6].

The measured arrival directions of UHECR are shedding light into the complex picture. A dipole anisotropy of 6.5% for energies above 8 EeV was discovered with 5.2σ significance pointing to $(l, b) = (233°, -13°)$ in galactic coordinates, 125° away from Galactic Center [7]. Evidence for extragalactic sources is now compelling as it is reasonably aligned to the excess dipole in the galaxy distribution [7]. Interestingly a 3.9σ significance has also been reported for correlations of arrival directions with the locations of Star Burst Galaxies (SBG) at angular scales $\sim 10°$ and CR energies above 40 EeV [8]. At these energies composition inference is lacking because of the limited duty cycle of the FD. Clearly measurements are closing in on the UHCER sources, but more data are needed.

Neutrinos provide complementary information. Interactions of UHECR in the sources or during propagation produce EeV neutrinos such as the *cosmological neutrinos* resulting from their GZK-interactions with the CMB. Having no charge, neutrinos point to their production sites even when they are shielded by large amounts of matter. Their energy spectrum is also related to cosmic ray composition at the highest energies, which is largely unknown. As extraterrestrial neutrinos of several PeV have already been detected [9], extending the spectrum measurement to the EeV region is most natural to effectively constrain the production mechanisms of UHECR. Their detection would probe neutrino interactions at energies orders of magnitude above accelerator experiments. Information about flavor ratios besides constraining the production mechanisms may even signal new physics.

[1]Note that the Telescope Array (TA) [3] is about a fourth of the size of the Observatory and has been running for less time.

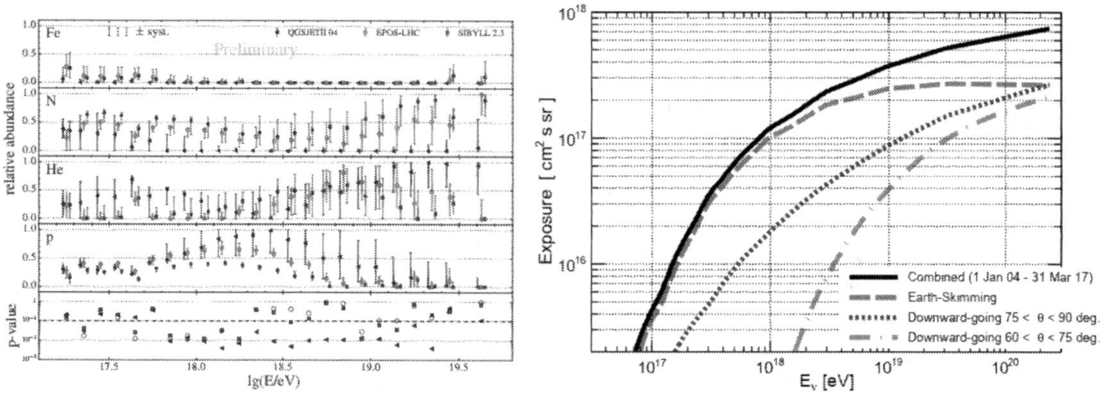

Figure 1. Left: Composition measured at the Pierre Auger Observatory using $X_{\max}$ as a function of energy. Right: Total exposure (solid) as a function of neutrino energy and contributions of the ES (dashes), DGH (dots) and DGL (dot dashes) searches at Earth assuming equal fluxes for all neutrino flavors.

2 Neutrino bounds with the Pierre Auger Observatory

It has been known for long that the Pierre Auger Observatory can also be used to detect inclined neutrino interactions deep in the atmosphere [11] and nearly horizontal *Earth-skimming* (ES) tau neutrinos [12]. If an ES tau neutrino has a charged current interaction near the surface of the Earth, the tau lepton will often exit to the atmosphere and decay in flight, producing a nearly horizontal EAS. This channel is only sensitive to tau neutrinos mostly between ~ 0.1 and 25 EeV, where cosmological neutrinos are expected. Deep inclined showers produced by down-going (DG) neutrinos and by ES tau-neutrinos can be distinguished from showers due to cosmic rays that interact in the upper layers because their electromagnetic component reaches the particle detectors. In contrast, the electromagnetic component of inclined cosmic ray showers gets absorbed after $\sim$ 800 g cm^{-2} (shower maximum) and only the muon component reaches ground level. These showers, which are regularly detected and reconstructed [13], consist of $\sim$ 20 to 200 GeV energy muons that travel nearly parallel to the shower axis and undergo few interactions before reaching the ground. As a result they accumulate small time delays and deposit the energy in a short burst, while deep neutrino showers produce signals with a much broader time spread because of the abundance of electrons and photons.

The time structure provides the basis for neutrino identification [11]. The signal of the SD stations of the Observatory is digitized in 25 ns time bins allowing a detailed measurement of the time profile of the shower front. Extensive simulations of neutrino events have been made varying the energy, zenith and azimuth angles, depth of interaction and neutrino impact parameter and, in the ES case, the tau lepton decay altitude. The response of the SD is also simulated with standard software. The search strategy involves a pre-selection of inclined events followed by a selection of neutrino candidates. An optimized variable is chosen, studying signal and background distributions after applying the same selection. A cut is made to separate simulated neutrinos from the background of cosmic showers.

The selection process is optimized splitting it in three categories: (1) ES selection requires at least three stations, a high eccentricity of the triggered area on the ground, an apparent average speed of the signal close to c and a small spread. The selection of DG neutrinos requires at least 4 stations and is further split in High (DGH) and Low (DGL) zenith angle ranges. (2) DGH selection for zenith angles

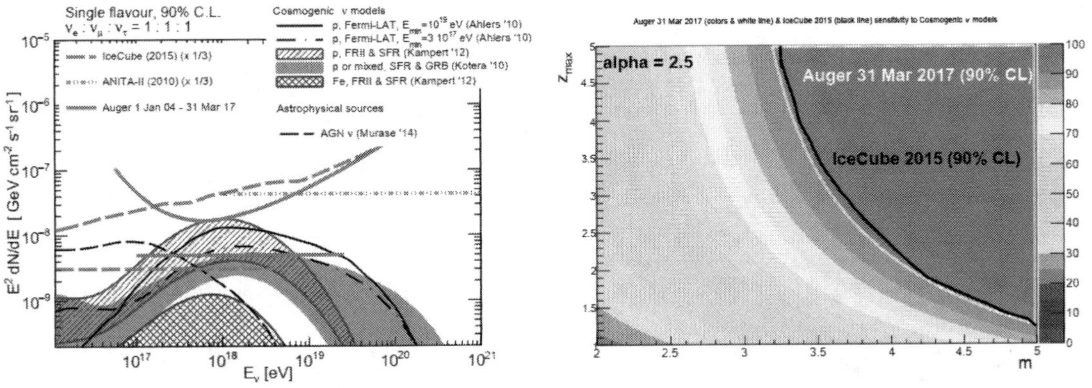

Figure 2. Left: Integral upper limit (at 90% C.L.) for the normalization, k, of a diffuse flux $dN/dE_\nu = kE^{-2}$ of single flavor neutrinos (straight red line), and differential upper limit (see text) compared to those from ANITAII [15] and IceCube [16] and several predictions (cosmogenic [17–19], astrophysical [20].) All flavors are assumed to have equal fluxes. Right: Constraints on parameter space for cosmological neutrinos in proton models ($dN/dE \propto E^{-2.5}$) as a function of m and $z_{\rm max}$ [29] (see text). Colored areas represent levels of exclusion. The region above the black (white) line is excluded at 90% C.L. by IceCube [16] (Auger) data.

exceeding 75° after a plane fit reconstruction must satisfy less stringent cuts on the same ES variables. (3) DGL selection for zenith angles in the 58.5° to 76.5° range [14].

The discriminating variable is related to the width of the digitized signal trace namely the Area over Peak (AoP). For the ES selection it is the average AoP of the triggered stations ($\langle$AoP$\rangle$) [14] and for the DG cases up to ten variables related to the AoP of 4 or 5 early or central stations are combined in a Fisher discriminant approach [14]. The cut value is chosen using a small fraction of the data which is assumed to be made up of cosmic ray showers and demanding a background rate of one event every 50 years. No neutrino candidates have been found.

The effective exposure as a function of neutrino energy convolved with a given neutrino flux spectrum, gives the expected number of events. It is obtained as the integral of the interaction and detection (trigger, selection and identification) probabilities over surface area, solid angle and time. For ES tau-neutrinos the probability of the tau-lepton exiting the Earth is calculated using a code that includes regeneration. It is then multiplied the probabilities of the tau decaying (a standard exponential) and by the detection probability. Both depend on decay position and this variable must be integrated too. For the DG selection the neutrino flavor and the type of interaction must be considered carefully to calculate interaction and detection probabilities. The time integral is performed sampling the effective probabilities over array configurations chosen every three days. The searches presented here span data collected from 1 January 2004 to 31 March 2017. Results are displayed in Fig. 1-right, assuming all flavors have equal fluxes. The ES channel dominates below ~ 50 EeV.

The 90% C.L. limits for a diffuse isotropic flux of single flavor neutrinos are shown in Fig. 2, compared to results from other telescopes. A more restrictive bound is obtained integrating a conventional $k\,E^{-2}$ neutrino spectrum, giving $k < 5 \times 10^{-9}$ GeVcm^{-2}s^{-1}, and plotted between 0.1 to 25 EeV, where 90% of the events are expected. The differential bound, obtained integrating in bins of 0.5 in $\log_{10} E_\nu$, illustrates the good SD sensitivity for detecting EeV neutrinos. Several predictions for cosmogenic neutrinos from models that address the UHECR origin are excluded at the 90% C.L., particularly those assuming proton primaries and attributing strong redshift evolution to the sources [18, 19]. Exclusion plots can be made as function of relevant parameters, namely the source evolution (m), the maximum

DOI: 10.1051/epjconf/201716501057

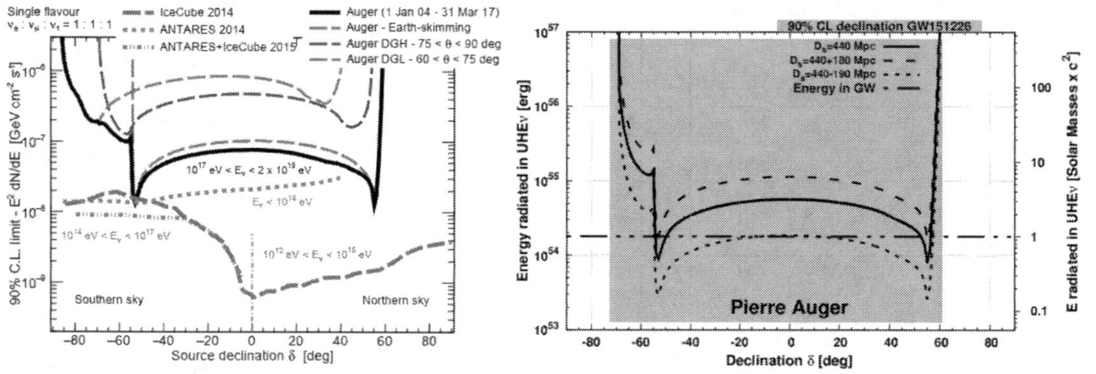

Figure 3. Left: Upper limits at 90% C.L. for a single flavor neutrino flux $dN/dE_\nu = k^{PS} E_\nu^{-2}$ as a function of δ for ES, DGH and DGL searches as labeled, compared to those of for IceCube [22], ANTARES [23] and a combination of both [24] (equal flavor scenario). Right: Constraints on the radiated energy in UHE neutrinos (per flavor) from the source of GW151226 as a function of δ [26]. Energies above the solid line are excluded at the 90% C.L.. The calculation is done for a distance $D_s = 410$ Mpc (dashed lines correspond to reported distances at the 90% C.L.). The inferred energy radiated in gravitational waves, $E_{\rm GW} \simeq 1.9 \times 10^{54}$ erg [27], is also displayed (dot-dashed line). The shaded region indicates the 90% C.L. region for the event location.

redshift ($z_{\rm max}$), the spectral index (α) and the maximum energy of the UHECR flux ($E_{\rm max}$) [28], using a conservative analytical approximation of the cosmogenic neutrino flux [29]. An example for a spectral index of 2.5 and $E_{\rm max}$ = 300 EeV is shown in Fig. 3.

3 Point source searches and correlations

Bounds on point sources as a function of declination (δ), obtained integrating the exposure in right ascension, are displayed in Fig. 3 [21]. The ES exposure displays strong peaks because this channel is very effective at converting the tau neutrinos into exiting tau leptons for directions close to the horizontal. The bulk of the ES sensistivity is for zenith angles between 90° and 92.5° and the sky coverage is $-54.5° < \delta < 59.5°$. The DG search extends it South to $-84.5°$.

The discovery of gravitational wave events reported by the Advanced Ligo Collaboration has triggered an intense program of targeted searches for coincidence events. A search has been made for EeV neutrinos in the Observatory data correlated with black-hole binary coalescences GW150914, GW151226, GW170104 and LVT151012 (candidate) [27]. Assuming a steady flux during a day after the event, motivated by an upper limit to the duration of GRB afterglows [26], a bound on the EeV neutrino emission has been obtained. The directionality of the neutrino exposure makes the search dependent on event position (only known as a likelihood plot over a broad region) and the results are given in terms of declination (δ). The neutrino exposure averaged over a day covers a large fraction of the sky and has thus large overlap with all the events, as exemplified by the shadowed band in Fig. 3 for event GW151226, for which practically the whole 90% C.L. declination band is covered. The 90% C.L. bound, obtained for an E^{-2} spectrum and converted into a bound on the energy fluence radiated in EeV neutrinos, is shown in Fig. 3 for GW151226. Similar bounds for GW150914 are somewhat stronger while for GW170104 they are weaker because it is further away [26].

Correlations between directions of UHECR detected with the Auger Observatory and TA and those of the neutrinos detected with IceCube have also been searched for. Angular correlations between neutrino and UHECR pairs have been compared to expectations from isotropy (as a function of separation angle). A stacking method (fixing both the neutrino and the UHECT directions) has been also used to maximize the likelihood varying the fraction of correlating events. Plausible assumptions about angular dispersion are required to account for direction tolerances. First reports pointed to interesting correlations [30] that have nevertheless become weaker in a recent update of the analysis [31].

Searches for anisotropies of UHECR are revealing precious information about their production sites but more data are needed. The Auger Observatory is effective to constrain models searching for diffuse neutrino fluxes and for point sources as part of a multi-messenger approach. As data from the Observatory continue to be analyzed and combined with other experiments we will get closer to deciphering the elusive origin of the most energetic particles in the Universe.

References

[1] A. Aab *et al.* Phys. Rev. **D 90**, 122005 (2014).
[2] E. Zas J. Phys. Conf. Ser. 632, no. 1, 012104 (2015).
[3] Abu-Zayyad, T. *et al.*; Nucl.Instrum.Meth. A **689** (2013) 87-97.
[4] J. Abraham *et al.* Phys. Rev. Lett. **101**, 061101 (2008).
[5] R. U. Abbasi *et al.*, Astropart. Phys. **64**, 49 (2015).
[6] V. de Souza for the Pierre Auger and TA Collaborations; Proc. of ICRC 2017, PoS (2017) 522.
[7] A. Aab *et al.* (in press) Science **358**, (2017).
[8] G.U. Giaccari for the Pierre Auger Collaboration; Proc. of ICRC 2017, PoS (2017) 483.
[9] M. G. Aartsen *et al.* Phys. Rev. Lett. **113**, 101101 (2014).
[10] A. Aab *et al.* Nuc. Inst. Meth. A **789**, 172 (2015).
[11] K. S. Capelle *et al.*; Astropart. Phys. **8**, 321 (1998).
[12] X. Bertou *et al.*; Astropart. Phys. **17**, 183 (2002).
[13] A. Aab *et al.*; JCAP **1408**, 019 (2014).
[14] A. Aab *et al.* [Pierre Auger Collab.], Phys. Rev. D **91**, 092008 (2015).
[15] P. W. Gorham *et al.* [ANITA Collab.], Phys. Rev. D **85**, 049901(E) (2012).
[16] M. G. Aartsen *et al.* [IceCube Collab.] Phys. Rev. Lett. **117**, 241101 (2016).
[17] M. Ahlers *et al.*, Astropart. Phys. **34**, 106 (2010).
[18] K. -H. Kampert, M. Unger, Astropart. Phys. **35**, 660 (2012).
[19] D. Allard *et al.*, JCAP **10**, 013 (2010).
[20] K. Murase, Y. Inoue, and C. D. Dermer, Phys. Rev. D **90**, 023007 (2014).
[21] P. Abreu *et al.*, Astrophys. J. **755**, L4 (2012).
[22] M.G. Aarsten et al., Astrophys. J. **796**, 109 (2014).
[23] S. Adrian-Martinez et al. [ANTARES Collab.], Astrophys. J. Lett. **786**, L5 (2014).
[24] S. Adrian-Martinez et al., Astrophys. J. **823**, 65 (2016).
[25] K. Kotera and J. Silk; Astrophys. J. **823**, no. 2, L29 (2016).
[26] A. Aab *et al.* [Pierre Auger Collab.], Phys. Rev. D **94**, 122007 (2016).
[27] B. P. Abbott *et al.* [LIGO Scientific Collab. and Virgo Collab.], Phys. Rev. Lett. **116**, 061102 (2016); *ibid.* **116** 241103 (2016); *ibid.* **118** 221101 (2017).
[28] M. G. Aartsen *et al.* Phys. Rev. Lett. **117**, 241101 (2016).
[29] S. Yoshida and A. Ishihara, Phys. Rev. D **85**, 063002 (2012).
[30] M.G. Aartsen *et al.*; JCAP **1601**, 037 (2016).
[31] I. Al Samarai and G. Golup; Proc. of ICRC 2017, PoS (2017) 961.

And so it all began: A personal tribute to the man behind the scientist

Marialuisa Aliotta[1,*]

[1] *SUPA, School of Physics and Astronomy, University of Edinburgh, EH9 3FD Edinburgh, United Kingdom*

Abstract. At the time I began my scientific career as a PhD student under the supervision of Claudio Spitaleri, the Trojan Horse Method was still in its infancy. Like with any new-born idea, it took time and passion and effort to plant the early seeds that would eventually develop into a now well-established method in nuclear astrophysics research. Here, I offer my own recollection of those early years as a personal homage to Claudio's unique mix of human traits that shaped our professional relationship for many years since.

1 Preamble

If you work in the field of Nuclear Astrophysics and hear about a Trojan Horse Method (THM) experiment, you immediately identify it with the group in Catania, and more specifically with the group of Prof. Claudio Spitaleri. Yet, a little less than two decades ago, almost no-one in the field would have heard about the Trojan Horse Method. And those few who had, would probably have not concealed their skepticism about it. It was an indirect method based on a quasi-free breakup mechanism that under appropriate conditions could be used for nuclear astrophysics studies otherwise difficult to access with direct approaches [1].

On paper, the method was brilliant in its simplicity: much like Ulysses had managed to overcome the walls of Troy by hiding his men and himself within the belly of a huge wooden horse brought to the city as a gift [2, 3], so could a little particle overcome the repulsive Coulomb barrier of a targeted foe by being brought into it hidden inside a suitable Trojan horse nucleus. This way, the reaction of astrophysical interest could be studied unhindered by the Coulomb suppression and essentially devoid of any electron screening effect, a still-puzzling phenomenon that plagues direct measurements at extremely low energies.

It just seemed too good to be true! Surely, there must have been some hidden issues that were not immediately obvious at a first glance? And indeed some there were. Firstly, the method did not allow for a measurement of the absolute cross section of any nuclear reaction and as such relied on the existence of direct data, at least at high energies, for normalisation purposes. Secondly, it had to fulfil a number of strict validation tests. For example it was necessary to distinguish (and only select) the quasi-free mechanism events from those generated by any other process (*e.g.* sequential decay) leading to the same particles in the final state. Thirdly, it required making sure that the other particle (effectively the Trojan horse companion) did not take part in the interaction and left the scene

* e-mail: m.aliotta@ed.ac.uk

of crime as a spectator. Of course this called for a non-trivial theoretical formalism and that too was rather complex and still in its infancy.

The method, and indeed its name, had already been proposed by Gerard Baur in 1986 [4] as a possible way to study reactions of astrophysical interest without the challenges posed by low-energy direct measurements. Yet, somehow, the suggestion had not been picked up until the late-mid 1990s. By that time, Claudio Spitaleri had already been working on break-up mechanisms for almost twenty years along with Marcello Lattuada, Francesco Riggi, Domenico (Mimmo) Vinciguerra, Djuro Miljanic, Mile Zadro, and others. By then, Claudio had realised that the idea proposed by Baur had a lot of potential and decided to pursue it at maximum speed. Luckily for me, this also coincided with the beginning of my PhD in 1995.

2 Chance encounters

I had graduated in physics in 1993 with a thesis on the use of radioactive ion beams for the study of the ^{19}Ne(p,γ)^{20}Na reaction in the context of nova explosions. Mine was the first undergraduate thesis in nuclear astrophysics at the University of Catania. Indeed, the field itself was probably very little known there at that time. In fact, I had found out about it almost accidentally as I attended a seminar given by a nuclear physicist from my home university, Prof Mimmo Vinciguerra, a few months before deciding on the topic of my thesis. In his seminar, Vinciguerra was reporting about new prospects to study nuclear reactions of interest to explosive astrophysical scenarios thanks to the possibility of creating and accelerating short-lived unstable nuclei for the first time. This opportunity was just emerging. At the Centre de Recherche du Cyclotron in Louvain-la-Neuve, Belgium, people had come up with the idea of coupling two existing cyclotrons to first produce and then re-accelerate radioactive nuclei to energies of astrophysical interest. This – Vinciguerra said in his seminar – would open up exciting new opportunities for research in Nuclear Astrophysics, a relatively new branch of science focussed on understanding the origin of the elements and the evolution of stars and galaxies. I got fascinated and quickly decided that this would become the topic of my thesis. Vinciguerra agreed to supervise me, jointly with Prof Aldo Zappalà from the Astronomical Observatory in Catania, and as part of my undergraduate thesis work I went to Belgium to take part in the ^{19}Ne(p,γ)^{20}Na experiment.

In the summer of 1993, a few months after my graduation, another chance encounter unfolded that would eventually prove to be crucial not only for my career, but also, as I will explain, for the dissemination of the THM beyond its initial restricted playground. At a summer school on Exotic Nuclei in Belgium, I happened to meet Prof Claus Rolfs, a world-leading figure in nuclear astrophysics and author of the book, *Cauldrons in the Cosmos* [5], by many regarded as the 'bible' in the field. I approached him during a coffee break and asked whether I might spend time at his institute in Germany. He replied that as long as I had funds of my own, I would be very welcome. So, I decided to apply for a scholarship from my university to spend a year abroad and moved to Bochum in 1994 at the Lehrstuhl für Ionenstrahlen of the Ruhr-Universität Bochum, Germany, led by Claus Rolfs. By the time I returned to Catania a year later I knew I wanted to do a PhD in Nuclear Astrophysics and I wanted it to be on an experimental nuclear physics project. I just did not know on what. The initial plan had been for me to continue working with Zappalà and Vinciguerra, so I started taking part in some experiments by the nuclear physics group at the Laboratori Nazionali del Sud (LNS) of the Istituto Nazionale di Fisica Nucleare (INFN) in Catania. That is when I met Claudio for the first time.

One day I was sitting on my own in the PhD students' office, feeling slightly unsure about what to do and which direction to take. Unexpectedly, Claudio came along and started asking me about my plans and stressing how important it was that I began working on a specific project as soon as possible. I guess the very fact that he showed interest in me, at a time where I was feeling somewhat under the weather, and the fact that he seemed to have a better idea about what I should do than I did, got me interested in the Trojan Horse Method and persuaded me to go for a change of project and supervisor.

It was soon after our initial chat that I started telling Claudio about my experience in Germany and how much I had enjoyed being in Rolfs' group. He showed great interest in finding out more about what I had done in Germany and it became clear that he too would have liked to meet the man. He did not know yet, but chance was around the corner once again, ready to offer another opportunity. In those days, Rolfs was spending a sabbatical in Naples as part of the NaBoNa collaboration with Prof Filippo Terrasi and his group. No sooner did I mention this to Claudio that he decided we should all go to Naples to visit him in person! Of course, by 'we' I mean just the three of us: Claudio, myself and Silvio Cherubini, Claudio's first PhD student who had just completed his doctorate. That's all who we were as far as the nuclear astrophysics group in Catania was concerned.

Claudio's quick decision to leave for Naples as soon as feasible summarises much of the attitude he has always had towards his work and was, as I would learn later, a typical trait of the enthusiasm and passion he poured into any new adventure he would embark on. So, without wasting time, I got in touch with Rolfs and asked whether it was ok for us to fly to Naples to meet him. Needless to say, he was delighted and quickly agreed to the plan with his usual, jovial cheerfulness. I do not remember much of that trip, except for a brief moment that still sticks indelible in my memory and that again epitomises much of Claudio's personality. We had just left the Physics Department in Naples after meeting with Rolfs and discussing with him about the THM. As we were walking towards our hired car, Claudio suddenly stopped in the middle of the path – tall eucalyptus trees all around us – looked at Silvio and me and, branding his finger mid air, with a tone required of solemn occasions, said: "Remember this day?! One day, maybe ten or twenty years from now, when we will have established a big group and everyone will know about us, you will think back at today and say "That's when it all began!"".

3 The early days of ASFIN

Back at the lab in Catania, one of my first tasks was to write a research proposal that would form the basis of my PhD project. The idea was to use the THM to investigate the elastic scattering of $^{12}C+^{4}He$ as a validity test [6] towards the study of the $^{12}C(\alpha,\gamma)^{16}O$ reaction, arguably one of the most important reactions in the field. Perhaps, you might think that tackling a simpler problem in nuclear astrophysics would have been wiser for someone approaching this field with an indirect method that was still very much in its infancy. But again, Claudio's personality says a lot about this choice. As he would often remind me for years: "Si t'ha mbriacari, mbriacati di vinu bonu!", a Sicilian phrase which roughly translates as "If you wish to get drunk, do so with quality wine!", an apt mantra to his philosophy of life. And so it was that we started to apply for funds under a new project name: ASFIN, simply standing for AStroFIsica Nucleare. That acronym still survives today as a highly successful research line within INFN.

Yet, if we wanted to turn the THM method from a mere idea into a well-established tool for nuclear astrophysics studies, we needed much theoretical support and a more sophisticated formalism. In those days, a German professor from the Ludwig-Maximilian Universität in Munich used to spend extended research visits at the Laboratori Nazionali del Sud di Catania. That professor was Hermann Wolter, who had recently collaborated with his post-doc Dr Stefan Typel precisely on matters closely related to the THM. Claudio discussed things with him and soon agreed to send me on a research visit to Munich. That again marked the beginning of many years of close collaboration which took the THM to a whole new ground.

As part of my PhD, Claudio also suggested that I should spend a period of six months back in Bochum. In part, this decision was dictated by the desire to test some new monolithic ΔE-E silicon

 DOI: 10.1051/epjconf/201716502001

Figure 1: (Left panel) Two ladies at the DTL: Aurora Tumino (front right) and I (front left) with a bunch of flowers given to us by our German colleagues (behind us). (Right panel) Is someone sabotaging our experiment behind our backs? Aurora thinks this is funny!

detectors that were being developed in Catania as a collaboration between LNS and STMicroelectronics. But that of course, much like a Trojan horse, was just a pretext: more importantly Claudio was exploiting the opportunity as a way to further establishing and consolidating a closer collaboration with Rolfs. Soon after I arrived in Bochum for the second time, I was asked to give a group seminar about the Trojan Horse method and this quickly became a hot topic of discussion both within the group and outside it, with other close collaborators of Claus. A seed had been planted to disseminate the idea further.

My second time in Bochum was simply fantastic. Every member of the group, all males as it happened, were incredibly eager to offer their support. Probably this was further helped by the fact that now two girls were around at the Dynamitron Tandem Laboratorium (DTL) in Bochum (Figure 1): one was me, the other was Aurora Tumino (a young student at that time working with Marcello Lattuada in Catania) whom Claudio had decided to involve in our scientific adventures.

As for Claudio, he was always there with us in the front line. I still remember the day before our first experiment in Bochum! I hadn't slept in 36 hours trying to get everything ready for beam time. Before reinforcements arrived from Catania, I had been in charge of preparing every aspect of the experiment, from testing the detectors, assembling the electronics, and testing the data acquisition system, which proved to be the hardest part back in the days of a multi-parameter box that no-one had used before. I was exhausted and worried that something might go wrong. Just as we were about to receive beam on target, Claudio sat by my side at the acquisition desk, to support and encourage me: all will be fine. The experiment, or rather the test, proved somewhat inconclusive, but it was enough for us to have started putting another foot on the international context.

4 The scientist, the mentor, the man

It would be impossible here to recall the many trips and adventures we shared throughout the years that followed; the places we visited; the experiments we performed; the laughter and the companionship I experienced during the time of my PhD. Those of us who were around in those days may still remember when I shamelessly mingled with the guests of a wedding banquet in the foyer of Schoenbrunn Palace in Vienna, at the margin of a conference Claudio had been invited to; or the time during the 1996 Nuclei in the Cosmos conference in Notre Dame when Silvio and I were desperately

Figure 2: (Left panel) Claudio, Aurora and I at a bus stop somewhere in Europe. (Right panel) Claudio and Claus (right) enjoying wine at a conference dinner.

looking for a shoe shop only to end up calling for a taxi from the desk of a down-town bank; or still the night at a restaurant in Louvain-la-Neuve when the waiter brought a Dame blanche dessert to the table when in fact Silvio had called for an 'ambulance' for Claudio who was feeling unwell; or the time at the Steakhouse Argentinien in Bochum where I suggested to Claus and Claudio that we should start a summer school in Nuclear Astrophysics, now famously known within the community as the 'St Tecla' School.

We travelled extensively, always establishing new connections and planting the seeds for new collaborations and friendships that last to this day (Figure 2). Behind every trip and every new initiative was Claudio's unwavering enthusiasm, his desire to go always a step beyond and make his dream come true even in the face of personal tragedy.

Today, the ASFIN group, or Claudio's group (Figure 3), or the Trojan Horse Method, are one and the same. Even by modern standards and metrics, the THM has made for a truly successful story: over 300 papers published to date (with some 120 submitted in the last five years alone); more than 150 invited talks at international conferences; international collaborations with over 70 scientists from more than 30 different countries. The legacy is huge.

As a scientist, Claudio has had the ability to look ahead where no-one else had dared; to stick with his intuition that the method had potential; and to do all he could to make it widely recognised and established. This he has accomplished thanks to his ability to constantly enthuse new recruits and engage established scientists, not least several theoreticians who have further developed the method to adapt it to resonant reactions and, more recently, to reactions with radioactive nuclei [7]. Over the years, the Catania group has steadily increased in size and many students have taken their first steps under Claudio's protective wing towards now-established research careers.

As a mentor and supervisor, Claudio was always present to support and encourage; to challenge and reward. He constantly engaged me in decisions about the group's next moves; involved me in meetings with senior scientists; and generally made me feel I was his peer, not just his PhD student. Of course, this was not out of insecurity or indecision, but rather out of a deep rooted desire to share his vision for the future he had in mind. It wouldn't be fair to claim that working with Claudio has

Figure 3: The ASFIN group in Catania (as of 2010. From left to right: Stefano Romano, Gabor Kiss, Silvio Cherubini, Marisa Gulino, Nellí Puglia, Claudio Spitaleri, Aurora Tumino, Gianluca Pizzone, Roberta Spartà, Marco La Cognata, Giuseppe Rapisarda (photo credit: G. Pizzone).

always been easy. We have certainly had discussions at times and occasionally strong disagreement, but there has always been a deep mutual respect and the deliberate intention to find new ways forward.

Yet for me, the most important trait of Claudio's personality has always been his humanity. And this is what I have cherished the most throughout the years. More than a supervisor, Claudio has been like a father figure. Someone I could go to when things, professionally and personally, had taken an unexpected turn, like when I left Germany to take up a permanent lectureship in the UK in 2001 or when my entire family moved away from Sicily just a few months afterwards. Throughout the many changes in my life and some moments of darkest despair, he has remained a constant presence and a reference point: someone I could get back to for advice or encouragement at every step along the way. For this, I am immensely grateful.

5 Afterword

As I sign off this contribution, I would like to take the opportunity to express my deepest gratitude to my many colleagues in Catania, who have been and continue to be some of my closest, dearest friends: Aurora Tumino and Stefano Romano, for teasing me in their truly unique way that makes me laugh with tears in my eyes every time we are together; Silvio Cherubini and Gianluca Pizzone (the 'before' and 'after' me), the former for constantly arguing with me about physics, the latter for taking much of the advice I had to offer as he followed in my own footsteps; and all the 'new' recruits, most of whom I never really worked with but who have always welcomed me back to the group making me feel one of them. To Marco La Cognata, Livio Lamia, Roberta Spartà, Giuseppe Rapisarda, Maria

Letizia Sergi, Marisa Gulino, Sebastiana (Nellí) Puglia, Luca Guardo, and to all the new students in the group, my greatest wish for you is that you may enjoy plentiful and successful careers, paying tribute to the wonderful legacy that Claudio has left for us all.

Today, I am always pleased and proud to be able to fly back to Catania, my home at heart, and to see how much things have moved forward since the days of my PhD. It is always a pleasure to see Claudio and notice how nothing has changed of his indomitable enthusiasm and passion for his work.

For me, the success of a scientist is measured not by the number of papers he or she has published, or by the astronomical personal h-index he or she might have achieved, but by the knowledge that he or she has left an imprint, a legacy, something for others to live on. This, Claudio, is what you have accomplished. If you ever look back, rejoice in the awareness that yours has been a career truly well spent.

References

[1] C. Spitaleri *et al.*, Physics of Atomic Nuclei, **74** (2011) 1725.
[2] Homer, Odyssey, 8th century BC.
[3] Virgil, The Aeneid, 29-19 BC.
[4] G. Baur, Physics Letter B **178**, 135 (1986).
[5] C.E. Rolfs and W.S. Rodney, *Cauldrons in the Cosmos* (The University of Chicago Press, 1988).
[6] C. Spitaleri, M. Aliotta, *et al.*, European Physical Journal **A7** (2000) 181.
[7] C. Spitaleri, *et al.*, European Physical Journal **A52** (2016) 77.

EPJ Web of Conferences **165**, 02002 (2017) DOI: 10.1051/epjconf/201716502002
NPA8 2017

Nuclear clustering and the electron screening puzzle

C.A. Bertulani[1,⋆] and *C.* Spitaleri[2,3,⋆⋆]

[1] *Department of Physics and Astronomy, Texas A&M University-Commerce, Commerce, Texas 75429, USA*
[2] *Department of Physics and Astronomy, University of Catania, Catania, Italy*
[3] *INFN-Laboratori Nazionali del Sud, Catania, Italy*

Abstract. Electron screening changes appreciably the magnitude of astrophysical nuclear reactions within stars. This effect is also observed in laboratory experiments on Earth, where atomic electrons are present in the nuclear targets. Theoretical models were developed over the past 30 years and experimental measurements have been carried out to study electron screening in thermonuclear reactions. None of the theoretical models were able to explain the high values of the experimentally determined screening potentials. We explore the possibility that the "electron screening puzzle" is due to nuclear clusterization and polarization effects in the fusion reactions. We will discuss the supporting arguments for this scenario.

1 Introduction

Thermonuclear reactions in stars occur at ultra-low energies around the Gamow energy, E_G , of the order of few keV to 100 keV, much below the Coulomb barrier [1, 2]. The experimental determination of the fusion cross sections for these reactions, $\sigma_b(E)$, requires the use of either direct or indirect techniques [3, 4]. In the Sun, most nuclear reactions involve charged particles and non-resonant reactions [2, 5]. As the relative energies between "bare" nuclei decrease, their cross sections σ_b(E) decrease exponentially and it is more appropriate to express the cross section in terms of the astrophysical S-factor, defined as

$$\sigma_b(E) = \frac{1}{E} S(E) \exp\left[-2\pi\eta(E)\right], \tag{1}$$

where $1/E \propto \pi\lambda\!\!\!^{-2}$ accounts for the quantum mechanical area $\pi\lambda\!\!\!^{-2}$ within which the reaction can occur, with $\lambda\!\!\!^{-}$ being the reduced wavelength and $\exp(-2\pi\eta)$ is a function which approximately accounts for the tunneling probability through the barrier. The Sommerfeld parameter $\eta(E)$ is given by

$$\eta(E) = \frac{Z_1 Z_2 e^2}{\hbar v} = \frac{Z_1 Z_2 \alpha}{v/c}, \tag{2}$$

where Z_1, Z_2 are the nuclear charges, $v = \sqrt{2E/\mu}$ the initial relative velocity between the nuclei, and $\alpha = e^2/\hbar c$ is the fine-structure constant. With this definition, the astrophysical S-factor becomes a much smoother function of E than $\sigma_b(E)$ allowing for a more reliable extrapolation from the high

⋆e-mail: carlos.bertulani@tamuc.edu
⋆⋆e-mail: spitaleri@lns.infn.it

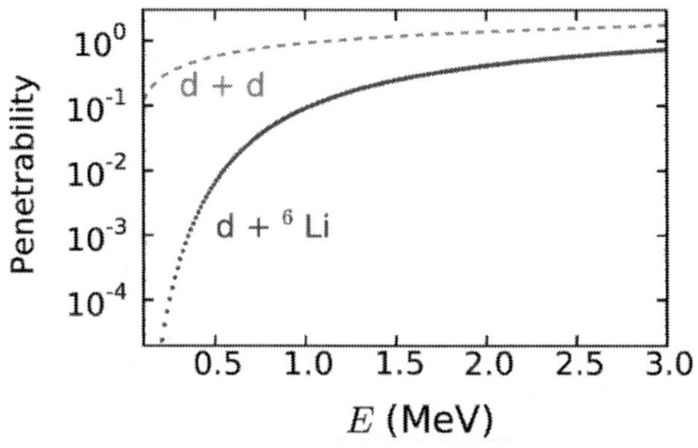

Figure 1. Barrier penetrabilities for d + d and for d + ^{6}Li reactions as a function of the relative motion energy.

energy values, where measurements (direct or indirect) have been carried out, to the low energies around Gamow peak [2].

The presence of electrons in atoms increases the tunneling probability, also increasing the measured (screened) cross-section σ_s. Free electrons in stars also induce and increase of the cross sections [1, 6–8]. One can define an enhancement factor $f_{lab}(E)$ in laboratory experiments by means of [1, 7, 8],

$$f_{lab}(E) = \frac{\sigma_s(E)}{\sigma_b(E)} = \frac{S_s(E)}{S_b(E)} \simeq \exp\left[\pi\eta \frac{U_e^{(lab)}}{E}\right], \tag{3}$$

where $U_e^{(lab)}$ is the electron screening potential, with the last step valid for $U_e \ll E$. This enhancement factor has been studied experimentally, but it was found out that in many cases the measured values extracted for $f_{lab}(E_G)$ are in significant disagreement with atomic-physics models [1, 8]. In this proceedings contribution we report a novel approach to explain this puzzle with nuclear clustering effects, as recently shown in Ref. [9].

In the Fock space, a properly anti-symmetrized nuclear wavefunction can be written as

$$|\psi_{nucleus}\rangle = \alpha\,|\psi_A\rangle + \beta\,|\psi_a\psi_B\rangle + \gamma\,|\psi_c\psi_D\rangle + \cdots, \tag{4}$$

where $|\psi_A\rangle$ is a single-cluster wave function of A nucleons, $|\psi_a\psi_B\rangle$ is a cluster (a+B) nuclear wave function and (α, β, γ) are spectroscopic amplitudes for such configurations. Since the Coulomb penetrability changes for each cluster configuration, fusion probabilities are effectively reduced, increasing the cross sections and leading to a similar effect as that of electron screening. We illustrate this with one example. Experiments show that the ^{6}Li + ^{6}Li $\rightarrow 3\alpha$ cross section is orders of magnitude larger than expected from barrier penetrabilities based on a non-clusterized ^{6}Li nucleus [10]. But ^{6}Li has a non-negligible probability to be found as $d + \alpha$ cluster structure. With this organized structure, fusion is enhanced as the Coulomb barrier for deuterons with ^{6}Li is reduced. The cross section can be written as

$$\sigma_L = C_{66}P_{^6\mathrm{Li}+^6\mathrm{Li}} + C_{26}P_{\mathrm{d}+^6\mathrm{Li}} + C_{46}P_{^4\mathrm{He}+^6\mathrm{Li}} + \cdots, \tag{5}$$

where C_{66}, C_{26}, etc, are variables that include the energy dependence not included in the tunneling probabilities for the ^{6}Li + ^{6}Li, d + ^{6}Li, etc, configurations. Certainly, $P_{^6\mathrm{Li}+^6\mathrm{Li}}/P_{\mathrm{d}+^6\mathrm{Li}} \rightarrow 0$ as $k \rightarrow 0$, and the second term in Eq. (5) will be dominant as $E \rightarrow 0$. The energy dependence of these variables are much weaker than of the barrier penetrabilities and therefore cluster-like structures will enhance

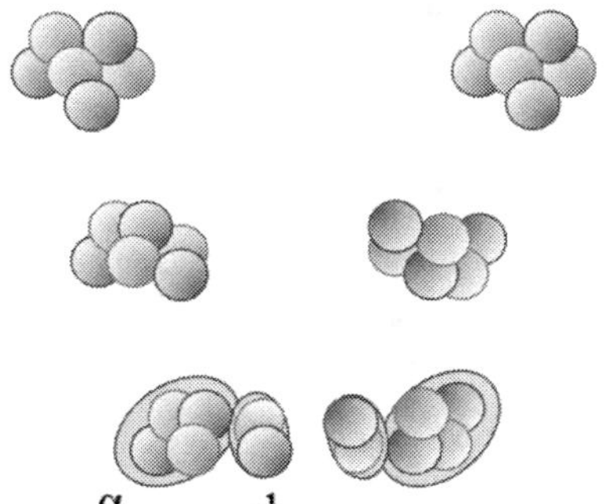

Figure 2. Schematic view of polarization and orientation as nuclei with large probabilities for cluster-like structures approach each other, shown here for ^{6}Li + ^{6}Li.

the cross sections by many orders of magnitude. Moreover, if the deuterons within the ^{6}Li nuclei approach each other and directly fuse to α particles, it would increase enormously the cross section for the reaction, as shown in Figure 1. That is what one observes with the cross section for ^{6}Li + ^{6}Li $\rightarrow 3\alpha$ [10–14]. It is likely that the deuterons within ^{6}Li tunnel through the barrier individually to form the α-particles, leaving the other two α-particles behind. The enhancement is probably helped by polarization, because ^{6}Li nuclei align so that the two deuterons can be located closer to each other during the tunneling process, thus reducing the effective barrier. The alpha particles stay farther away from each other due to polarization and alignment, as schematically shown in Figure 2. An average over all configurations leads to a reduced effective Coulomb repulsion. This was demonstrated using a simple cluster model for ^{6}Li in Ref. [9].

The model described in Ref. [9] clearly proves that the action of both polarization and orientation, always enhances fusion in the dicluster-dicluster configuration compared to the non-cluster configuration. Polarization is remnant of the Oppenheimer-Phillips effect [15, 16], which has been used before to quantify deuteron polarization effects in the Coulomb field of a nucleus in reactions induced by deuterons. But little can be said about the amplitudes C_i in Eq. (4). There is no robust theory with a predictive power for the clustering amplitudes in nuclei, despite intensive theoretical efforts beyond the naïve shell-model [17]. The nuclear clustering theory also needs to account for cluster pre-formation amplitudes, which needs to be tackled with in ab-initio models (see, e.g., Ref. [18–20]). But even without information about pre-formation factors, or spectroscopic amplitudes for clustering configurations, the mere assumption of the existence of clusters in nuclei will lead to an effective enhanced tunneling when averaged over all orientations [9].

Typical cases of experimentally studied thermonuclear reactions are shown in Table 1 and Figure 3. We claim that in most of these cases, clusterization fusion enhancements have already been observed. Our argument is based on the fact that a clear separation exists for $Z = 1$ nuclei reactions with nuclei without cluster sub-structures, and those with cluster-like nuclei. There exists a discrepancy between the adiabatic limit of the screening potential, U_e^{adlim}, and its experimental value, U_e^{exp}, but the disagreement is larger when a cluster structure is expected. The values displayed in Table 1 favor a nuclear clustering solution for the electron screening problem at the thermonuclear energies of astrophysical interest. To summarize, the evidences are: (a) Reactions involving non-cluster-like nuclei with a ($Z = 1$) nucleus display an agreement, within experimental errors, with the adiabatic theory of the screening energy [6] (e.g., the d + d and d + p reactions), (b) Reaction cross sections of cluster-

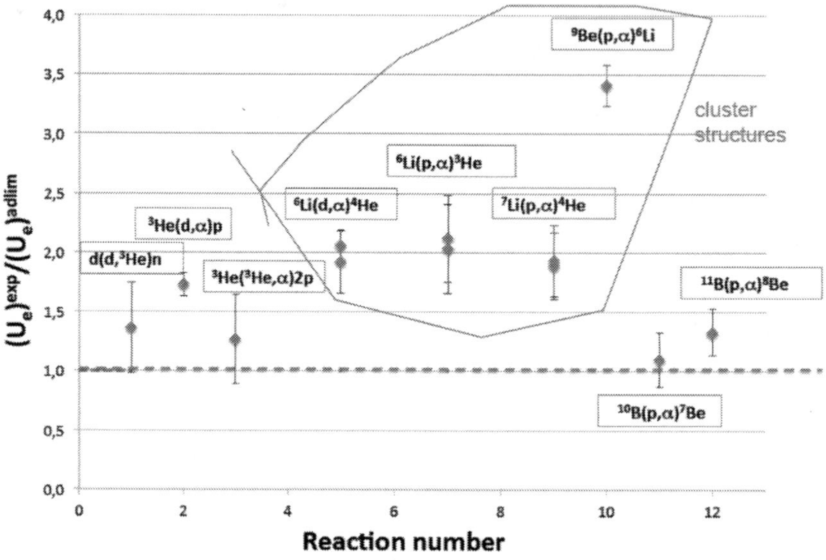

Figure 3. Ratio of experimentally determined electron screening potential U_e^{exp} and the adiabatic limit U_e^{adlim} for the reactions listed in Table 1. The reactions involving cluster-like nuclei are encircled by a sketchy line.

Table 1. Experimentally determined electron screening potentials, U_e^{exp}, and the adiabatic limits, U_e^{adlim}.

	Reaction	U_e^{adlim} (eV)	U_e^{exp} (eV)	Note	Ref.
[1]	^{2}H$(d,t)^1$H	14	19.1±3.4		[16, 21]
[2]	^{3}He$(d,p)^4$He	65	109±9	D_2 gas target	[22]
[3]	^{3}He$(d,p)^4$He	120	219±7		[22]
[4]	^{3}He(^{3}He,2p)^{4}He	240	305±90	compilation	[2]
[5]	^{6}Li$(d,\alpha)^4$He	175	330±120	H gas target	[23]
[6]	^{6}Li$(d,\alpha)^4$He	175	330±49		[23, 24]
[7]	^{6}Li$(p,\alpha)^3$He	175	440±150	H gas target	[23]
[8]	^{6}Li$(p,\alpha)^3$He	175	355±67		[23, 25, 26]
[9]	^{7}Li$(p,\alpha)^4$He	175	300±160	H gas target	[23]
[10]	^{7}Li$(p,\alpha)^4$He	175	363±52		[23, 26, 27]
[11]	^{9}Be$(p,\alpha_0)^6$Li	240	788±70		[28, 29]
[12]	^{10}B$(p,\alpha_0)^7$	340	376±75		[30, 31]
[13]	^{11}B$(p,\alpha_0)^8$Be	340	447±67		[30, 32]

like nuclei with $Z = 3$ and $A = 6,7$ (e.g., p + ^{6,7}Li) are more than a factor 1.5 larger than expected using the theoretical U_e^{adlim}, (c) Experimental values of U_e^{exp} for reactions with $Z = 4$ cluster nuclei clearly disagree with the adiabatic approximation. And this discrepancy increases with nuclei with more evidenced cluster structures (e.g., p + ^{9}Be and p + ^{10,11}B).

We conclude that the electronic screening problem might be due to a nuclear structure effect in the form of polarization and orientation of nuclei during the fusion process and we propose that new studies be carried out to confirm our claims. Candidates for such a search can involve reactions with ^{6}Li or ^{7}Li. For example, one could envisage studies with ^{6}Li + ^{6}Li, ^{7}Li + ^{7}Li, ^{7}Li + ^{6}Li , ^{9}Be + ^{3}He, ^{9}Be + ^{7}Li at thermonuclear energies. For more details and explicit description of calculations mentioned in this proceedings contribution, see Ref. [9]. There are other experiments which extracted different electron screening potentials that were not mentioned here. The electron screening potentials as obtained from out theory are independent of the target environment, and we do not expect that all experimental data on electron screening will be explained.

We have benefited from useful discussions and a collaboration with A. Vitturi and L. Fortunato. This work has been partially supported by the Italian Ministry of University MIUR under the grant "LNS-Astrofisica Nucleare (fondi premiali)" and the U.S. NSF Grant No. 1415656, and U.S. DOE grant No. DE-FG02-08ER41533.

References

[1] C. Rolfs, W. Rodney, *Cauldrons in the Cosmos* (The University of Chicago, 1988).
[2] E.G. Adelberger, al., Review of Modern Physics **195**, 83 (2011).
[3] C.A. Bertulani and A. Gade, Physics Reports **485**, 195 (2010).
[4] R. Tribble, C.A. Bertulani, M. La Cognata, A. Mukhamedzhanov, and C. Spitaleri, Rep. Prog. Phys. **77**, 106901 (2014).
[5] C.A. Bertulani and T. Kajino, Prog. Part. Nucl. Phys. **89**, 56 (2016).
[6] H. Assenbaum, K.L. Langanke, C. Rolfs, Z. Phys. A **327**, 461 (1987).
[7] G. Fiorentini, R. Kavenagh, C.Rolfs, Z. Phys. A **350**, 289 (1995)
[8] F. Strieder, C. Rolfs, C. Spitaleri, P. Corvisiero, Naturwissenschaften **88**, 461 (2001)
[9] C. Spitaleri, C.A. Bertulani, L. Fortunato, A. Vitturi, Phys. Lett. B **755**, 275 (2016).
[10] M. Lattuada, F. Riggi, D. Vinciguerra, C. Spitaleri, G. Vourvopulos, D. Miljanić, Z. Phys. A-Atomic Nuclei **330**, 183 (1988)
[11] D. Manesse, M. Coste, C. Lemeille, L. Marquez, N. Saunier, Nucl.Phys. **55**, 433 (1964)
[12] L. Gadeken, E.Norbeck, Phys.Rev. C **6**, 1172 (1972)
[13] M. Lattuada, F. Riggi, D. Vinciguerra, C. Spitaleri, D. Miljanić, Z. Phys. A-Atomic Nuclei **328**, 497 (1987)
[14] C. Spitaleri, A. Tumino, M. Lattuada, S. Tudisco, M. Lattuada, F. Riggi, D. Miljanić, M. Milin, N. Soic, Phys.Rev.C **91**, 024612 (2015)
[15] J. Oppenheimer, M. Phillips, Phys. Rev. **500**, 48 (1935)
[16] U. Greife, F. Gorris, M. Junker, C. Rolfs, and D. Zahnow, Z. Phys. A **107**, 351 (1995).
[17] M. Freer, H. Horiuchi, Y. Kanada-En'yo, D. Lee, and U.-G. Meißner, arXiv:1705.06192 (2017).
[18] P. Navratil, S. Quaglioni, Phys. Rev. C **83**, 044609 (2011).
[19] T. Suhara, N. Itagaki, J. Cseh, and M. P?oszajczak, Phys. Rev. C **87**, 054334 (2013).
[20] Chang Xu, G. Röpke, P. Schuck, Zhongzhou Ren, Y. Funaki, H. Horiuchi, A. Tohsaki, T. Yamada, Bo Zhou, Phys. Rev. C **95**, 061306(R) (2017).

[21] A. Tumino, R. Spartá, C. Spitaleri, A.M. Mukhamedzhanov, S. Typel, R. Pizzone, E. Tognelli, V. Burian, V. Kroha, Z. Hons et al., The Astrophysical Journal **96**, 785 (2014).

[22] M. Aliotta, F. Raiola, G. Gyrky, A. Formicola, R. Bonetti, C. Broggini, L. Campajola, P. Corvisiero, H. Costantini, A. D'Onofrio et al., Nucl. Phys. A **690**, 790 (2001).

[23] S. Engstler, G. Raiman, C.Angulo, U.Greife, C.Rolfs, U.Schroeder, E. Somoraj, B.Kirch, K. Langanke, Z. Phys. A **187**, 471 (1992).

[24] A. Musumarra, R.G. Pizzone, S. Blagus, M. Bogovac, P. Figuera, M. Lattuada, M. Milin, D. Miljanic, M.G. Pellegriti, D. Rendic, C. Rolfs, N. Soic, C. Spitaleri, S. Typel, H.H. Wolter, M. Zadro, Phys. Rev. C **64**, 068801 (2001).

[25] L. Lamia, C. Spitaleri, R.G. Pizzone, E. Tognelli, A. Tumino, S. Degl'Innocenti, P. G.Prada Moroni, M. La Cognata, L. Pappalardo, M. L.Sergi, Ap. Journal **768**, 65 (2013).

[26] J. Cruz, H. Luis, M. Fonseca, Z. Fulop, G. Gyurky, F. Raiola, M. Aliotta, K.U. Kettner, A.P. Jesus, J.P. Ribeiro, F.C. Barker, C. Rolfs, J. Phys. G, Nucl. Part. Phys. **35**, 014004 (2008).

[27] L. Lamia, C. Spitaleri, M. La Cognata, R.G. Pizzone, A.&A. **541**, A **158** (2012).

[28] D. Zahnow, C. Rolfs, S. Schmidt, H.P. Trautvetter, Z. Phys. A **359**, 211 (1997).

[29] Q.G. Wen, C.B. Li, S.H. Zhou, Q.Y. Meng, J. Zhou, X.M. Li, S.Y. Hu, Y.Y. Fu, C. Spitaleri, A. Tumino et al., Phys. Rev. C **78**, 035805 (2008)

[30] C. Angulo, S. Engstler, G. Raiman, C.Rolfs, W.H. Schhulte, E. Somorjai, Z. Phys. A **231**, 345 (1993).

[31] C. Spitaleri, L. Lamia, S. M.R.Puglia, S. Romano, M. La Cognata, V. Crucillá, R. G. Pizzoneand G. G. Rapisarda, M. L. Sergi, M. Gimenez Del Santo, N. Carlin et al., Phys. Rev. C **90**, 035801 (2014).

[32] L. Lamia, C. Spitaleri, V. Burjan, N. Carlin, S. Cherubini, V. Crucillá, M. Gameiro Munhoz, M. Gimenez Del Santo, M. Gulino, Z. Hons et al., Journ. Phys. G **39**, 015106 (2012).

EPJ Web of Conferences **165**, 02003 (2017) DOI: 10.1051/epjconf/201716502003
NPA8 2017

Neutron-captures in Low Mass Stars and the Early Solar System Record of Short-lived Radioactivities

Maurizio Busso[1,2,⋆], *Diego* Vescovi[1,2], *Oscar* Trippella[1,2], *Sara* Palmerini[1,2], *Sergio* Cristallo[3,2], and *Luciano* Piersanti[3,2]

[1] *Department of Physics and Geology, Via A. Pascoli, 06123 Perugia, Italy*
[2] *I.N.F.N., Section of Perugia, Via A. Pascoli, 06123 Perugia, Italy*
[3] *I.N.A.F., Observatory of Abruzzo, Teramo, Italy*

Abstract. Noticeable improvements were recently introduced in the modelling of n-capture nucleosynthesis in the advanced evolutionary stages of giant stars (Asymptotic Giant Branch, or AGB, stars). Two such improvements are closely linked together and concern the introduction of non-parameterized, physical models for extended mixing processes and the adoption of accurate reaction rates for H- and He-burning reactions, including the one for the main neutron source $^{13}C(\alpha,n)^{16}O$. These improvements profited of a longstanding collaboration between stellar physicists and C. Spitaleri's team and of his seminal work both as a leader in the Nuclear Astrophysics scenario and as a talent-scout in the recruitment of young researchers in the field. We present an example of the innovative results that can be obtained thanks to the novelties introduced, by estimating the contributions from a nearby AGB star to the synthesis of short-lived ($t_{1/2} \leq 10$ Myr) radioactive nuclei which were alive in early Solar System condensates. We find that the scenario indicating an AGB star as the source of such radioactivities, discussed for many years by researchers in this field, appears now to be no longer viable, when the mentioned improvements of AGB models and nuclear parameters are considered.

1 Introduction

Computations of slow neutron capture nucleosynthesis (the so-called *s*-processing) in the final evolutionary phases of red giant stars (the AGB stages) were for many years based on parameterized approaches. In particular, several years ago it was recognized [1–3] that the main neutron source had to be identified with the $^{13}C(\alpha,n)^{16}O$ reaction, starting from an abundance of ^{13}C necessarily much larger than the equilibrium concentration left for this nucleus by the occurrence of H-burning in a shell. Fresh ^{13}C had to be synthesized locally; its abundance was usually varied parametrically [3, 4] in order to derive constraints on its production from observations. This last was suggested to occur during the recurrent inward extensions of the envelope, known collectively under the name of *Third Dredge-up* or TDU. Even in absence of a model, it was assumed that at TDU deep mixing phenomena could inject protons into the He-rich layer in concentrations suitable to interact with the abundant ^{12}C, producing the required ^{13}C.

⋆e-mail: maurizio.busso@pg.infn.it

Today we deal with a scenario noticeably improved with respect to those early attempts. The upgrades would not have been possible, if not for the longstanding collaboration in Nuclear Astrophysics issues between research groups studying stellar physics and the team founded and headed by Claudio Spitaleri, who provided for many years a wealth of new accurate estimates for crucial reaction rates obtained by the ASFIN collaboration through the indirect experimental method known as *Trojan Horse*. He also helped in various ways in the recruitment of bright young researchers in the field. In this last role he has become known worldwide especilly for his Santa Tecla school, one of the most successful and rewarding initiatives in the whole Nuclear Astrophysics panorama.

Among the rates which were provided by Claudio and his collaborators, and that are now adopted in our nucleosynthesis computations, we remind first of all the one for (α,n) captures on ^{13}C, i.e. the neutron source itself [5, 6]. Moreover, of high relevance are the cross sections for p-induced reactions on intermediate-mass isotopes, which must be considered as they hamper the possibility of those nuclei to capture neutrons or α particles. Among them we remind (p,α) reactions on ^{15}N [7], on ^{17}O, ^{18}O [8, 9] and on ^{19}F [10, 11]. More recently, ASFIN made available also the rate for (α,p) captures on ^{19}F, occurring during the thermal instabilities of the He-shell in direct competition with both (α,γ) and (α,n) captures on ^{18}O and ^{22}Ne [12]. Finally, neutron captures have started to become available from the collaboration, see e.g. the case of ^{17}O(n,α)^{14}C [13], which is important for the neutron absorption from the daugther nucleus ^{14}N.

Another crucial ingredient, as important as the availability of high-quality nuclear physics inputs, was the formulation of physical models for the non-convective mixing processes affecting the inner border of the convective envelope in AGB stars.

Recently, models for deep mixing phenomena have been explored by various groups in the aim of overcoming the uncertain parameterizations and provide a more quantitative knowledge of AGB nucleosynthesis. Among such attempts, we recall the work by [14], exploring the effects of gravity waves and Kelvin-Helmoltz instabilities. This work revisited previous suggestions by [15, 16] and others. Another approach was pursued by [17, 18], on the basis of revisions of the physics at the convective border; similar attempts were also made by [19]. Concerning the study of low-metallicity AGB stars and the peculiar coupling of proton injection and nucleosynthesis occurring in them, see also [20].

A rather different paradigm was discussed by [21, 22] and by [23]. In it, an extended penetration of protons into the He-rich layers is induced as a byproduct of Magneto-hydrodynamical (MHD) effects, namely the buoyancy of magnetized bubbles, formed deep below the TDU inner border and forced to emerge into the envelope by their extra pressure provided by the magnetic field. This buoyancy pushes down protons for mass conservation: see details of the original MHD simulations in [24, 25] and, especially, [26].

Although it is still premature to judge which one of the above-mentioned suggestions will prove to be more promising, they all point to the same direction, i.e. the need of overcoming the era of free parameterizations, passing to more physical types of modelling. Then the complexity of stellar plasma phenomena makes it probable that more than one mechanism be at play in AGB stars.

In this contribution we shall in particular consider the approach by [22] that was already shown to be able to produce extended ^{13}C reservoirs and to induce neutron captures capable of reproducing not only the solar-system distribution of *s*-elements, but also the more stringent constraints [27–29] that are posed by the isotopic composition of presolar SiC grains: see e.g. [30]. As an example of the further results that can be obtained in this scenario, we consider here the possible contributions of an intermediate-mass star to the production of the short-lived ($\tau \leq 20$ Myr) radioactivities (SLR) that were found to be alive in the first condensates of the Early Solar System (ESS). The complementary mixing scenario suggested by [17, 18] will be considered in a forthcoming paper.

For computing the detailed nucleosynthesis pattern produced by an AGB star we had to adopt an extended set of cross sections for $p-$ and α-capture reactions on light and intermediate nuclei; this includes many more processes, in addition to those mentioned above. They were taken from the compilation by [31], having again Claudio Spitaleri as a coauthor. Concerning the neutron captures induced by the (α,n) neutron sources active in the He-layers of an AGB star, we followed the recommendations by [32] and [33]. Initial abundances were chosen to be solar, and were adopted from [34].

2 An Attempt to Explain ESS Radioactivites with an IMS Star.

Starting from the pioneering work by [35], suggesting the presence of live ^{129}I in pristine Solar System materials, the inventory of radioactive nuclei demonstrated to be alive in the ESS grew over the years considerably. Recent discussions of the present-day ideas on the early concentrations of those nuclei were presented by [36] and by [37]. The emerging situation is summarized in Table 1 (see the quoted papers for the original experimental works on which the Table is based).

Table 1. Short-Lived Nuclei in the ESS

Rad.	Ref.	Mean Life (Myr)	$(N_P/N_I)_{ESS}$ *at* $t = 0$
^{10}Be	^{9}Be	2.0	$\sim 5 \cdot 10^{-4}$
^{26}Al	^{27}Al	1.03	$(5-5.2) \cdot 10^{-5}$
^{36}Cl	^{35}Cl	0.43	10^{-4}
^{41}Ca	^{40}Ca	0.15	$\geq 1.5 \cdot 10^{-8}$
^{53}Mn	^{55}Mn	5.3	$6.7 \cdot 10^{-5}$
^{60}Fe	^{56}Fe	3.75	$(2-4) \times 10^{-7}$
^{107}Pd	^{108}Pd	9.4	$5.9(\pm 2.2) \cdot 10^{-5}$
^{129}I	^{127}I	23	$1.0 \cdot 10^{-4}$
^{135}Cs	^{133}Cs	3,3	$(1-2) \cdot 10^{-4}$
^{146}Sm	^{144}Sm	148	$1.0 \cdot 10^{-2}$
^{182}Hf	^{180}Hf	12.8	$9.7 \cdot 10^{-5}$
^{205}Pb	^{204}Pb	22	$(1-2) \cdot 10^{-4}$
^{244}Pu	^{232}Th	115	$3 \cdot 10^{-3}$
^{247}Cm	^{232}Th	23	$(1-2.4) \cdot 10^{-4}$

We adopted the model of a 5 $M_\odot$ star of solar metallicity from the calculations of full stellar models performed with the Schwarzschild criterion for convection, as part of the input material for the review by [38]. The star has a lifetime slightly longer than 100 Myr and was computed with a Reimers-like mass loss, using the value $\eta = 3$ for the free parameter of the formula. Figure 1 illustrates a sketch of the evolution in time of the stellar structure; there the C-O core, the He-rich buffer and the overlying convective envelope with its recurrent inward extensions (TDU) are shown. The H-rich buffer below the convective envelope is barely appreciable as its mass is smaller than the one of the He-rich zone by more than a factor of 10.

Inspection of the physical structure of the model reveals that conditions suitable for obtaining the analytical solution of MHD equations discussed by [26] exist in two thin layers sited, respectively, above the H-burning shell and below the inner border of convection at TDU. These conditions include a density that drops rapidly as a negative power-law of the radius, a temperature that decreases more

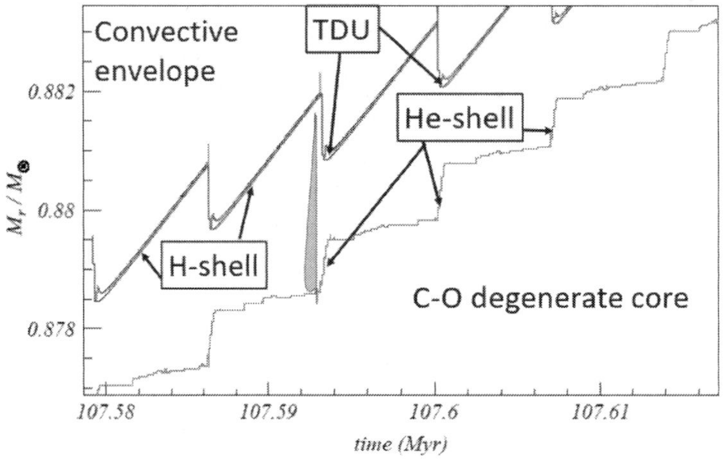

Figure 1. The basic structure of our model star during the thermally pulsing AGB phase around the 5th thermal pulse (shaded in grey). Note how thin is the H-rich layer below the convective envelope.

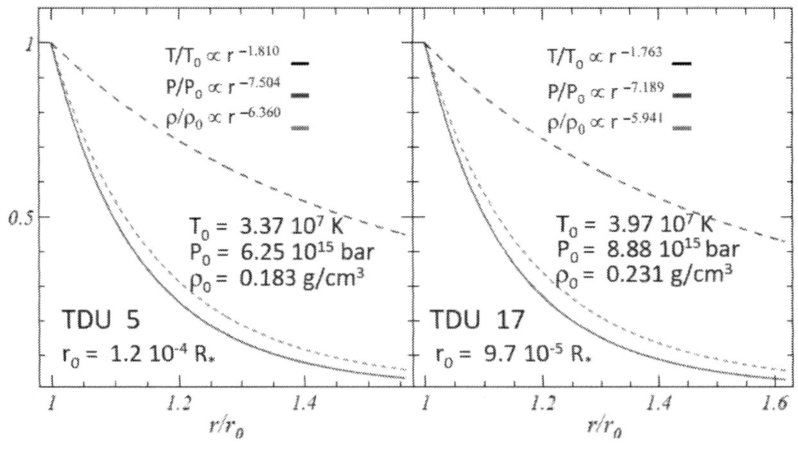

Figure 2. The physical conditions, in terms of temperature, density and pressure, of the layers in the He-rich zone where a quasi-ideal MHD condition holds, so that one expects that a circulation of material occurs for magnetic buoyancy, inducing the penetration of protons from the envelope and the formation of a ^{13}C pocket. The mass of such a pocket is smaller than for a star of 1.5 $M_{\odot}$ by a factor from 28 to 35, depending on the pulse number.

rapidly than $1/r$, a small kinematic viscosity and a large value for the magnetic Prandtl number. These conditions imply that magnetic buoyancy easily induces mixing there. Above the H-burning shell this yields the synthesis of some ^{26}Al, but this production is not very efficient, due to the thin radiative H-rich layer. In the He-zone, a ^{13}C pocket is formed. This last is established over a very small mass region (smaller by some 30 times than in low mass stars). Nevertheless, some neutron captures do occur in such a pocket, adding their effects to the *s*-processing induced by the ^{22}Ne source [2] in the convective region before a TDU episode. We therefore find that a condition like the one advocated by [39], in which an AGB star was assumed to host extramixing phenomena in the radiative H-rich zone, without the formation of a ^{13}C pocket, cannot occur.

This fact is fatal for the possibility of an AGB star (of any mass) to explain the ESS radioactivities. Indeed, ^{26}Al is produced only moderately by deep mixing processes and any ^{13}C burning in the He-layers invariably leads to a ratio ^{107}Pd/^{26}Al by far too high than required to account for the extinct radioactivities. Furthermore, the limited number of neutrons available from ^{13}C burning strongly favors Pd and Cs over Hf and Pb; also the production ^{60}Fe is excessive and a global solution is impossible.

An example of such finding is shown in Table 2, where we applied, to the envelope composition at the end of the evolution of our model star, the procedure described in [36, 37, 39] to infer the concentrations of some radioctive nuclei in a forming star close in time and space to the mass losing 5 $M_\odot$ AGB star we modelled. As the Table shows, if we try to fit Al and Ca all the other nuclei are not explained.

TABLE 2.
Our Predictions for Short Lived Nuclei in the ESS.

	5.0 $M_\odot$,	$Z = Z_\odot$ $f_0 = 2.56 \times 10^{-2}$	Measured or extrapolated
Parent *P*	Index *I*	$(N_P/N_I)_{\Delta_1}$ $\Delta_1 = 0.65$ Myr	at $t = \Delta_1$
^{26}Al	^{27}Al	$5.05 \cdot 10^{-5}$	$(5 - 5.2) \cdot 10^{-5}$
^{41}Ca	^{40}Ca	$1.08 \cdot 10^{-8}$	$\geq 1.5 \cdot 10^{-8}$
^{60}Fe	^{56}Fe	$1.15 \cdot 10^{-5}$	$(2 - 4) \cdot 10^{-7}$
^{107}Pd	^{108}Pd	$8.1 \cdot 10^{-4}$	$5.9(\pm 2.2) \cdot 10^{-5}$
^{135}Cs	^{133}Cs	$7.6 \cdot 10^{-4}$	$(1 - 2) \cdot 10^{-4}$
^{182}Hf	^{180}Hf	$1.24 \cdot 10^{-5}$	$9.7 \cdot 10^{-5}$
^{205}Pb	^{204}Pb	$1.89 \cdot 10^{-6}$	$(1 - 2) \cdot 10^{-4}$

We remind that recently [36] failed to explain the ESS radioactivities, even considering more massive AGB stars, producing ^{26}Al from Hot Botton Burning. If those results and our ones will be confirmed, then this would put an end to the idea that a chance encounter with an AGB star might be at the origin of the isotopic anomalies of pristine meteorites induced by the decay of short lived radioactivities. If this were the case, then the most promising alternatives might reside either in the sequential pollution of the parent molecular cloud by various massive stars, as in the model by [40] or in some revision of a single SN contributions, e.g. in the new scenarios of rotating massive stars [41].

References

[1] Iben, I., Jr. and Renzini, A. 1983, ARA&A 21, 271
[2] Busso, M., Gallino, R., Wasserburg, G.J. 2001, ARAA 37, 239
[3] Gallino, R., Arlandini, C., Busso, M., et al. 1998, ApJ 497, 388
[4] Käppeler, F., Gallino, R., Bisterzo, S. et al. 2011, Rev. Mod. Phys. 83, 157
[5] La Cognata, M., Spitaleri, C., Trippella, O., et al. 2013, ApJ 777, 143

[6] Trippella, O., La Cognata, M. 2017, ApJ 845, 191
[7] La Cognata, M., Goldberg, V.Z., Mukhamedzhanov, A.M. et al. 2009, Phys. Rev. C 80, 12801
[8] La Cognata, M., Spitaleri, C., Mukhamedzhanov, A.M, et al. 2010 ApJ 708, 796
[9] Palmerini, S., Sergi, M.L., La Cognata, M. et al. 2013, ApJ 764, 128
[10] La Cognata, M., Mukhamedzhanov, A.M., Spitaleri, C., et al. 2011, ApJ 739, L54
[11] Indelicato, I., La Cognata, M., Spitaleri, C., et al. 2017, ApJ 845, 191
[12] Pizzone, R.G., D'Agata, G., La Cognata, M., et al. 2017, ApJ 836, 57
[13] Guardo, G., Spitaleri, C., Lamia, L. et al. 2017, Phys. Rev. C. 95, 25807
[14] Battino, U., Pignatari, M., Ritter, C., et al. 2016, ApJ 827, 30
[15] Denissenkov, P.A., and Tout, C.A. 2003, MNRAS 340, 322
[16] Talon, S. 2008, in EAS Publication Series vol 32, ed. C. Chabonnel and J.-P. Zahn, p. 81
[17] Cristallo, S., Piersanti, L., Straniero, O., et al. 2011, ApJS 197, 17
[18] Cristallo, S., Straniero, O., Piersanti, L., Gobrecht, D. 2015 ApJS 219, 40
[19] Pignatari,. M., Herwig, F., Hirschi, R. et al. 2016, ApJS 225, 24
[20] Stancliffe, R.J., Dearborn, D.S.P., Lattanzio, J.C., et al. 2011, ApJ 742, 2
[21] Trippella, O., Busso, M., Maiorca, E., et al. 2014, ApJ 787, 41
[22] Trippella, O., Busso, M., Palmerini, S., et al. 2016, ApJ 818, 125
[23] Palmerini, S., Trippella, O., and Busso, M., 2017a, MNRAS 467, 1193
[24] Busso, M., Wasserburg, G.J., Nollett, K.M., and Calandra, A. 2007, ApJ 671, 802
[25] Nordhaus, J., Busso, M., Wasserburg, G.J., Blackman, E.G., and Palmerini, S. 2008, ApJ 684, L29
[26] Nucci, M.C., and Busso, M. ApJ 2014, 787, 141
[27] Liu, N., Gallino, R., Bisterzo, S., et al. 2014a, ApJ 786, 66
[28] Liu, N., Gallino, R., Bisterzo, S., et al. 2014b, ApJ 788, 163
[29] Liu, N., Savina, M.R., Gallino, R. et al. 2015, ApJ 803, 12
[30] Palmerini, S., Trippella, O., Busso, M., Vescovi, D., et al. 2017b, GCA in press.
[31] Adelberger, E. G., García, A., Robertson, R. G., et al. 2011, Rev. Mod. Phys. 83, 195.
[32] Dillmann, I., Heil, M., Käppeler, F., et al., 2006, AIP Conference Proceedings 819, 123
[33] Dillmann, I. 2014, in *Proceedings of the XIII Symposium Nuclei in the Cosmos*, http://pos.sissa.it/cgi-bin/reader/conf.cgi?confid=204, id.57
[34] Lodders, K. and Palme, H. 2009, MPSS, 72, 5154
[35] Reynolds, J. H. 1960, Phys. Rev. Lett. 4, 351.
[36] Wasserburg, G. J., Karakas, A. I, and Lugaro, M. 2017, ApJ 836, 126
[37] Busso, M. 2018, in *Astronomy with Radioactivities* (Springer: in press)
[38] Straniero, O., Domínguez, I., Cristallo, S., Gallino, R. 2003, PASA 20, 389
[39] Wasserburg, G. J., Busso, M., Gallino, R., Nollett, K. M. 2006, Nucl. Phys. A, 777, 5
[40] Gounelle, M. and Meynet, G. 2012, A&A 545, 4
[41] Limongi, M, & Chieffi, A. 2015, in Chemical and dynamical evolution of the Milky Way and the Local Group, Proc. Sedem Conf. id 15

EPJ Web of Conferences **165**, 02004 (2017) DOI: 10.1051/epjconf/201716502004
NPA8 2017

α-cluster resonances in light nuclei

Vladilen Goldberg[1,⋆], *Alexander* Volya[2], *Aliya* Nurmukhanbetova[3], *Dosbol* Nauruzbayev[3,4], and *Grigoriy* Rogachev[1]

[1]*Cyclotron Institute, Texas A&M University, College Station, TX 77843-3366, USA*
[2]*Department of Physics, Florida State University, Tallahassee, Florida 32306, USA*
[3]*National Laboratory Astana, Nazarbayev University, Astana, 010000, Kazakhstan*
[4]*Saint Petersburg State University, Saint Petersburg, 199034, Russia*

Abstract. Thick target inverse kinematics technique was combined with Time of Flight method to study resonance reactions induced by heavy ions at low energy, to minimize background and to identify various possible nuclear processes in extended target. The ^{17}O, ^{20}Ne spectrum, the cluster and nucleon spectroscopic factors were calculated using cluster-nucleon configuration interaction model.

1 Introduction

The phenomenon of alpha clusterization is well known in light 4N nuclei (^{8}Be, ^{12}C, ^{16}O...). In particular, it remarkably manifests itself as twin bands of an alternate parity with large reduced α particle widths. Because of the importance of this phenomenon in nuclear structure and due to large abundance of helium in the Universe, the properties of α-cluster states are also an important factor for the understanding of nuclear processes in stars. The astrophysical important reactions proceed often trough the states which are very close to the α particle decay threshold or even through subthreshold states. The small cross sections for corresponding energies well below the Coulomb barrier cannot be measured in laboratories. To calculate these cross sections one needs to know the "interaction" of the cluster states with the states around because strong α-cluster states can provide α width to the states that are closer to the region of astrophysical interest through configuration mixing [1]. The problems of a description of α-cluster states using models based on the shell model conception are well discussed (see [2] and references there). Twenty years ago C. Spitalery just began to develop the ideas of the Trojan Horse Method (THM). He also actively investigated new experimental approaches to study nuclear structure; these appeared as a result of the technological developments in the acceleration of rare beams.

New insight in the problem of the relationship between the single particle and cluster degrees of freedom can be obtained through the experimental studies of the α-cluster states in N≠Z nuclei. In N≠Z nuclei, the nucleon decay threshold is usually below that for α particle (opposite to the 4N nuclei case), and the penetrability factors do not inhibit the nucleon decay from the states in question. Therefore, data on the decay properties of the α-cluster states in N≠Z nuclei might give insight in the relation between the single particle and cluster degrees of freedom. Moreover, the properties of the

⋆e-mail: goldberg@comp.tamu.edu

alpha cluster states in proton and neutron-rich nuclei can be compared using the isospin conservation in the reaction with mirror nuclei. However, any investigation of resonance interaction of a radioactive nucleus with helium seemed very difficult. Indeed, radioactive nuclei cannot be used as a target, and rare beams cannot be used to study a broad energy region with good energy resolution using small controlled changes of the energy as it was in conventional measurements. The first work to study alpha cluster states in N≠Z nuclei using ^{18}Ne rare beam (and a comparison with the ^{18}O data) was made using a novel Thick Target Inverse Kinematics (TTIK) method [3] in collaboration with Prof. Spitaleri's group [4]. It showed exciting and unexpected differences in the excitation functions of mirror reactions ^{18}O(α,α) and ^{18}Ne(α,α). This work is considered as a breakthrough in the new field.

In this report we discuss some details of the experimental work on resonance scattering at DC-60 cyclotron in Astana, Kazakhstan and the results of a comparison of the prediction cluster-nucleon configuration interaction model (CNCIM) [5] with the experimental data.

2 Experimental method and results

The DC-60 cyclotron [6] accelerates heavy ions up to the 1.9 MeV/A energy. The TTIK technique [3] is used in the measurements. The incoming ions are slowed down and stopped in a helium (or hydrogen) gas of the extended target. The light recoils, α particles and protons produced in a scattering event are detected. They hit a Si detector array located in forward hemisphere. Together with the amplitude signal, the Si detectors provided for a fast signal. This signal together with a "start" signal from RF of the cyclotron was used for the Time-of-Flight measurements. The combined E-TF

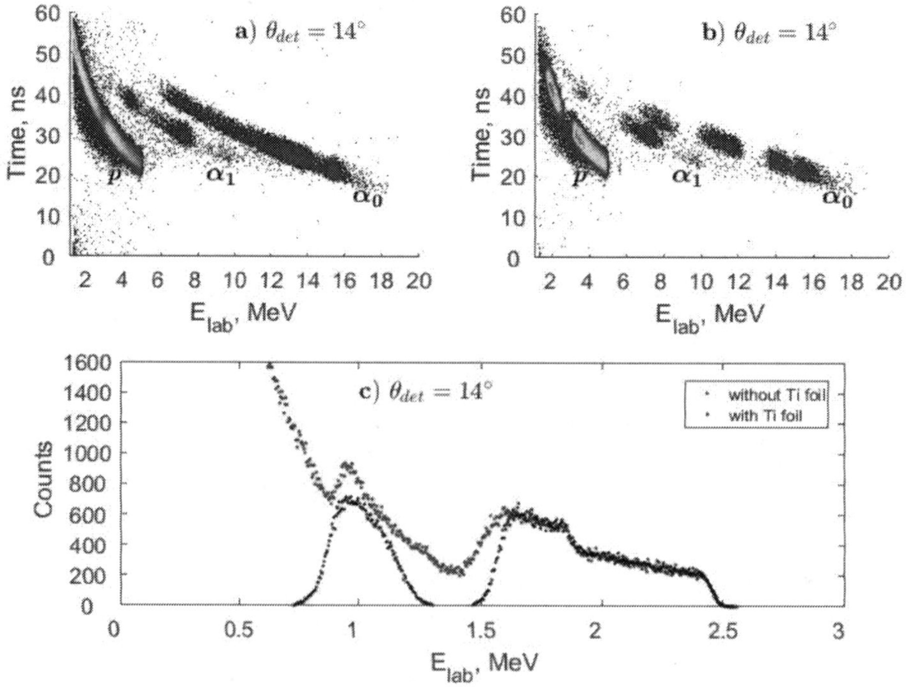

Figure 1. E-TF spectra of protons and α particles from the interaction of ^{15}N with hydrogen (a) without and (b) with the foil in the gas, and (c) energy spectra of protons from the ^{15}N(p,p) reaction.

measurements enable us to minimize the background and to separate elastic and inelastic scattering [7]. The TTIK approach provides for a continuous excitation function as a result of the slowing down of the beam. The best energy resolution of the method is reached at zero degrees (180° c.m.), and it is typically 25 keV c.m.

More details of the experimental technique are given in Ref.[7]. There is still a source of an uncertainty in TTIK measurements especially when resonance data are practically absent (the α+^{17}O case); it is beam energy loss in the gas at low energy [8]. To address this issue we developed a simple device to place a thin (2 um) Ti foil at different distances from the entrance window. The foil can be placed along the beam path or taken away during the experiment without cycling vacuum.

Fig.1.a,b demonstrates Energy vs Time (E-T) spectra for p, α_0 and α_1 events corresponding to ^{15}N(p,p), ^{15}N(p,α)^{12}C$_{g.s.}$ and ^{15}N(p,α)^{12}C* reactions without and with the foil on the path of the ^{15}N beam in hydrogen gas target. Fig.1.c represents a projection of the p, events on to E axis. A rapid change of the yield when the beam crosses the foil (Fig.1.b) enables us to check the energy loss of heavy ions in the gas. As a result, the 5.81 MeV energy loss of ^{15}N in hydrogen can be obtained with a precision of 120 keV, and the time when the beam reaches the foil can be obtained with precision of 0.45 ns. An analysis of the obtained results showed that the calculations [9] satisfactory fit the observations (see more details elsewhere).

Fig.2. shows a comparison of our data with the ^{17}O(p,p) results of Ref.[10]. As seen in Fig.2, our approach to the evaluation of the energy loss of heavy ions in the hydrogen gas results is in good agreement with the energy calibration obtained in the classical measurements. The cross section differences are probably related to the angular dependence of the cross sections.

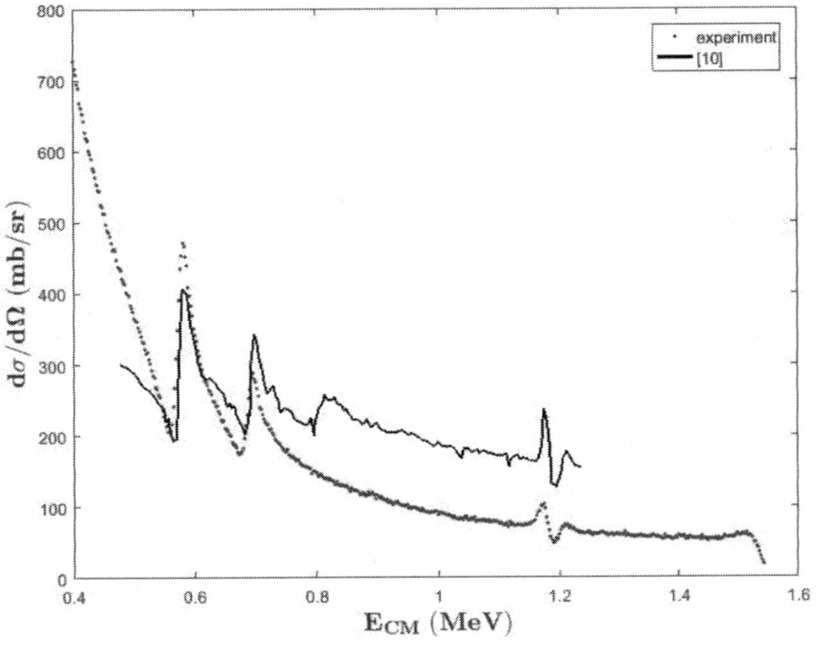

Figure 2. The ^{17}O(p,p) elastic scattering excitation function: the blue line is our experimental data at 180° c.m. and black line is data of Ref. [10] at 161° c.m.

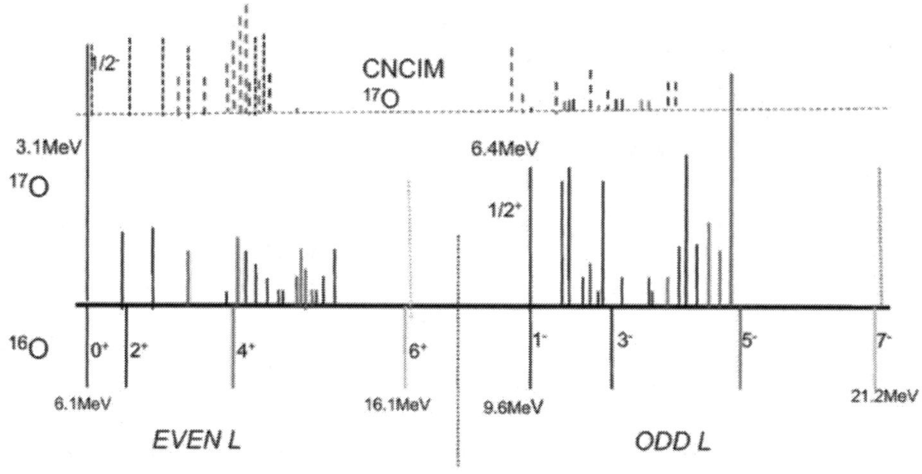

Figure 3. α-cluster structure in ^{17}O, comparison with predictions of CNCIM model.

The experimental elastic scattering excitation functions were analyzed using multilevel multichannel R matrix code [11]. We have obtained data on alpha and nucleon widths of the resonances. Below we report results on the alpha particle decay properties of the nuclear states. While it was found that all strong α-cluster states have nucleon widths, these will be analyzed elsewhere. Fig.3 shows the α-cluster Spectroscopic Factors (SF) for the ^{17}O states obtained as a result of the excitation functions analysis [12] in comparison with the predictions CNCIM model. The length of the sticks is proportional to the value of SF.

We characterize the α-cluster properties of the states above the α-particle decay threshold by SF $= \gamma_\alpha = \Gamma_{\alpha\,\mathrm{exp}}/\Gamma_{\alpha\,\mathrm{cal}}$, where $\Gamma_{\alpha\,\mathrm{cal}}$ is the α-particle width calculated in the α-core potential. To calibrate the potential, the SFs were first calculated for the well-known α-cluster states in ^{16}O. The Woods-Saxon potential was used to calculate the limit ($\Gamma_{\alpha\,\mathrm{cal}}$) for the width of the single-particle states. First we fit the widths of known α-cluster states, 1^- and 4^+, in ^{16}O so that $\gamma_\alpha = \Gamma_{\alpha\,\mathrm{exp}}/\Gamma_{\alpha\,\mathrm{cal}} \sim 1.0$. The real part of the potential was changed to fit the binding energy of the states. The radius of the potential was chosen to be R = $r_0 \times 12^{1/3}$; the Coulomb potential was taken into account as a charge sphere potential with R_{Coul} = R. We made our first calculations (1) with r_0 = 1.31 fm and the diffuseness a = 0.65 fm, then we set r_0 = 1.23 fm (2), and we finally performed a third set of calculations (3) with r_0 = 1.23 fm and a = 0.6 fm. The results are summarized in Table I. The γ_α calculations for the states in investigated nuclei (^{17}O and ^{20}Ne) are made with the final (third) set of parameters, because these parameters provide for a better description of the widths of the known ^{16}O α-cluster states.

As it seen in Fig.3, the α-cluster strength is distributed over many states in ^{17}O, quite different from the ^{16}O case (one could expect that the number of states in ^{17}O would be just two times higher

Table 1. α + ^{12}C levels in ^{16}O

^{16}O level	$\Gamma_{\alpha\,\mathrm{exp}}$ keV [13]	$\gamma_\alpha(1)$; -V_0 MeV	$\gamma_\alpha(2)$; -V_0 MeV	$\gamma_\alpha(3)$; -V_0 MeV
1^-, 9.58 MeV	420 ± 20	0.70; 138.2	0.72; 150.0	0.84; 158.5
4^+, 10.36 MeV	26 ± 3	0.68; 125.3	0.88; 139,6	1.21; 143.2

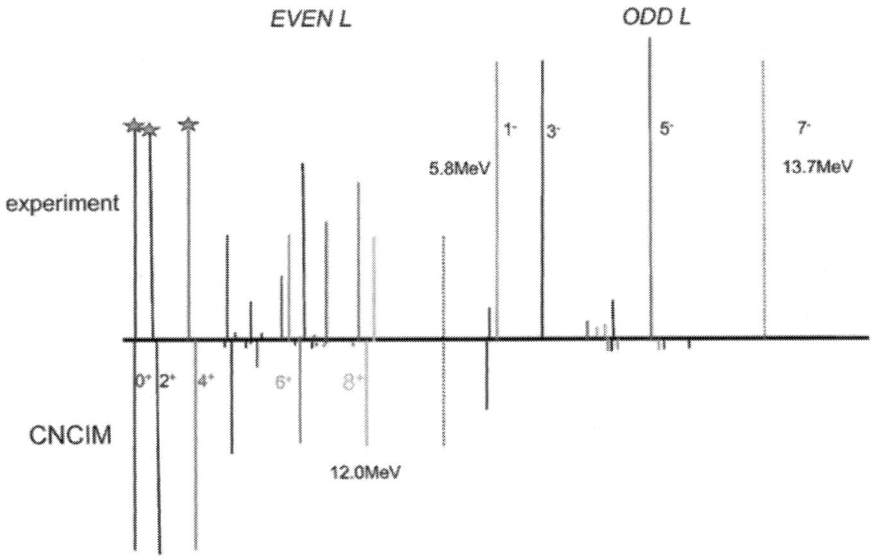

Figure 4. α-cluster structure in ^{20}Ne, comparison with predictions of CNCIM model.

than in ^{16}O). The comparison with CNCIM calculations shows that the theoretical model is able to generate strong α-cluster SF which are close to the known values for the strong states. It is true at relatively small excitation energy. At higher excitation energy, there is evident disagreement which cannot be corrected by inclusion of *f2p* shells.

Fig.4 shows a spectrum of the α-cluster states in ^{20}Ne in comparison with the prediction of the CNCIM model. The model describes well the properties of the famous α-cluster ground state band up to 8^+ level (experimental SFs for the states below the α particle decay threshold in ^{20}Ne obtained from the α-cluster transfer reactions). However, the model generates SF which are too small for the positive parity states with a larger number of nods of the cluster wave functions [14]. Similar drastic disagreement is evident for the levels belonging to the cluster band of negative parity.

3 Summary

1. TTIK approach greatly improves the possibilities to study resonance interaction of rare isotopes with helium and hydrogen at low energies.

2. A new development of shell model, CNCIM model, succeeded in the prediction of large SF observed experimentally for strong α-cluster states in light nuclei. An evident achievement of the model is a detailed description of the ground state band in ^{20}Ne. However, the model predictions for the states at higher excitation energy are in evident disagreement with the experimental data. The reasons of this disagreement are not evident.

3. We hope that the analysis of these widths will help to understand relationship between two degrees of freedom. There is a CNCIM characterization of ^{20}Ne at low energies by 7 MeV excitation energy allowing to predict reasonably well the excitation energy of the states and their cluster and single-particle properties. The disagreement between theoretically predicted and experimentally spectra is observed at broad resonances of high energies.

Acknowledgments

This work was supported by Ministry of Education and Science of the Republic of Kazakhstan (grant number #0115PK03029 "NU-Berkeley", 2014-2018; grant number #0115PK022465, 2015-2017). This material is also based upon work supported by the U.S. Department of Energy Office of Science, Office of Nuclear Physics under Grants No DE-FG02-93ER40773 and No. DE-SC0009883.

References

[1] M.L.Avila, G.V.Rogachev, V.Z.Goldberg, E.D.Johnson, K.W.Kemper, Yu.M.Tchuvil'sky, A.S.Volya Phys.Rev. C **90**, 024327 (2014)

[2] B. A. Brown and W. A. Richter, Physical Review C **74**,34315 (2006).

[3] K. Artemov, O.P. Belyanin, A.L. Vetoshkin, R. Wolski, M.S. Golovkov, V.Z. Goldberg, M. Madeja, V.V. Pankratov, I.N. Serikov, V.A. Timofeev, Sov. J. Nucl. Phys. **52**, 408 (1990)

[4] V.Z.Goldberg, G.V.Rogachev, W.H.Trzaska, J.J.Kolata, A.Andreyev, C.Angulo, M.J.G.Borge, S.Cherubini, G.Chubarian, G.Crowley, P.Van Duppen, M.Gorska, M.Gulino, M.Huyse, P.Jesinger, K.-M.Kallman, M.Lattuada, T.Lonnroth, M.Mutterer, R.Raabe, S.Romano, M.V.Rozhkov, B.B.Skorodumov, C.Spitaleri, O.Tengblad, A.TuminoPhys.Rev. C **69**, 024602 (2004)

[5] A. Volya and Y. M. Tchuvil'sky, Physical Review C **91**, 44319 (2015)

[6] B. Gikal, S. Dmitriev, P. Apel et al., Physics of Particles and Nuclei Letters 5(7) (2008) **642**

[7] A. K. Nurmukhanbetova, V. Z. Goldberg, D. K. Nauruzbayev, G. V. Rogachev, M. S. Golovkov, N. A. Mynbayev, S. Artemov, A. Karakhodjaev, K. Kuterbekov, A. Rakhymzhanov, Z. Berdibek, I. Ivanov, A. Tikhonov, V. I. Zherebchevsky, S. Y. Torilov, and R. E. Tribble, Nuclear Instruments and Methods in Physics Research Section A: Accelerators, Spectrometers, Detectors and Associated Equipment **847**, 125 (2017).

[8] Helmut Paul. Nuclear Stopping Power and It's on the determination of electronic stopping Power. AIP Conf. Proc. **1525**, 309 (2013);

[9] J. F. Ziegler and J. P. Biersack, in Treatise on Heavy-Ion Science: Volume 6: Astrophysics, Chemistry, and Condensed Matter, edited by D. A. Bromley (Springer US, Boston, MA, 1985), pp. 93–129.

[10] J. C. Sens, F. Rietsch, A. Pape, and R. Armbruster, Nuclear Physics A **199**, 232 (1973).

[11] E. D. Johnson. The Cluster Structure of Oxygen Isotopes. The Florida State University, 2008.

[12] N.A.Mynbayev, A.K.Nurmukhanbetova, V.Z.Goldberg, M.S.Golovkov, G.V.Rogachev, V.N.Dzyubin, M.V.Koloberdin, I.A.Ivanov, R.E.Tribble J.Exper.Theo.Phys. **119**, 663 (2014); Zh.Eksp.Teor.Fiz. **146**, 754 (2014)

[13] D. R. Tilley, C. M. Cheves, J. H. Kelley, S. Raman, and H. R. Weller, Nuclear Physics A **636**, 249 (1998).

[14] D. K. Nauruzbayev, V. Z. Goldberg, A. K. Nurmukhanbetova, M. S. Golovkov, A. Volya, G. V. Rogachev, and R. E. Tribble, Physical Review C **96**, 14322 (2017).

EPJ Web of Conferences **165**, 02005 (2017) DOI: 10.1051/epjconf/201716502005
NPA8 2017

At the origins of the Trojan Horse Method

Marcello Lattuada

Dipartimento di Fisica ed Astronomia, Via S. Sofia 64, 95123 Catania, Italy
Istituto Nazionale di Fisica Nucleare- LNS, Via S. Sofia 62, 95123 Catania, Italy

Abstract. During the seventies and eighties a long experimental research program on the quasi-free reactions at low energy was carried out by a small group of nuclear physicists, where Claudio Spitaleri was one of the main protagonists. Nowadays, *a posteriori*, the results of these studies can be considered an essential step preparatory to the application of the Trojan Horse Method (THM) in Nuclear Astrophysics.

Everything began in the seventies, with the publication on Nuclear Physics A of a paper reporting the results of an experiment carried out by a group of Japanese researchers on the three alpha-particles produced in the ^{3}He + ^{9}Be reaction at incident energy of 4 MeV [1]. The coincidence detection of two of the outgoing alpha particles showed that, in addition to the usual sequential decay through states of ^{8}Be, another mechanism contributed to the spectra, giving place to a bump centered at the minimum momentum of the undetected alpha. Angular correlations and energy sharing spectra showed that the shape of this bump was consistent with a contribution of the ^{5}He(^{3}He, α)^{4}He virtual reaction occurring in the frame of a quasi-free process. In this process the undetected alpha cluster α_s acts as spectator of the process and keeps the momentum it owns at the time of the interaction, as schematized in fig. 1.

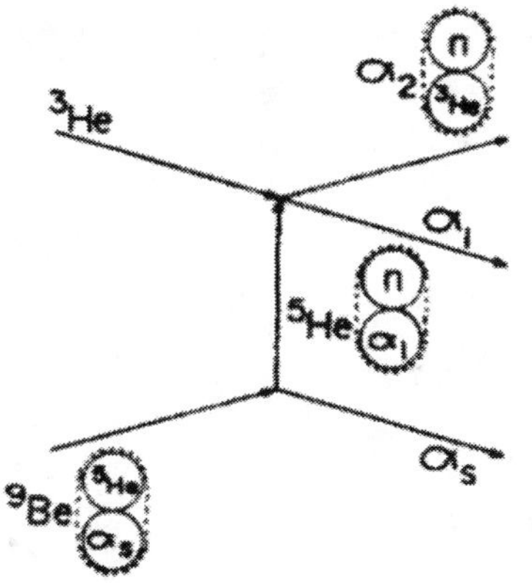

Fig. 1. Schematic picture of the ^{3}He + ^{9}Be quasi-free reaction mechanism (from [1]).

In this representation the reaction proceeds through two independent steps corresponding to two virtual processes: the virtual decay of one of the particles in the entrance channel (^{9}Be into ^{4}He and ^{5}He in this case) and the virtual reaction of one of the decay products (^{5}He) with the other participant (^{3}He). No one of the two processes conserves the energy separately but of course they conserve the energy on the whole.

The dominant s-wave motion of the ^4He and ^5He clusters in ^9Be produces a maximum in the $^9Be(^3He, \alpha\,\alpha)^4He$ spectra in correspondence of the minimum spectator momentum, according to the Impulse Approximation (IA) predictions, which for the triple differential cross section gives a simple formula:

$$\frac{d^3\sigma}{dE_1\, d\Omega_1\, d\Omega_2} \approx KF\, \mathbf{\Phi}^{\,2}(\boldsymbol{p}_s)\, \frac{d\sigma}{d\Omega} \qquad (1)$$

Here $\mathbf{\Phi}^2(\boldsymbol{p}_s)$ is the squared momentum distribution of the spectator particle, that in the Plane Wave IA (PWIA) is simply given by the Fourier transform of the cluster wave function, $d\sigma/d\Omega$ is the half-off-energy-shell cross section of the virtual two body reaction and KF is a factor proportional to the phase space. Actually the use of the IA requires a high momentum transfer and the Q-value of the three body reaction guarantees that this requirement was satisfied, indeed. On the other hand the use of the PWIA instead of the more realistic Distorted Wave IA (DWIA) approach was justified since it has been shown that in many cases the PWIA satisfactorily approximates the DWIA in the region close to the minimum spectator momentum. Of course the PWIA analysis cannot provide any information on the absolute value of the cross section.

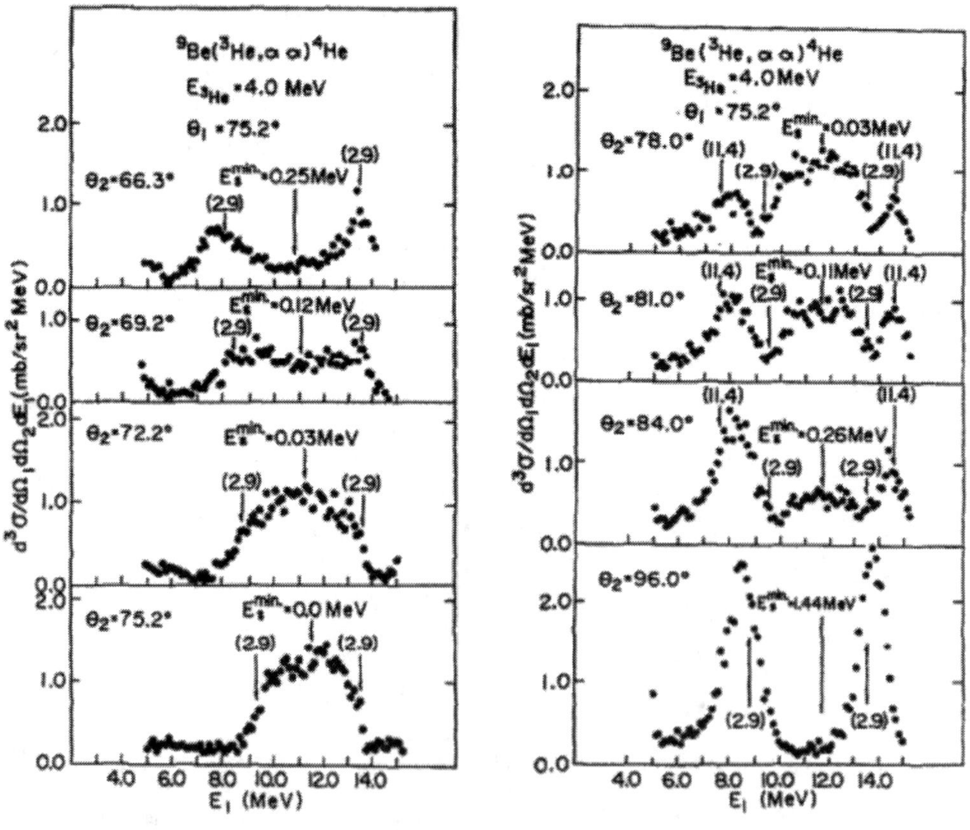

Fig. 2. Alpha-alpha coincidence spectra projected on the energy of one of the detected alpha-particles from the $^9Be(^3He, 2\alpha)^4He$ reaction for various angle pairs (from [1])

The question arose whether such contribution was present also at lower energy and/or in other similar systems. Actually the occurrence of a quasi-free scattering like (p, pα) or (α,

DOI: 10.1051/epjconf/201716502005

2α) in the interaction of light nuclei had already been found in earlier years for some other light systems even at energy of some tens of MeV, but on the other hand very few were the systems showing the contribution of quasi-free reactions at so low energy.
At that time a small single stage Van de Graaff, with a maximum voltage of about 3 MV, was operating at the Catania Physics Department. So the experiment was repeated with this accelerator at the sub-barrier energy of 2.8 MeV with a set of detectors covering the phase space regions where the minimum ^{4}He spectator momentum assumes different values, with the purpose of following the evolution of the differential cross section [2]. In this experiment and in experiments performed later at different incident energies [3] the appearance of a quasi-free reaction contribution well described by Eq. 1 was again confirmed and coherent shapes of the momentum distribution were extracted from all the coincidence spectra.
In addition the pole character of the mechanism was tested with a later experiment, applying the Treiman-Yang criterion to the ^{3}He+ ^{9}Be reaction [4]. The isotropy of the Treiman-Yang distribution, even if it is a necessary but not sufficient condition, represented a further support to the independence of the two poles.
According to a standard procedure, the quasi-free data were analyzed in two steps. First a phase space region where the relative energy and center of mass angles of the two interacting particles are appreciably constant was selected to deduce the experimental momentum distribution using the Eq.1. Then, dividing the three body cross section by the experimental momentum distribution and by the kinematical factor, it was possible to deduce, again from Eq.1, the virtual two body cross section as a function of angle and energy. Of course no comparison with the free cross section could be done in the case of the ^{5}He(^{3}He, α)^{4}He reaction, due to the unbound nature of ^{5}He.
Other light systems were experimentally studied in the following years (see for instance [5-8] and references therein), possessing in the entrance channel projectile or target with a marked ground state cluster structure, like ^{6}Li and ^{2}H and indication of a quasi-free mechanism was found in many cases. The case of the ^{6}Li(^{6}Li, 2α)^{4}He reaction [9] was particularly interesting since the α-d momentum distribution in ^{6}Li could be measured within the same experiment by selecting alternatively the events corresponding to the quasi-free mechanism with the spectator either in the target or in the projectile. The coincidence cross section, divided by the proper kinematical factor, for events corresponding to the two quasi-free mechanisms, showed similar resonant behavior peaked at the same energy as reported in Fig. 3.

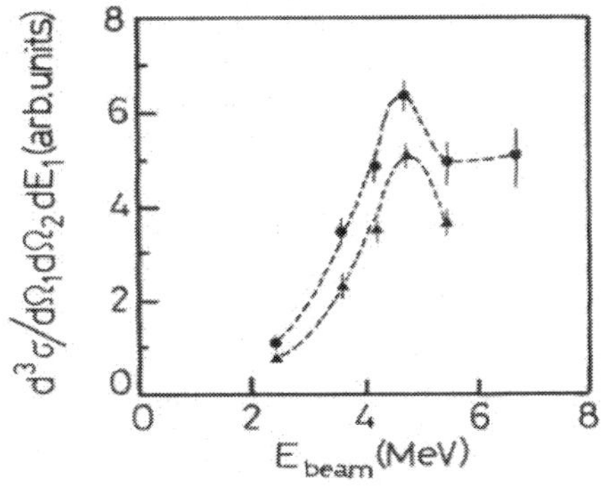

Fig. 3. Excitation function of the ^{6}Li(^{6}Li, 2α)^{4}He reactions at around zero of the spectator α–particle momentum, either in the target (points) or in the projectile (triangles).The cross sections are divided by the proper kinematical factor. Figure from [9].

But the most clear indication came from two experiments which allowed the extraction of the two body $^1H(^6Li, \alpha)^3He$ and $^1H(^7Li, \alpha)^4He$ cross sections down to sub-barrier energies, from the three body $^2H(^6Li, \alpha^3He)n$ and $^2H(^7Li, 2\alpha)n$ reactions respectively. The agreement of the THM excitation functions with the free reaction data was impressive, probably also due to the absence of Coulomb interaction between the spectator neutron and the participants. The resonances corresponding to excitation of known states of ^7Be (at 7.2 MeV) [10] and ^8Be (at 19.9 and 22.2 MeV) [11] in the ^1H $(^6Li, \alpha)^3He$ and $^1H(^7Li, \alpha)^4He$ reactions respectively were fairly reproduced by the THM data. The comparison between free and quasi-free data is shown in fig. 4 where the fits to the free reaction data are reported as continuous curves, whilst the points represent the virtual reaction cross sections.

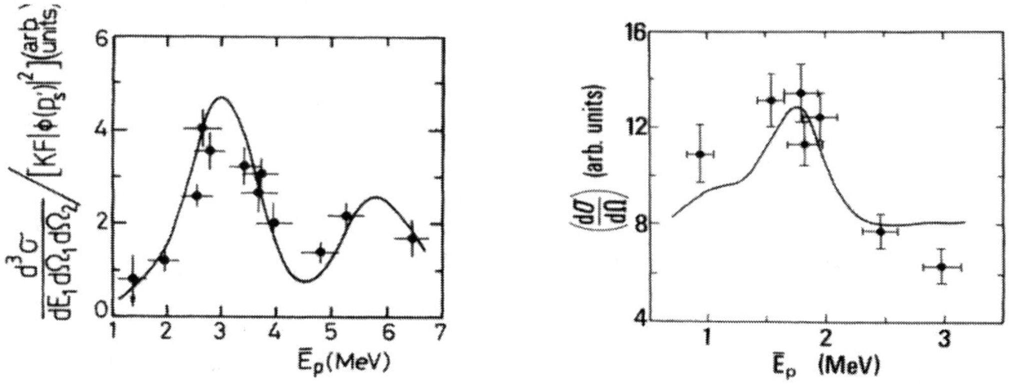

Fig. 4. Two body $^1H(^7Li, \alpha)^4He$ (left, from [10]) and $^1H(^6Li, \alpha)^3He$ (right, from [11]) excitation functions extracted from the $^2H(^7Li, 2\alpha)n$ and $^2H(^6Li, \alpha\ ^3He)n$ three body reactions respectively.

It was a clear, experimental confirmation that one can indirectly measure a two body reaction cross section even at very low interaction energy by choosing a suitable three body reaction at energy above the Coulomb barrier and selecting appropriate kinematical regions where the quasi-free contribution is favored. Thanks to the *high* projectile energy, the entrance channel relative energy can exceed the Coulomb barrier. Once the barrier has been overcome, the virtual reaction can occur even at very low relative energies, resulting from the energy balance where the deuteron binding energy and the n-p relative motion inside the deuteron are taken into account. Then the first important implication of the success of the method was the possibility of an indirect measurement of astrophysical reaction cross sections whose low interaction energy often makes it difficult, and sometimes prohibitive, a direct experimental approach. Moreover another interesting perspective on the side of nuclear physics is that the quasi-free mechanism can be regarded as a unique tool for the study of reactions with an unbound nucleus in the entrance channel.

The possibility of using the deuteron as virtual source of low energy protons or, more interesting, neutrons opened new opportunities not only in nuclear astrophysics but also in nuclear physics. For this reason and also for its low binding energy and its well-known momentum distribution, the deuteron can be considered as the ideal Trojan Horse nucleus.

Claudio Spitaleri firmly believed in the use of quasi-free reactions in nuclear astrophysics [12] and devoted his efforts to the involvement of a quite large number of researchers in international collaborations working on this topic. The result is that important improvements have been introduced in these years, from the point of view of the THM formalism as well as from the point of view of the use of more sophisticated experimental set-ups. Nowadays the THM has an important role as a tool for the indirect study of nuclear astrophysical processes, together with the Coulomb dissociation and the ANC method [13]. Around Claudio Spitaleri, a group of young researchers grew in Catania attracted by the

simplicity and wide perspectives of this powerful tool and by the enthusiasm for the studies of nuclear astrophysics transmitted by him.
It is difficult to mention the many collaborators and colleagues who were involved with Claudio Spitaleri in these studies, without forgetting anybody. So big is the net of collaborations developed on this subject. Nevertheless a special mention is deserved by some of them who, for different, and sometimes sad, reasons could not attend this NPA8 special session, but have contributed in different ways and times to the study of the quasi-free reaction mechanism and its relevance in the development of the THM.
Mimmo Vinciguerra must be remembered for bringing into the group his previous experience in the field of quasi-free scattering. Gerhard Baur gave an enlightening contribution with the introduction of the original concept of the Trojan Horse. Herman Wolter contributed with his competence to the development of the THM formalism. The continuous encouragement of Claus Rolfs along with his decisive push for the foundation of the Santa Tecla European Summer School on Experimental Nuclear Astrophysics cannot be forgotten. A deep gratitude has to be expressed to Djuro Miljanic for sharing his wide knowledge of the light nuclei interaction and for thirty-five years of collaboration and friendship.

References

1. J. Kasagi, T. Nakagawa, N. Sekine, T.Tohei, H. Ueno, Nucl. Phys. A **239**, 233 (1975)
2. N. Arena, D. Vinciguerra, F. Riggi, C. Spitaleri, Lett. N. Cimento **17**, 231 (1976)
3. M. Lattuada, F. Riggi, C. Spitaleri, D. Vinciguerra, D. Miljanic, M. Zadro, Yao Jinzhang, Nucl. Phys. A **458**, 493 (1986)
4. P.G. Fallica, M. Lattuada, F. Riggi, C. Spitaleri, C.M. Sutera, D. Vinciguerra, Phys. Rev. C **24**, 1394 (1981)
5. M. Lattuada, F. Riggi, S. Barbarino, C. Spitaleri, D. Vinciguerra, Lett. N. Cimento **20**, 193 (1977)
6. M. Lattuada, F. Riggi, C. Spitaleri, Vinciguerra, Lett. N. Cimento **33**, 433 (1982)
7. M. Lattuada, F. Riggi, C. Spitaleri, D. Vinciguerra, C.M. Sutera, Il N. Cimento A **71**, 429 (1982]
8. S. Blagus, C, Blyth, G. Calvi, O. Karban, M. Lattuada, D. Miljanic, F. Riggi. C. Spitaleri, Z. Phys. A – Atomic Nuclei **337**, 297 (1990)
9. M. Lattuada, F, Riggi, D. Vinciguerra, C. Spitaleri, D. Miljanic, Z. Phys. A – Atomic Nuclei **328**, 497 (1987)
10. M. Zadro, D. Miljanic, C. Spitaleri, G. Calvi, M. Lattuada, F. Riggi, Phys. Rev. C **40**, 181 (1989)
11. G. Calvi, M. Lattuada, D. Miljanic, F. Riggi, C. Spitaleri, D. Vinciguerra, Phys. Rev. C, **41**, 1848 (1990)
12. C. Spitaleri, Proc. Fifth Hadronic Physics Winter Seminar, Folgaria–TN, Italy (Ed. World Scientific, Singapore) (1990)
13. R. Tribble C. A. Bertulani, M. La Cognata, A. M. Mukhamedzhanov C. Spitaleri, Rep. Prog. Phys. **77**, 106901(2014)

Coulomb dissociation

another "Trojan Horse"

Tohru Motobayashi[1,★]

[1] *RIKEN Nishina Center, 2-1 Hirosawa, Wako, Saitama 351-0198, Japan*

Abstract. Coulomb dissociation is discussed in connection with the Trojan-Horse method and other quasi-free reactions.

1 ^{1}H(d,^{2}He)n reactions - a Quasi Free Reaction Process

The Trojan-Horse method relies on the "quasi-free reaction" mechanism involved in a direct reaction with three particles in the final state. Claudio Spitaleri and his group in Catania have developed this method to a level at which reliable astrophysical reaction rates are extracted. They often adopt a simplest approximation for the quasi-free process, PWIA (Plane Wave Impulse Approximation), with successful results [1].

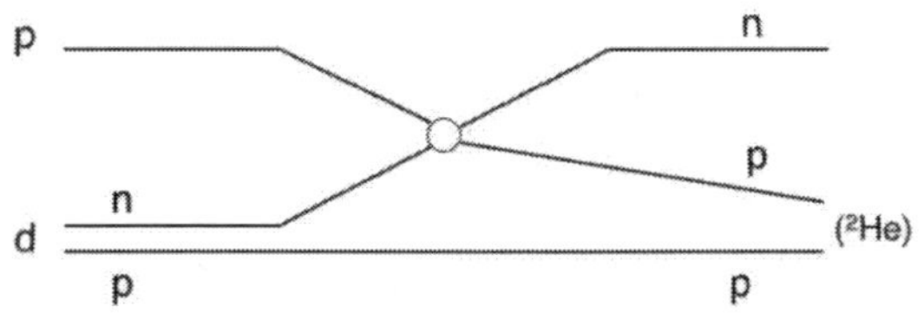

Figure 1. Diagram for the quasi-free process for the ^{1}H(d,^{2}He)n reaction. The three-body final state is dominated as a final-state interaction by the strong $^1S_0$ enhancement in the proton-proton interaction.

The ^{1}H(d,^{2}He)n reaction at an intermediate energy may also be dominated by the quasi-free reaction process, where the two-body process in question is the charge exchange scattering by the neutron in the projectile deuteron and the target proton (see Fig. 1). The symbol ^{2}He here denotes a system of two protons in their unbound singlet S state, with a strong enhancement at around 400 keV relative energy. The p-n charge-exchange could be viewed as a process exchanging a (virtual) pion, and thus the entire process is analogous to the Trojan-Horse process with the pion as the "soldier". The data taken at Saturne in late 1980s using 200 MeV polarized deuteron beams were analyzed with PWIA [2]. The tensor analyzing powers and even the absolute magnitudes of the differential cross sections have been successfully explained by assuming a n+p→p+n quasi-free process at small momentum transfers. In PWIA calculations, charge exchange nucleon-nucleon amplitudes are employed to predict

★e-mail: motobaya@riken.jp

spin-dependent observables [3]. That was our surprise, because nucleon rearrangement reactions in this energy regime are not always well described by such a simple treatment. The major purpose of this study was to establish the reaction mechanism, with its practical application to the measurements of the deuteron tensor polarization. It should be noted that the quasi-free assumption works well only at small momentum-transfers, in which charge exchange or pion-exchange dominates the n+p interaction. That is similar to the Trojan-Horse applications, where the kinematical conditions are selected to assure the quasi free mechanism.

2 Coulomb Dissociation

Many radiative proton-capture reactions play key roles in astrophysical hydrogen burning [4]. Their direct measurements are often affected by the small cross sections. Difficulty of preparing radioactive targets could be the serious concern for the cases involving unstable nuclei. The Coulomb dissociation provides an indirect method for determination of radiative capture reactions [5, 6]. In a Coulomb dissociation experiment, the residual nucleus A of the reaction B(x,γ)A bombards a high-Z target as lead, and it is excited in the Coulomb field to its unbound state that decays to the B+x channel. The Coulomb dissociation process is regarded as absorption of a virtual photon, A(γ,x)B, the inverse of the radiative capture of interest, and therefore the dissociation yield can be related to the relevant capture cross section.

Figure 2 schematically shows the Coulomb dissociation process. When a projectile nucleus A is passing the vicinity of a high Z nucleus, the projectile feels the presence of the electric field generated by the high Z nucleus. The time-dependent amplitude of the electric field (shown in the bottom-left of Fig. 2) can be decomposed in electromagnetic waves with different frequencies, and hence is equivalent to the exposition of nucleus A to photons with different energies. The photon spectrum is obtained by the Fourier transform of the impulse. With fast beams of $\approx$50 MeV/nucleon or higher, the maximum energy of the virtual photons can cover the range of excitation energy of the nucleus A in most of the cases of astrophysical interest [5].

Application of the Coulomb dissociation process to astrophysical radiative capture has been discussed by Baur, Bertulani and Rebel [5] for the first time, and some review articles are available [6, 7]. The photo-absorption is an inverse process of the radiative capture, their cross sections are related by the detailed balance as,

$$\sigma_{\mathrm{A}+\gamma\to\mathrm{B}+x} = \frac{(2j_{\mathrm{B}}+1)(2j_{\mathrm{x}}+1)}{2(2j_{\mathrm{A}}+1)}\frac{k^2}{k_\gamma^2}\sigma_{\mathrm{B}+x\to\mathrm{A}+\gamma}. \tag{1}$$

The spins of A, B and x are denoted by j's, respectively. The symbols k and k_γ denote the wave numbers in the B+x channel and A+γ channel, respectively. Due to the large phase space volume for the B+x channel, the photo-absorption cross section is enhanced. For example the wave-number ratio k^2/k_γ^2 in Eq. 1 is calculated to be about 1000 for the case of ^{7}Be(p,γ)^{8}B reaction. This enables experimental studies on explosive burning with beams of unstable nuclei or radioactive isotope (RI) beams, the intensity of which is expected to be low.

The first experimental result with RI beams was obtained at RIKEN on the ^{13}N(p,γ)^{14}O reaction, relevant to the hot CNO cycle nuclear burning [9]. This work was published in 1991, almost at the same time and on the same subject as the experiment at Leuvain la Neuve [10] using a radioactive ^{13}N beam for a direct measurement. The Coulomb dissociation experiment used fast RI beams of ^{14}O at 87.5 MeV/nucleon. The results of these two experiments agreed with each other. The higher statistics of the Coulomb dissociation experiment demonstrate the above mentioned advantage of the method.

The solar-neutrino production process ^{7}Be(p,γ)^{8}B has been studied by the Coulmob dissociation method in several laboratories [11–13]. In the first experiment at RIKEN, radioactive ^{8}B nuclei were

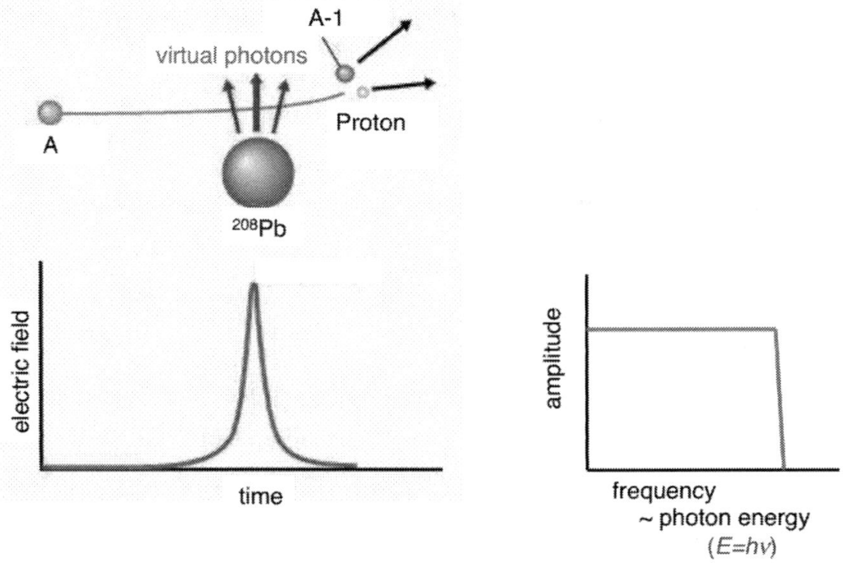

Figure 2. Schematic picture of the Coulomb dissociation. The left and right lower panels show the electric force felt by the projectile as a function of time and its Fourier transform, respectively. This figure is originally from ref. [8]

produced by the ^{12}C+^{9}Be fragmentation reaction with a 92 MeV/nucleon ^{12}C primary beam, then analyzed and transported to a ^{208}Pb secondary target. The energies and directions of the fragments were measured to obtain the p-^{7}Be relative energy event-by-event, with a ΔE-E plastic scintillator hodoscope [7, 11].

Figure 3 shows the astrophysical S_{17}-factors obtained in the experiments at RIKEN together with the results from several direct (p,γ) measurements. Our Coulomb dissociation data agree well with the results by Filippone *et al.* [19] and Vaughn *et al.* [17]. Direct data taken in 2000s by Junghans *et al.* [22] and Baby *et al.* [23] (not shown in the figure), are slightly larger than the ones shown by the theoretical curves, which are normalized to the second Colomb dissociation data. The comparison between the results by the direct capture and Coulomb dissociation provided useful information on this important reaction, since the two methods are independent, and their sources of possible systematic uncertainties are different [7].

The Coulomb dissociation is also a process with a three body final state. As indicated in Fig. 2, the virtual photon induces the breakup of the nucleus A. Therefore the Coulomb dissociation may be another analogue of Trojan-Horse reaction with the "soldier" of the virtual photon. In contrast to the Trojan-Horse applications, the process is theoretically treated often as inelastic excitation of the nucleus A to its excited state above the particle decay threshold. Another difference lies in the

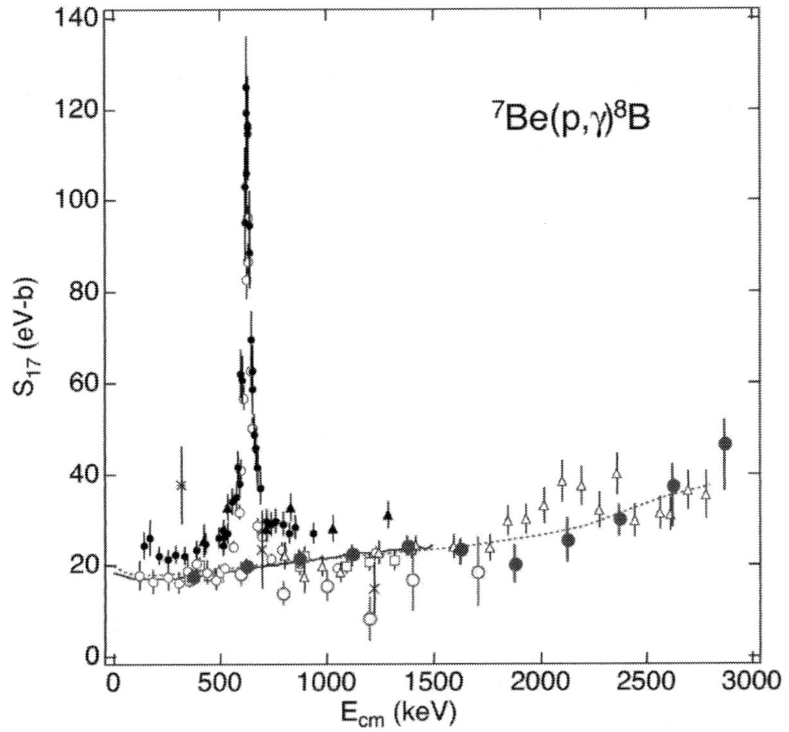

Figure 3. Astrophysical S_{17}-factors for the ^{7}Be(p,γ)^{8}B reaction extracted from the first (large open circles in purple) [11] and second (large solid circles in blue) [12] experiments at RIKEN. Direct (p,γ) data by Kavanagh [14] (crosses), Parker [15] (closed triangles) Kavanagh *et al.* [16] (closed circles), Vaughn *et al.* [17] (open triangles), Wiezorek *et al.* [18] (asterisk), and Filippone *et al.* [19] (open circles) are also shown. The energy dependent S_{17} factors theoretically predicted by Barker and Spear [20] (solid curve) and Descouvemont and Baye [21] are shown. They are normalized to the data from the second Coulomb dissociation experiment.

incident energy. The Coulomb dissociation method is applied at relatively high energy to make the virtual photon capable to excite the nucleus to its unbound state. Both the methods have a common feature of yield enhancement. The Trojan-Horse method takes this advantage to mostly overcome the low cross section of the relevant reaction of astrophysical interest, while the Coulomb dissociation is often applied to a capture reaction involving a short-lived nucleus, the beam intensity of which is very low.

3 Summary and Remarks

Two examples of the reactions having three particles in the final states are considered. As well as the success for various Trojan Horse reactions thoroughly studied by Claudio Spitaleri and his Catania group, PWIA well describes the ^{1}H(d,^{2}He)n reaction data including the differential cross sections and tensor analyzing powers. The Coulomb dissociation is another indirect method for studying astrophysical reactions. Its mechanism can be regarded as absorption of a virtual photon, which plays a similar role of the "soldier" in a Trojan Horse reaction. Thus the inverse process, radiative capture, can be studied by the Coulomb dissociation. Results of Coulomb dissociation studies agree with relevant direct data, demonstrating its usefulness. It should be noted that the Coulomb dissociation can indirectly study radiative capture, often of importance in nucleosynthesis, while the Trojan Horse method can indirectly access particle transfer reactions of astrophysical interest.

These successful applications of simple theoretical treatments to various types of reactions with three-body final state lead me a "feeling": the quasi-free mechanism or first-order Coulomb excitation can naturally find a way to particle-unbound final state, providing a simple and reliable tool for analysis. On the other hand, population of discrete bound-states requires more kinematically restricted conditions and may allow for complicated mechanism to be involved. That is only a prejudice, but we may thank this favorable situation.

The author of this article conducted several Coulomb dissociation experiments in the period when Spitaleri and his group were vigorously studying and establishing the Torojan Horse method. We appreciate their long experimental efforts and careful analysis to extract the relevant cross sections of astrophysical interest.

References

[1] C. Spitaleri *et al.*, Phys. Rev. C **60**, 055802 (1999); S. Typel and G. Baur, Annals Phys. **305**, 228 (2003); C. Spitaleri *et al.*, Eur. Phys. J. A **52**, 77 (2016)

[2] T. Motobayashi *et. al.*, Phys. Lett. B **233**, 69 (1989); S. Kox *et. al.*, Nucl. Phys. A **556**, 621 (1993)

[3] D.V. Bugg and C. Wilkin, Phys. Lett. B **152**, 37 (1985)

[4] E.M. Burbidge, G.R. Burbidge, W.A. Fowler, and F. Hoyle, Rev. Mod. Phys. **29**, 547 (1957); C. Rolfs and W. Rodny, *Cauldrons in the Cosmos* (University of Chicago Press, 1988)

[5] G. Baur, C.A. Bertulani and H. Rebel, Nucl. Phys. A **458**, 188 (1986).

[6] G. Baur and H. Rebel, J. Phys. G **20**, 1 (1994); Ann. Rev. Nucl. and Part. Sci. **46**, 321 (1996).

[7] T. Motobayashi, Nucl. Phys., A **693**, 258 (2001).

[8] T. Motobayashi and H. Sakurai, Prog. Theor. Exp. Phys. **2012**, 03C001 (2012).

[9] T. Motobayashi *et al.*, Phys. Lett. B **264**, 259 (1991).

[10] P. Decrock *et al.*, Phys. Rev. Lett. **67**, 808 (1991).

[11] T. Motobayashi *et al.*, Phys. Rev. Lett., **73**, 2680 (1994)

[12] T. Kikuchi *et al.*, Phys. Lett. B **391**, 261 (1997)

[13] N. Iwasa *et al.*, Phys. Rev. Lett. **83**, 2910 (1999); B. Davis *et al.*, Phys. Rev. Lett. **86**, 2750 (2001) F. Schümann *et al.*, Phys. Rev. Lett. **90**, 232501 (2003);

[14] R.W. Kavanagh, Nucl. Phys. **15**, 411 (1960)

[15] P.D. Parker, Phys. Rev. **150**, 851 (1966)

[16] R.W. Kavanagh, T.A. Tombrello, J.M. Mosher and D.R. Goosman, Bull. Am. Phys. Soc. **14**, 1209 (1969)

[17] F.J. Vaughn, R.A. Chalmers, D. Kohler, and L.F. Chase, Jr., Phys. Rev. C **2** 1657 (1970).

[18] C. Wiezorek, H. Kräwinkel, R. Santo, and L. Wallek, Z. Phys. A **282**, 121 (1977)

[19] B. Filippone, S.J. Elwyn, C.N. Davids, and D.D. Koetke, Phys. Rev. Lett., **50**, 412 (1983); Phys. Rev. C **28**, 2222 (1983).

[20] F.C. Barker and R.H. Spear, Astrophys. J. **307**, 847 (1986)

[21] P. Descouvemong and D. Baye, Nucl. Phys. A **567**, 341 (1994)

[22] A.R. Junghans *et al.*, Phys. Rev. Lett. **86**, 041101 (2002) .

[23] L.T. Baby *et al.*, Phys. Rev. Lett. **90**, (2003) 022501; Phys. Rev. **C 67** 065805 (2003).

EPJ Web of Conferences **165**, 02007 (2017) DOI: 10.1051/epjconf/201716502007
NPA8 2017

"Other" indirect methods for nuclear astrophysics

Livius Trache

IFIN-HH, Bucharest-Magurele, Romania

Abstract. In the house of Trojan Horse Method (THM), I will say a few words about "other" indirect methods we use in Nuclear Physics for Astrophysics. In particular those using Rare Ion Beams that can be used to evaluate radiative proton capture reactions. I add words about work done with the Professore we celebrate today. With a proposal, and some results with TECSA, for a simple method to produce and use isomeric beam of ^{26m}Al.

1 Introduction

As this is an invited lecture in the special session dedicated to prof. Claudio Spitaleri on the occasion of his (formal) retirement, I will treat it just like that! I mean, I will keep the promise of the title and abstract, but I will insist on work done together with Claudio and his people. It will obviously be difficult to put in 6 pages the content of 35 slides, including images worth 1000 words each! In all cases I will enumerate, rather than fully describe or justify, the most common indirect methods that we use in nuclear astrophysics. In this age when nuclear physics for astrophysics (NPA) and the rare ion beams (RIB) are spearheading most nuclear physics programs, the combination of the two is done through indirect methods (so far at least). That is, by using measurements at laboratory energies of tens or even hundreds or MeV/nucleon to obtain data that allow us to evaluate cross sections and/or reaction rates for nuclear processes that take place in stars at energies of tens – hundreds of keV/nucleon, or even lower. I cannot pretend in such a short time (during the conference) and space (in the Proceedings) to cover exhaustingly all indirect methods and all what was done so far with them. Therefore I will be very selective and subjective and in all cases I will use examples from works I participated to and which, to one extent or another, can be related to the honouree of the session. In particular in Sec. 3 and 4, where I will present TECSA and one interesting use of it, respectively.

I met Professore Spitaleri sometime last century! And as many of you here, I met him first through a paper of his before meeting him in person. That first encounter was the origin of many works we did, or we participated to, together, either at Texas A&M University, in Catania, or at IFIN-HH in Bucharest. Certainly others in this session and this room know him better and for longer, and they already have spoken or will, and said already moving words about him and his works. Repeating those will only seem formal. But I say that it is important to tell from the beginning that when I think and talk about Claudio I always have in mind also his group of people. A group that he raised in Catania, at the University and/or at LNS, which most probably represent his best and most enduring achievement of the career. Congratulations for that!

2 Indirect Methods in Nuclear Astrophysics

It is well known that we need to use indirect methods in NA basically for two reasons:

a) Stars are "cold" and the corresponding cross sections for the nuclear reactions in stars or stellar environments are very small when charged particles are involved, due to the Coulomb barriers;
b) Most of the reactions in stars involve radioactive species, which do not exist in nature, but can be produced in laboratory.

Many measurements in NPA are not direct measurements (= reactions between partners at stellar energies or close to them), are therefore indirect, but one cannot always classify easily them in a few types. Therefore the list of types of indirect methods may be subjective. Mine reads like this:

A. Coulomb dissociation
B. Transfer reactions (the ANC method)
C. Breakup (nuclear) of loosely bound nuclei
D. Resonance spectroscopy: β-decay, β-delayed proton-decay, transfer reactions, resonant elastic scattering, etc.
E. Trojan Horse Method.

2.1 Coulomb Dissociation

The study of a radiative capture reaction can be replaced by measuring its inverse, the photo-dissociation of the final nucleus. Coulomb dissociation (CD) is the process in which photo-dissociation is produced by virtual, rather than real photons [1]. One of the initiators of CD is in the audience, can better describe it, several other talks were describing it. I want only to comment on the fact that CD is sometimes called "the first indirect method in NA". I think that we had other indirect methods before. For example the study of weak interactions that allowed Bethe to evaluate the first step in the pp-chains. Or even, the mass measurements that allowed Sir A. Eddington to suggest that nuclear forces are at the origin of the energy production in Sun and stars!

2.2 Transfer reactions (the ANC method)

Transfer reactions were crucial in obtaining information about the structure of nuclei, in particular about their single particle, fermionic, structure. For over two decades now it was demonstrated and is recognized that one can use peripheral one-proton [2] or even one-neutron [3] transfer reactions to extract information about the peripheral behaviour of the radial wave function of the least bound nucleon and from it to evaluate the direct (non-resonant) component of the radiative proton capture reactions at low energies. The method consists in measuring cross sections for the transfer of a proton in nucleus-nucleus reactions, which allow the determination of the Asymptotic Normalization Coefficient (ANC) through comparison with DWBA calculations. It is applicable if the reaction is peripheral and one-step, and if good optical model potentials are available. The group at Texas A&M University was instrumental in proposing using and demonstrating its validity. It was applied to stable and radioactive beams, as well [4, 5, 6 and references therein]. It was extended to radiative alpha capture too (examples in [7, 8]).

DOI: 10.1051/epjconf/201716502007

2.3 Breakup (nuclear) of loosely bound nuclei

The nuclear breakup of loosely bound nuclei can also be peripheral. In a reaction of the type Y→X+p (breakup of a proton) at intermediate energies, from the shape of the measured momentum distribution of the core one can extract information about the quantum numbers of the proton orbital involved and from the cross section one can extract the ANC by comparison with appropriate calculations. Which ANC in turn can be used to evaluate the radiative capture cross section for the reaction X(p,γ)Y. We applied it successfully in several cases [9, 10]. I would argue that in many cases its most significant contribution was that cross sections (or astrophysical S-factor) where determined in reactions of crucial importance through a method that implied totally different systematic errors than those determined directly. Like in the case of the $S_{17}(0)$-factor), where we could extract $S_{17}(0)$=18.7(19) eVb [9], in good agreement with the best direct measurements for the $^{7}Be(p,\gamma)^{8}B$ reaction, of crucial importance for the understanding of the solar neutrino puzzle.

2.4 Resonance spectroscopy: β-decay, β-delayed proton-decay, transfer reactions, elastic scattering, etc.

In several cases reactions at stellar energies are proceeding through, and are even dominated by resonances. The reaction is a two-step process in this case: a meta-stable state in the compound nucleus is populated in the entrance channel, which later decays to the final configuration. In these situations, the reaction rates get sizable contributions from the resonances that are located approximately in the Gamow window of the reaction. The contribution of each isolated narrow resonance is entirely dependent on two parameters: the location of the resonance (its energy) and the resonant strength. If both (or at least one, the location for example) can be determined by the spectroscopy of the metastable state that constitutes the resonance, then we can evaluate the contribution to the reaction rate. In many cases it is easier to determine the position of the resonance, and this is important as it affects exponentially the rate. Any spectroscopic method that would then allow to find these resonances and characterize them, is an indirect method. One with an obvious connection is the beta-delayed proton decay: if conditions allow it ($Q_\beta > S_p$ and β-decay selection rules allow for populating same states relevant in proton capture), the metastable state may decay by proton emission; measured, it can determine its location, quantum numbers and decay branchings. There are two experimental problems we encounter, though: (1) the low proton energies relevant in NA are difficult to measure experimentally and (2) the lower the energy of protons, the lower the proton branching (due to a wider Coulomb barrier to be penetrating out). The combinations of these factors lead us to try first to detect βp-decay using very thin Si detectors (65 and 45 microns, see [11]), not quite satisfactory, as there was still a large beta-induced background at low energies. Another idea was to use a gas detector, based on a novel type of detector with Micromegas as amplifier. This was much better, detectors named ASTROBOX1 and ASTROBOX2 gave good results [12, 13] and the idea is already spreading (see these Proceedings).

2.5 Trojan Horse Method

This is a Catania speciality and Claudio's role here is crucial! You have heard much about it, I do not believe I have anything to add at this point. Except, may be, a comment: it was proposed about 30 years ago. And it produces excellent results for more than two decades, but no other group was so far able to adopt it fully and produce independent experimental results! One should do that in the next future, for the good of the recognition of the method.

And also say that it is the method that gives results that look closer to those of direct measurements: look in these Proceedings at S-factor graphs going down to E=0, and even under the threshold! Plus its first use with radioactive beams for the study of the $^{18}F(p,\alpha)^{15}O$ reaction [14, 15].

3 TECSA – Texas-Edinburgh-Catania Silicon Array

At some point in the cooperation with Claudio and his group, we realized that we can put together a device by employing existing Texas A&M facility (MARS) and manpower, complex multi-channel electronics from the Edinburgh group (PJ Woods, T. Davidson) and Catania's YY1 Si detectors. TECSA was built without any extra funding and a number of experiments were done at TAMU with it, some being "firsts" in their class. The commissioning was a $^{14}C(d,p)$ experiment in inverse kinematics, using ^{14}C radioactive beam accelerated by the K500 superconducting cyclotron onto a CD_2 target [16, 17]. In Figure 1 bellow, I show online data: the right panel lets you understand the geometry of the array, the left one shows the population of ground and first excited state in ^{15}C (with good hints on angular distributions).

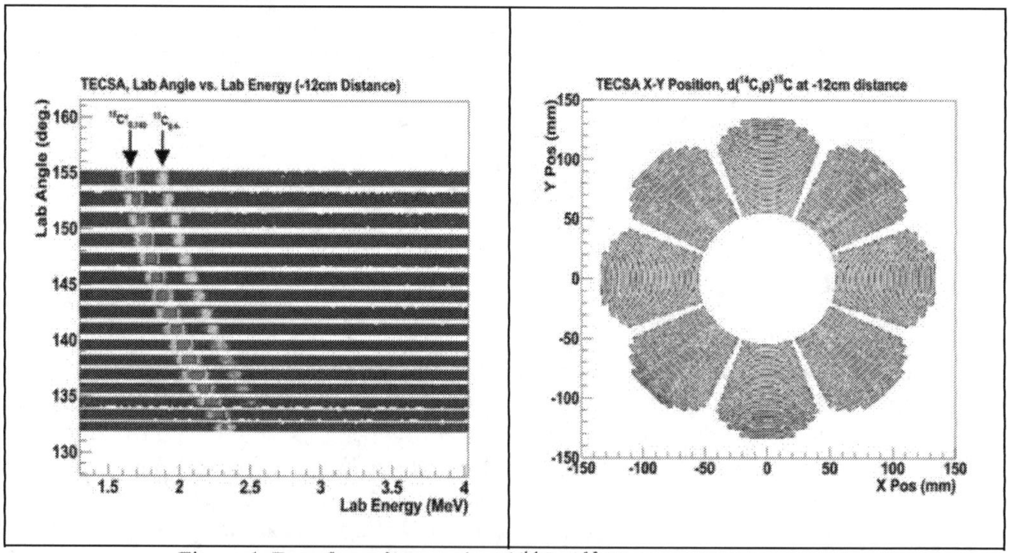

Figure 1. Data from the reaction $d(^{14}C,p)^{15}C$ taken with TECSA.

TECSA was also used for the study of the first THM using radioactive beam: d(18F,p)n15O.

4 ^{26m}Al beam production: proposal and demonstration

I will treat specially this experiment, which was the last before we dismantled TECSA in the spring of 2012. It is about obtaining am isomeric beam using a production mechanism, rather than a separator.

The isotope ^{26}Al is unstable with a long half-life of 0.72 million years for its $J=5^+$ ground state and has an isomeric state $J=0^+$ only 228 keV above with a much shorter half-life of 6.34 sec. The isotope is known to be important in nuclear astrophysics: the gamma-ray line $E\gamma$ = 1809 keV following the beta-decay of its long-lived ground state was the first discrete line to be detected by gamma-ray satellite telescopes. This was a proof of on-going

DOI: 10.1051/epjconf/201716502007

nucleosynthesis in the Universe. The origin of freshly produced ^{26}Al is not well known, however, and the comparison of observations with nucleosynthesis model predictions is hindered by the lack of precise nuclear data for some reactions in the reaction networks used in calculations. For all temperatures below 2 GK, ^{26g}Al and ^{26m}Al behave like separate, different isotopes and are to be treated as such in calculations. For higher temperatures, only reached in some explosive scenarios, they will mix. Therefore, it is important to evaluate reactions on ^{26m}Al and those can only be studied experimentally with an isomeric beam. The two states, g.s. and isomer, cannot be separated by current mass separators, because resolutions of 228 keV at masses of about 26 GeV are not available at current accelerators. Different methods must be found. We tried the use of a selective reaction mechanism at the MARS separator of the Cyclotron Institute of Texas A&M University. We started with a ^{26}Mg primary beam of 16 MeV/nucleon from the superconducting cyclotron K500 bombarding a H_2 cryotarget. With it ^{26}Al could be produced by ^{26}Mg(p,n) reaction by two mechanisms: fusion-evaporation and charge-exchange. The first should not be selective and produce a mixing of g.s. ($J=5^+$) and isomeric state ($J=0^+$) around the velocity of the compound system, while charge-exchange should select the isomeric state through the GT transition: $(\nu d_{5/2}{*}\nu d_{5/2})0^+ \rightarrow (\nu d_{5/2}{*}\pi d_{5/2})0^+$ at velocities close to that of the projectile. By tuning MARS for the appropriate magnetic rigidities we did separate the two. When we tuned MARS for the fusion-evaporation products we obtained a pure ^{26}Al beam of $3{*}10^6$ pps with 33% ^{26m}Al content. When we changed the rigidity to that of the transfer reaction, we increased the ^{26m}Al content of the beam to 67%, but the intensity decreased 10 times to about $3{*}10^5$ pps. The intensities were sufficient in both cases to study (d,p) reactions in inverse kinematics using TECSA detection system. The comparison of the results show that the two beams can populate same or different excited states in the final nucleus ^{27}Al. Details of the experiment and the results were only presented at one meeting and published online in the report [18], but it occurred to me that the method works and could be used further (*Note added in Proceedings: it was pointed to me that in a paper published in August this year, an ANL group reported a similar idea [19]*). It follows from the experiment that the production of the isomeric beam using a selective population mechanism like charge-exchange is possible. A better isomeric purity can be obtained by increasing the bombarding energy of the primary beam to 25-40 MeV/nucleon. The isomer production through charge-exchange will be favored by the larger energy and at the same time the fusion-evaporation mechanism will lead mostly to ^{25}Al through (p,2n). Also the rigidity separation between the two components will increase, favoring a better separation.

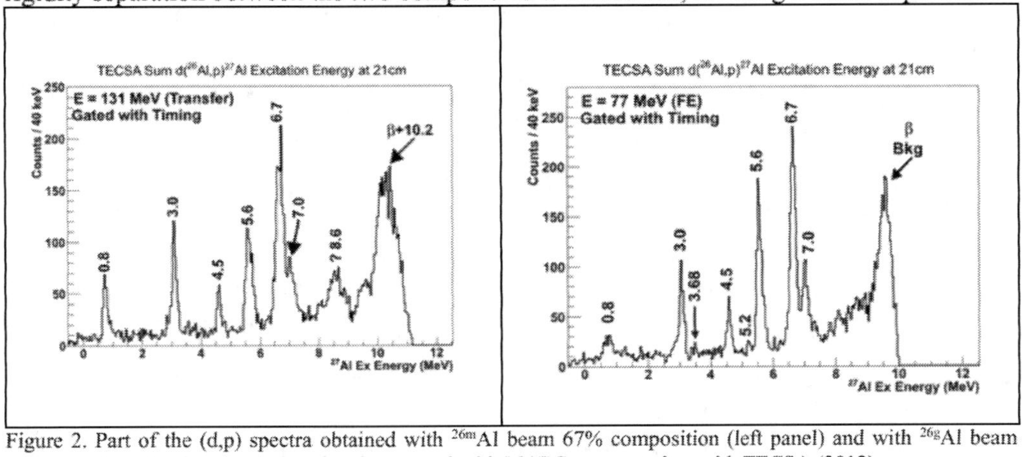

Figure 2. Part of the (d,p) spectra obtained with ^{26m}Al beam 67% composition (left panel) and with ^{26g}Al beam (right). Radioactive beams produced and separated with MARS, spectra taken with TECSA (2012).

5 Laudatio

I want to close by a few words as *laudatio* for Claudio! Best *laudatio* in this case is to enumerate a few of his achievements. Trojan Horse Method was invented, demonstrated, used, spread to over 10 laboratories where Claudio and his group had experiments. Nuclear astrophysics has become a recognized mark of Catania laboratories. Claudio is appreciated and has a large number of friends in the global nuclear astrophysics community. He is one of the initiators of the St. Tecla European School on Experimental Nuclear Astrophysics (ESENA, 9 editions so far) and of the European Network of Nuclear Astrophysics Schools (ENNAS).

But the most important achievement is in my opinion that he has formed and educated a solid group of young, talented and ambitious people. He formed and lead them with "a velvet glove on an iron fist". They have competence and dynamism and they have jobs! They teach to a large part in Sicily, work in Catania and all over the world's laboratories, some are in the rest of the world. Claudio was instrumental in these!

Congratulation Claudio and good health in the years that come and success in all activities you'll chose to embrace!

The author acknowledges that many of the above have been learned from working with many in the audience, in particular with the Texas A&M University group at the Cyclotron Institute and with the Catania group. Financial support through the grants NAIRIB C02-FAIR and PN 16 39 01 02 of the Romanian Ministry of Research and Innovation made this conference participation possible.

References

1. G. Baur, C. A. Bertulani and H. Rebel, Nucl. Phys. **A458**, 188 (1986)
2. A. M. Mukhamedzhanov et al. Phys Rev C **56**, 1302 (1997)
3. L. Trache et al. Phys Rev C **67**, 062801(R) (2003)
4. Azhari et al. Phys. Rev. Lett. **82**, 3960 (1999)
5. A. Banu et al, Phys. Rev. C **79**, 025805 (2009)
6. R.E. Tribble et al., Rep. Prog. Phys. **77**, 106901 (2014)
7. C. R. Brune et al., Phys. Rev. Lett. **83**, 4025 (1999)
8. M. L. Avila et al., Phys. Rev. Lett. **114**, 071101 (2015)
9. L. Trache et al, Phys. Rev. Lett. **87**, 271102 (2001); Phys. Rev. C **69**, 032802 (2004)
10. A. Banu et al., Phys. Rev. C **84**, 015803 (2011)
11. A. Saastamoinen et al. Phys. Rev. C **83**, 045808 (2011)
12. E. Pollacco et al, Nucl. Instr. Meth. A **723**, 102 (2013)
13. A. Saastamoinen et al., Nucl. Instr. Meth. B **376**, 357 (2016)
14. BT Roeder et al., Nucl. Instr. Meth. A **634**, 71 (2011)
15. M. McCleskey et al. Phys. Rev. C **89**, 044605 (2014)
16. S. Cherubini et al. Phys. Rev. C **92**, 015805 (2015)
17. R.G. Pizzone et al., Eur. Phys. J. A, **52**:24 (2016)
18. L. Trache, in 4th Eurisol User Group Topical Meeting, July 1-3, 2013, Krakow, Poland, http://www.eurisol.org/usergroup/?p=407 , p. 101
19. S. Almaraz-Calderon et al., Phys. Rev. Lett. **119**, 072701 (2017).

EPJ Web of Conferences **165**, 02008 (2017) DOI: 10.1051/epjconf/201716502008
NPA8 2017

Theory of the Trojan-Horse Method – From the Original Idea to Actual Applications

Stefan Typel[1,2,⋆]

[1] *Institut für Kernphysik, Technische Universität Darmstadt, Schlossgartenstraße 9, D-64289 Darmstadt, Germany*
[2] *GSI Helmholtzzentrum für Schwerionenforschung GmbH, Planckstraße 1, D-64291 Darmstadt, Germany*

Abstract. The origin and the main features of the Trojan-horse (TH) method are delineated starting with the original idea of Gerhard Baur. Basic theoretical considerations, general experimental conditions and possible problems are discussed. Significant steps in experimental studies towards the implementation of the TH method and the development of the theoretical description are presented. This lead to the successful application of the TH approach by Claudio Spitaleri and his group to determine low-energy cross section that are relevant for astrophysics. An outlook with possible developments in the future are given.

1 Introduction

Cross sections of low-energy nuclear reactions are crucial ingredients in astrophysical models that describe the nucleosynthesis during Big Bang and stellar evolution, see, e.g., Ref. [1]. They have to be known with sufficient accuracy to predict the abunduncies of the produced elements. Because charged particles participate in the reactions, the cross sections are strongly suppressed by the Coulomb barrier at the low thermal energies of the astrophysical environment. Their direct measurement in the relevant range of energies, the Gamov window, is extremely difficult due to the low reaction rates and many experimental challenges. As an alternative, indirect methods have been proposed to extract the wanted cross sections from suitable surrogate reactions. An example is the Coulomb dissociation of nuclei into fragments with virtual photons to study the inverse radiative capture reactions [2, 3].

A particular indirect method that is applicable to astrophysical rearrangement reactions of the type

$$A + x \to C + c \tag{1}$$

was introduced by Gerhard Baur as a theoretical concept in 1986 [4]. In fact, he had suggested the idea already in his invited talk on 'Breakup processes in nuclear reactions' at the 1985 Varna International Summer School on Nuclear Physics [5]. Instead of measuring the cross section of reaction (1) at low energies directly in the laboratory, the breakup reaction

$$A + a \to C + c + b \tag{2}$$

⋆e-mail: s.typel@gsi.de

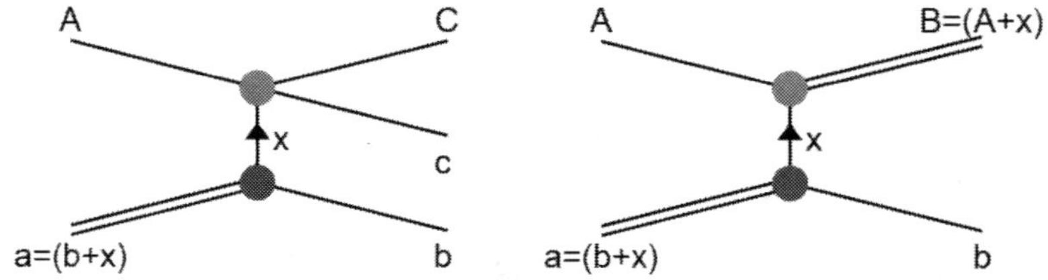

Figure 1. Diagrammatic representation of a transfer reaction to the continuum of the subsystem $C + c$ (left) and to a bound state of the subsystem B (right) with a Trojan-horse nucleus a and a spectator nucleus b.

is studied with energies in the initial state above the Coulomb barrier, see the diagram in the left panel of figure 1. The nucleus a is assumed to be well described as a bound state composed of two nuclear clusters b and x. Reaction (1) can be considered as a sub-process of (2) where x is the transfered particle and b acts as a spectator. Nucleus x is hidden inside a and brought across the Coulomb barrier to induce the reaction with A. This description clearly justifies to call nucleus a the Trojan horse, thus explaining the name of the method. It goes back to Homer's epic poems on the Trojan War and subsequent events. Actually, the capture of Troy with the subterfuge using the Trojan horse is mentioned only shortly in the Odyssey at two places, e.g., as '...κεκαλυμμενοι ιππω' [6]. A more detailed account of the story can be found in Virgil's Aeneid [7].

In order to extract the cross section of the astrophysically relevant reaction (1) from the reaction (2) measured in the laboratory, direct reaction theory is employed. Since the TH reaction is surface dominated, it is expected that the suppression of the cross section of the subprocess, which is astrophysically relevant, by the Coulomb barrier is reduced. Also electron screening will play no role due to the relatively high energies in the initial $A + a$ system. In the original proposal of Gerhard Baur, it was argued that the Fermi motion of x inside the TH a compensates the relative motion of x with respect to A and thus small relative energies in the $A + x$ system were accessible. This corresponds to exploring the tail of the $x + b$ wave function of relative motion in momentum space. However, in the actual application of the TH method by Claudio Spitaleri and his group it was realized that the reaction (2) has to be studied as a quasifree scattering process to exploit the virtues of the TH method. This corresponds to kinematical conditions covering the maximum and not the tail of the momentum distribution.

2 Theory of direct reactions

For a sucessfull application of the THM it is essential to utilise direct reaction theory to connect the cross sections of the TH reaction (2), depicted diagrammatically in the left part of figure 1, with that of the astrophysical reaction (1). The general cross section of a reaction with three particles in the final state can be written in nonrelativistic kinematics as

$$d\sigma = \frac{2\pi}{\hbar}\frac{\mu_{Aa}}{p_{Aa}}\frac{d^3 p_{Bb}}{(2\pi\hbar)^3}\frac{d^3 p_{Cc}}{(2\pi\hbar)^3}\left|T_{fi}\right|^2 \delta\left(E_{Bb} + E_{Cc} - E_{Aa} - Q\right) \tag{3}$$

where B denotes the $C + c$ system and

$$Q = (m_A + m_a - m_C - m_c - m_b)\, c^2 \tag{4}$$

is the Q value of the THM reaction (2). The mass of a system i is denoted by m_i and the relative momentum between i and j is represented by $\vec{p}_{ij} = \mu_{ij}\left(\vec{p}_i/m_i - \vec{p}_j/m_j\right)$ with the reduced mass $\mu_{ij} =$

$m_i m_j/(m_i + m_j)$. Actual cross sections that are measured are obtained from (3) by an integration over unobserved quantities taking the energy conservation, expressed by the δ function, into account.

The essential information on the reaction process is contained in the T-matrix element T_{fi} that is given in post formulation by

$$T_{fi} = \langle \phi_B \phi_b \exp\left(i\vec{p}_{Bb} \cdot \vec{r}_{Bb}/\hbar\right) |V_{Bb}| \Psi^{(+)}_{Aa} \rangle \tag{5}$$

with the full scattering wave function $\Psi^{(+)}_{Aa}$ in the initial state and the interaction V_{Bb} between the spectator b and the subsystem $B = C + c$ in the final state. The wave function ϕ_B is the full scattering wave function $\Psi^{(-)}_{Cc}$ of the $C + c$ system containing information on the scattering matrix elements of the reaction (1) and thus the cross section. If the final state of the the THM reaction was not a transfer to a continuum state $B = C + c$ but to a bound state $B = A + x$, depicted diagrammatically in the right panel of figure 1, the wave function ϕ_B would contain information on the spectroscopic factors or asymptotic normalisation coefficients of the nucleus B. The exact calculation of the T-matrix element (5) is practically impossible and approximations have to be introduced. Several choices are introduced frequently in the literature, e.g., the distorted-wave Born approximation (DWBA), the distorted-wave impulse approximation (DWIA), or the plane-wave impulse approximation (PWIA).

The fully differential cross section of the TH reaction (2) assumes a particular simple form if the PWIA is applied. In this case it can be written in a factorized form as

$$\frac{d^3}{dE_{Cc}d\Omega_{Cc}d\Omega_{(Cc)b}} = K \left|\Phi_a(\vec{Q})\right|^2 \frac{d\sigma^{\mathrm{HOES}}}{d\Omega_{Cc}} \tag{6}$$

with a kinematic factor K, and two factors that can be related to the two poles in the diagram of transfer reaction to the continuum, see the left panel of figure 1. The momentum distribution $|\Phi_a(\vec{Q})|^2$ of the transfered particle x inside the TH a depends on the momentum transfer $\vec{Q}$ to the spectator b and is derived from the ground state wave function of relative motion in momentum space Φ_a. This factor corresponds to the lower (blue) pole in the diagram. The last factor in eq. (6) is the differential cross section for the astrophysical reaction (1), however, not the wanted on-shell one but a so-called half-off-shell cross section. The difference is mainly caused by two facts: (1) the transfered particle x is not on the energy shell, because the standard dispersion relation between its energy and momentum does not apply, (2) the suppression by the Coulomb barrier is reduced as compared to the on-shell cross section. This last feature makes the THM very attractive because it allows to reach extremely low energies in the reaction with reasonable cross sections in contrast to the original astrophysical reaction (1). The actual form of the cross section $\frac{d\sigma^{\mathrm{HOES}}}{d\Omega_{Cc}}$ and its relation to the cross section of reaction (1) depends on the approximations made in deriving the factorized form (6) of the TH cross section. E.g., applying the PWIA and the surface approximation, these cross sections differ by a simple penetrability factor depending on Coulomb wave functions at a certain radius. In less drastic approximations this relation might change leading to a modification of the energy dependence of the extracted astrophysical cross section.

3 Conditions and problems

The THM can be realised successfully in experiments if several precondition are met. First of all, a suitable TH nucleus a has to be selected that is well clustered. Typically, a nucleus like ^{2}H or ^{6}Li is chosen in order to be able to study reactions induced by neutrons, protons, deuterons or α particles. The momentum distributions $|\Phi_a(\vec{Q})|^2$ of these Trojan horses are well known and peak at zero momentum because they are described as s-wave ground states in cluster models. The kinematic conditions for the TH reaction also have to be selected carefully. On the one hand quasifree scattering condition have to be satisfied, corresponding to a zero momentum transfer to the spectator for the above-mentioned Trojan horses, and on the other hand the beam energy in the initial state should match the accessible range of energies of the astrophysical reaction. The separation of the quasifree reaction from other reaction mechanism, e.g., sequential processes, has to be checked experimentally.

In the analysis of a TH experiment, the factorization of the cross section as in eq. (6) is usually presupposed to be valid, however the quality of this approximation needs to be tested in improved theoretical calculations in the future. The relation of the half-off-shell cross section $\frac{d\sigma^{\text{HOES}}}{d\Omega_{Cc}}$ to the wanted on-shell cross section of the reaction (1) used in present applications of the THM has to be validated in refined approaches. The extracted cross sections have to be normalized to directly measured data at high energies, which are usually available, because the approximations applied in the derivation of the cross section (6) do not allow to obtain absolute values.

4 Experimental studies and applications of the TH method

The kinematic conditions for TH experiments can be realised easily at the Laboratori Nazionali del Sud (LNS) in Catania, e.g., with the Tandem accelarator providing beams with energies of a few MeV/nucleon, and at similar 'small-scale' reseach laboratories. In addition, the expertise of Claudio Spitaleri and his group matches perfectly with those required to perform successful TH experiments. They studied momentum distributions of clustered nuclei extensively [8, 9] and performed tests of the cross section factorization in PWIA [10]. Furthermore, many experiments investigating quasifree scattering and reactions were conducted [11–17]. Hence it is not a surprise that Claudio Spitaleri became interested in the TH method.

The first experimental attempts to apply the TH method to astrophysically important reactions started in the second half of the 1980s [18–21] but it remained unclear how to relate the extracted half-off-shell cross section $\frac{d\sigma^{\text{HOES}}}{d\Omega_{Cc}}$ to the actual astrophysical cross section. This situation triggered detailed theoretical considerations in Refs. [22, 23] extending the previous formulation by Gerhard Baur and simulations of actual experiments were performed. The application of the 'surface approximation' lead to an explicit form of the penetrability correction that proved to be decisive in the further development of the TH method. Extensive experimental applications of the TH method followed, see, e.g., the early studies [24–26], and realistic cross sections of astrophysically relevant reactions were obtained.

5 Further developments and outlook

Nowadays, the standard application of the TH method concerns non-resonant rearrangement reactions. But the scope of applications was extended to include the study of resonance properties, low-energy elastic scattering, or electron screening effects. The experimental progress is nicely summarized in the recent review [27]. The theoretical development including other indirect methods is discussed in the review [3]. For a further refinement of the TH method, more dedicated high-precision experiments are needed as well as more detailed calculations in order to check the validity of the approximations and the accuracy of the extracted cross sections of astrophysically relevant reactions. E.g., the explicit propagation of the transfered off-shell cluster x can be considered in an extended theoretical description of the TH reaction, leading to a two-step process [28] instead of the one-step description (5). Nevertheless, the final expression of the cross section assumes a similar product form as in eq. (6). A comparison of different theoretical approaches will be very worthwhile. On the experimental side, the use of TH nuclei *a* with clusters in a p-wave ground state and a maximum of the corresponding momentum distribution at nonzero momentum could broaden the range of applications of the TH method even further.

References

[1] E. G. Adelberger *et al.*, Rev. Mod. Phys. **83**, 195 (2011)
[2] G. Baur, C. A. Bertulani and H. Rebel, Nucl. Phys. A **458**, 188 (1986)
[3] R. E. Tribble *et al.*, Rept. Prog. Phys. **77**, 106901 (2014)
[4] G. Baur, Phys. Lett. B **178**, 135 (1986)
[5] G. Baur, Nuclear Energy **25**, 183 (1987)
[6] Homer, Odyssey **VIII**, 503
[7] Virgil, Aeneid **II**, 13 ff.
[8] S. Barbarino *et al.*, Phys. Rev. C **21**, 104 (1980)
[9] M. Lattuada *et al.*, Nuovo Cim. A **63**, 530 (1981)
[10] P. G. Fallica *et al.*, Phys. Rev. C **24**, 1394 (1981)
[11] M. Lattuada *et al.*, Phys. Rev. C **26**, 1330 (1982)
[12] M. Lattuada *et al.*, Phys. Rev. C **30**, 531 (1984)
[13] M. Lattuada *et al.*, Nucl. Phys. A **458**, 493 (1986)
[14] M. Zadro *et al.*, Nucl. Phys. A **474**, 373 (1987)
[15] M. Zadro *et al.*, Phys. Rev. C **40**, 181 (1989)
[16] G. Calvi *et al.*, Phys. Rev. C **41**, 1848 (1990)
[17] S. Blagus *et al.*, Phys. Rev. C **44**, 325 (1991)
[18] S. Cherubini *et al.*, Ap. J. **457**, 855 (1996)
[19] G. Calvi *et al.*, Nucl. Phys. A **621**, 139c (1997)
[20] C. Spitaleri *et al.*, Phys. Rev. C **60**, 055802 (1999)
[21] M. Aliotta *et al.*, Eur. Phys. J. A **9**, 435 (2000)
[22] S. Typel and H. H. Wolter, Few-Body Systems **29**, 75 (2000)
[23] S. Typel and G. Baur, Annals Phys. **305**, 228 (2003)
[24] C. Spitaleri *et al.*, Phys. Rev. C **63**, 055801 (2001)
[25] A. Musumarra *et al.*, Phys. Rev. C **64**, 068801 (2001)
[26] M. G. Pellegriti *et al.*, Nucl. Phys. A **688**, 543c (2001)
[27] C. Spitaleri *et al.*, Eur. Phys. J. A **52**, 77 (2016)
[28] C. A. Bertulani, M. S. Hussein and S. Typel, arXiv:1707.04563 [nucl-th]

AUTHOR INDEX

AUTHOR INDEX

AUTHOR INDEX

AUTHOR INDEX

AUTHOR INDEX

AUTHOR INDEX

AUTHOR INDEX